The Functional Interpretation of Logical Deduction

Advances in Logic

Series Editor: Dov M Gabbay FRSC FAvH
Department of Computer Science
King's College London
Strand, London WC2R 2LS
UK
dg@dcs.kcl.ac.uk

Published

Advances in Logic – Vol. 5

The Functional Interpretation of Logical Deduction

Ruy J G B de Queiroz • Anjolina G de Oliveira

UFPE, Brazil

Dov M Gabbay

King's College, UK

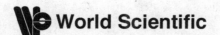 World Scientific

NEW JERSEY · LONDON · SINGAPORE · BEIJING · SHANGHAI · HONG KONG · TAIPEI · CHENNAI

Published by

World Scientific Publishing Co. Pte. Ltd.

5 Toh Tuck Link, Singapore 596224

USA office: 27 Warren Street, Suite 401-402, Hackensack, NJ 07601

UK office: 57 Shelton Street, Covent Garden, London WC2H 9HE

British Library Cataloguing-in-Publication Data
A catalogue record for this book is available from the British Library.

Advances in Logic — Vol. 5
THE FUNCTIONAL INTERPRETATION OF LOGIC DEDUCTION

Copyright © 2012 by World Scientific Publishing Co. Pte. Ltd.

ISBN-13 978-981-4360-95-1
ISBN-10 981-4360-95-3

Typeset by Stallion Press
Email: enquiries@stallionpress.com

Printed in Singapore.

To
Lucas and Pedro
and little Ira

Contents

Preface

The current volume arose out of a sequence of peer-reviewed scientific papers around a nonstandard perspective on the so-called Curry-Howard functional interpretation. Common to all those publications is the idea that the notion of proof and inference are indeed key to a theory of meaning for the language of mathematics, but not in the epistemological sense as prevalent in the so-called Dummett-Prawitz verificationist theory of meaning. Rather, the idea is to look at the role of proof and inference from a game-theoretic, or dialogue-theoretic, perspective, drawing on an interpretation of Wittgenstein's dictum that "meaning is use" where the role of (immediate) consequences of a sentence is key to fixing its meaning. Indeed, further to the connections between meaning and use, it seems useful to consider the (explanation of the immediate) consequences one is allowed to draw from a proposition as something directly related to its meaning/use. As a matter of fact, Wittgenstein's references to the connections between meaning and the consequences, as well as between use and consequences are sometimes as explicit as his celebrated 'definition' of meaning as use given in the *Investigations*. Very much in this spirit we attempt to draw on some of these references, elaborating on how an intuitive basis for the construction of a more convincing proof-theoretic semantics (than, say, assertability conditions semantics) for the mathematical language can arise out of this connection meaning/use/(explanation of the immediate) consequences.

At this point we should like to take the opportunity to thank our publishers for the way they have handled the whole process since signing of the contract, through to being patient enough to wait for the long overdue delivery of the typescript, and finally going through copy-editing with such a great sense of professionalism: Sarah Haynes and Catharina Weijman from Imperial College Press, and Steven Patt from World Scientific.

And, of course, it is never too much to express how grateful we are to our families for their patience, love and support.

Ruy J.G.B. de Queiroz
Anjolina G. de Oliveira
Dov M. Gabbay

Recife and London, August 2011

Overview

Introduction

The functional interpretation of logical connectives is concerned with a certain harmony between, on the one hand, a functional calculus on the expressions built up from the recording of the deduction steps (the *labels*), and, on the other hand, a logical calculus on the formulae. It has been associated with Curry's early discovery of the correspondence between the axioms of intuitionistic implicational logic and the type schemes of the so-called combinators of combinatory logic [Curry (1934)], and has been referred to as the *formulae-as-types* interpretation. Howard's extension of the formulae-as-types paradigm to full intuitionistic first-order predicate logic [Howard (1980)] meant that the interpretation has since been referred to as the 'Curry–Howard' functional interpretation. Although Heyting's intuitionistic logic [Heyting (1930, 1956)] did fit well into the formulae-as-types paradigm, it seems fair to say that, since Tait's intensional interpretations [Tait (1965, 1967)] of Gödel's *Dialectica* system of functionals of finite type [Gödel (1958)], there has been enough indication that the framework would also be applicable to logics beyond the realm of intuitionism. Ultimately, the foundations of a *functional* approach to formal logic are to be found in Frege's [Frege (1879, 1893, 1903)] system of 'concept writing', not in Curry, or Howard, or indeed Heyting.

In an attempt to account for some of the less declarative aspects of certain non-classical logics, in a way that those aspects could be handled directly in the object language, D. Gabbay has set up a novel research programme in his book on *Labelled Deductive Systems* [Gabbay (1994)]. The idea, which may be seen as the seeds of a more general alternative to the type-theoretic interpretation of two-dimensional logical systems (i.e. 'terms alongside formulas'), is that the declarative unit of a logical system is to be seen as a *labelled* formula '$t : A$' (read 't labels A'). From this perspective, a logical system is taken to be not simply a calculus of logical deductions on formulas, but a suitably harmonious combination of a functional calculus on the labels and a logical calculus on the formulas. A logic will then be defined according to the meta-level features of the conceptual norm that the logic is supposed to formalise: the allowable logical moves will then be 'controlled' by appropriate constraints on 'what has been done so far' (has the assumption been used at all; have the assumptions been used in a certain order; has the assumption been used more than once; etc.).

Here we wish to present a framework for studying the mathematical foundations of *Labelled Deductive Systems*. We could also regard it as an attempt at a reinterpretation of Frege's logical calculus where abstractors and functional operators work harmoniously alongside logical connectives and quantifiers. In other words, the functional interpretation (sometimes referred to as the Curry–Howard–Tait interpretation) can be viewed in a wider perspective of a labelled deductive system that can be used to study a whole range of logics, including some of which may not abide by the tenets of the intuitionistic interpretation (e.g. classical implicational logic, many-valued logics, etc.). The result is a labelled natural deduction system which we would like to see as a reinterpretation of Frege's 'functional' account of logic: it is as if the theory of functions of *Grundgesetze* is put together with the theory of predicates of *Begriffsschrift*, in such a way that a formula is true (valid) if and only if a deduction of it can be constructed where the label contains no free variable (i.e. its proof-construction is a 'complete' object, which means that the truth of the formula relies on no assumptions). The weaker the logic, the stricter are the ways by which assumptions may be withdrawn. Classical implicational logic, for example, will have a procedure for withdrawing implicational assumptions depending on the history of the deduction, which its intuitionistic counterpart will not have. So, we need to look for a paradigm for two-dimensional logical systems (terms alongside formulas) which can account for a general perspective on the harmony between withdrawal of assumptions in the logic and abstraction of variables in the functional calculus. We are beginning to find what seems to be a reasonable architecture for such a methodology underlying logical systems based on term calculi: *Grundgesetze* alongside *Begriffsschrift* [de Queiroz (1992)].

Labels and Gentzen's programme

In order to prove the *Hauptsatz*, which could not be done in the natural deduction calculi NJ and NK because of lack of symmetry in NJ and lack of elegance in NK, Gentzen went on to develop the 'logistic' calculi.

"In order enunciate and prove the *Hauptsatz* in a convenient form, I had to provide a logical calculus especially suited to the purpose. For this the natural calculus proved unsuitable. For, although it already contains the properties essential to the validity of the *Hauptsatz*, it does so only with respect to its intuitionist form, in view of the fact that the law of excluded middle, as pointed out earlier, occupies a special position in relation to these properties." [Gentzen (1935)] (Opening Section, §2)

A major improvement on Gentzen's original programme of analysis of deduction via analysis of connectives was put forward by D. Prawitz in his monograph on natural deduction [Prawitz (1965)]. The main features of Prawitz's framework can be summarised as follows:

- Definition of normalisation (i.e. the so-called 'reduction' rules) for NJ, therefore 'pushing' the *cut* principle down to the level of connectives, rather than the level of consequence relation; e.g.:

$$\cfrac{\Sigma_1 \qquad \cfrac{\begin{array}{c}[A]\\ \Sigma_2 \\ B\end{array}}{A \to B} \to \text{-}intr}{B} \to \text{-}elim \qquad \triangleright \qquad \begin{array}{c}\Sigma_1\\ [A]\\ \Sigma_2\\ B\end{array}$$

  where the Σ's (i.e. Σ_1, Σ_2) stand for whole deduction trees.
- Definition of (classical) *reductio ad absurdum*, i.e.:

$$\cfrac{\begin{array}{c}[\sim A]\\ \Lambda\end{array}}{A}$$

  where A is atomic and different from Λ, '$\sim$' stands for negation, and 'Λ' is the distinguished propositional constant for absurdity. With the addition of this rule to the intuitionistic system, Prawitz provided an inferential counterpart to Gentzen's special place for the axiom of the excluded middle.
- Proof theory is based on the *subformula principle*, which compromised the credibility of natural deduction systems (especially the full fragment, i.e. with $\vee$, $\exists$), on what concerned decision procedures.
- Little emphasis on the formulation of a proof theory for classical logic, perhaps due to the philosophical underpinnings of his [Prawitz (1977, 1980)] programme (joint with M. Dummett [Dummett (1975, 1980)]) on a language-based philosophical account of intuitionism.

Adding an extra dimension

The main features of a system of natural deduction where there is an additional dimension of labels alongside formulas can be summarised as follows:

- It is 'semantics driven': by bringing meta-level information back into the object-level, it is bringing a little of the semantics (i.e. names of individuals and dependency functions, names of possible worlds, etc.) into the proof-calculus.
- It retakes Gentzen's programme of analysis of logical deduction via an analysis of connectives (via introduction/elimination rules, and a distinction between assumptions, premises and conclusions), by introducing the extra-dimension (the *label*), which will take care of deduction in a more direct fashion than the sequent calculus. That is to say, the extra-dimension will take care of eventual dependencies among referents. In other words, the shift to the sequent calculus was motivated by the need for recovering symmetry (i.e. hypotheses and conclusions stand in opposition to each other in terms of polarity) and local control (for each inference rule, no need for side conditions or global control), but in fact the calculus can only do 'bookkeeping' on formulas, but not on individuals, dependency functions or possible worlds. The handling of *inclusive* logics, i.e. logics which also deal with empty domains, is much improved by the explicit introduction of individuals as variables of the functional calculus on the labels. Thus, a formula like '$\forall x P.(x) \rightarrow \exists x.P(x)$' is not a theorem in a labelled system, but its counterpart with the explicit domain of quantification '$D \rightarrow (\forall x^D.P(x) \rightarrow \exists x^D.P(x))$', to be interpreted as 'if the domain of quantification D is non-empty, then if for all x in D, P is true of x, then there exists an x in the domain such that P is true of it'.
- It recovers symmetry (non-existent in either Gentzen's NJ [Gentzen (1935)] or Prawitz's I [Prawitz (1965)]) by allowing a richer analysis of the properties of '$\rightarrow$', whose labels may play the rôle of *function* or *argument*.
- It recovers a 'connective-based' account of deduction, for reasons already mentioned above.
- It replaces the *subformula principle* by the *subdeduction principle*. With this proviso we shall avoid the complications introduced when the straightforward notion of *branch* is replaced either by more complex notions such as *path* or *track*. Cf. [Prawitz (1971)], where the notion of *path* is replacing the notion of *branch*, and [Troelstra and van Dalen (1988)], where the complicated notion of *track* has to be defined in order to account for the proof of the subformula property. As we shall see, the use of *subdeduction*, instead of *subformula* is especially useful for the so-called 'Skolem'-type connectives such as disjunction, existential quantifier and

propositional equality, just because their *elimination* rules may violate the *subformula* property of a deduction, while they will always respect the *subdeduction* property.

- From the properties of implication, it is easy to obtain a *generalised reductio ad absurdum* for classical positive implicational logic:

$$\frac{[x : A \to B] \\ b(x, \ldots, x) : B}{\lambda x.b(x, \ldots, x) : A} \begin{array}{|c|} \hline \text{`}A \to B\text{'} \\ \text{as minor \&} \\ \text{as ticket} \\ \hline \end{array}$$

- It incorporates the handling of first-order variables into the calculus, therefore dispensing with special conditions on *eigenvariables*.
- With the new dimension introduced, it is easier to check the connections between the proof procedures and the model-theoretic-based resolutions (e.g. Skolem's, Herbrand's) because variables for dependency functions (the Skolem functions) and for 'justification of equalities' (substitution) (the Herbrand functions) are introduced and manipulated accordingly in the functional calculus on the labels, yet without appearing in the logical calculus on the formulae.
- The *Hauptsatz* is recast in a more 'realisability'-like presentation, as in the Tait's [Tait (1965)] method: cut elimination is replaced by normalisation, i.e. main measure of redundancy is in the label.
- It recovers the 'continuation' aspect of classical logic, which, unlike Gentzen's [Gentzen (1935)] NJ or Prawitz's [Prawitz (1965)] C, the sequent calculus did manage to keep by allowing more than one formula to the right of the turnstile. Via a more careful analysis of what is at stake when 'new' branches have to be open in a proof tree whenever a Skolem-type connective is being eliminated, one observes the importance of reasoning about 'our proof so far' that we can do with a labelled system due to the added dimension. For example, the replacement of *subformula* by *subdeduction* would seem to facilitate the proof of decidability results via Gentzen-type techniques. It also makes it possible to define *validity* on the basis of *elimination* rules, something which is not easily done with plain systems. (Cf. [Prawitz (1971)] (p. 290), on the definition of validity through elimination rules breaking down the induction for the cases of disjunction and existential quantification.)
- Recovering duality by establishing that any label will either play the rôle of a *function* or that of an *argument*.
- The additional dimension is crucial in the establishment of a proof theory for equality, given that referents and dependency functions are handled

directly by the two-dimensional calculus.

- The definition of normal derivations becomes easier than the one given in [Prawitz (1971)] (II.3, p. 248), because it is to be based on the normality of the expression in the label, the latter containing the encoding of the steps taken, even if they involved Skolem-type connectives (those which may violate the subformula property, but which in a labelled system will not violate the subdeduction property).

Labels and computer programming

There are a number of features of *labelled* systems that can have significant benefits on the logical treatment of computer programming issues. Some of these features were already pointed out in P. Martin-Löf's [Martin-Löf (1982)] seminal paper, such as the connections between constructive mathematics and computer programming. It has also been pointed out that the essence of the correspondence between proof theory (i.e. logic as a deduction system) and computer programming are to be found in the so-called *conversion* rules such as those of the Curry–Howard isomorphism.[1]

Furthermore, developments on computational interpretations of logics have demonstrated that there is more to the connections between *labelled* proof theory and computer science, such as, e.g. the establishment of a logical perspective on computer programming issues like:

- iteration versus recursion
- potential infinity and lazy evaluation
- implementation of a type into another
- use of resources

---

[1]Cf.:

> "These equations [arising out of the Curry–Howard term-rewriting rules] (and the similar ones we shall have occasion to write down) are the essence of the correspondence between logic and computer science." [Girard *et al.* (1989)] (Section **Computational significance**, subsection **Interpretation of the rules**)

and:

> "The idea that a reduction (normalisation) rule can be looked at as a semantic instrument should prove a useful conceptual view that could allow the unification of techniques from theories of abstract data type specification with techniques from proof theory, constructive mathematics and λ-calculus." [de Queiroz and Maibaum (1990)] (p. 408)

- flow of control
- order of evaluation

While the first three topics were addressed in [de Queiroz and Maibaum (1990)], [de Queiroz and Maibaum (1991)], the remaining ones are dealt with in the functional interpretation (sequent-style) of linear logic given by S. Abramsky [Abramsky (1990)].

Labels and information flow

There has been a significant amount of research into the characterisation of the concept of *information flow* as a general paradigm for the semantics of logical systems. It has been said, for example, that Girard's [Girard (1987a)] *linear logic* is the *right* logic to deal with information flow.[2] It is also claimed that Barwise's [Barwise (1989)] *situation theory* is the most general mathematical formulation of the notion of information flow.

The approach generally known as the *construction-theoretic* interpretation of logic (cf. e.g. [Tait (1983)]), which underlies Gabbay's *Labelled Deductive Systems*, is more adequate for the formalisation of the concept of *information flow* and its applications to computer science. Some of the reasons for this claim could be enumerated as follows:

- It is neither a specific logic nor a specific semantical approach, but a general (unifying) framework where the integration of techniques from both proof theory and model theory is the driving force.
- It accounts for 'putting together' meta-language and object-language in a disciplined fashion.
- It can be viewed as an attempt to benefit from the devices defined in *Begriffsschrift* (i.e. connectives and quantifiers) on the one hand, and *Grundgesetze* (i.e. functional operations, abstractors, etc.) on the other hand, by having a functional calculus on the labels harmonised with a logical calculus on the formulas. In fact, by developing the basis of formal logic in terms of function and argument, Frege is to be credited as the real pioneer of the functional interpretation of logic, not Curry, Howard, or indeed Heyting.
- It is closer to the realisability interpretation than the (intuitionistic) Curry–Howard interpretation, thus giving a more general account of the paradigm 'formulas and the processes which realise them'. A formula is

---

[2]V. Pratt's contribution to 'linear-logic' e-mail list, Feb 1992.

a theorem if it can be proved with a 'complete object' (no free variable) as its label. The label can be thought of as the 'evidence' (the 'reason') for the validity of the formula. Thus, by appropriately extending the means by which one can 'close' the term labelling a formula one extends the stock of valid formulas;

Cf. [Lambek and Scott (1986)] (p. 47):

> "Logicians should note that a deductive system is concerned not just with unlabelled entailments or sequents $A \rightarrow B$ (as in Gentzen's proof theory), but with deductions or proofs of such entailments. In writing $f : A \rightarrow B$ we think of f as the 'reason' why A entails B."

In a paper which has appeared in the *JSL* [Gabbay and de Queiroz (1992)] (abstract in [Gabbay and de Queiroz (1991)]), we demonstrate how to extend the interpretation to various logics, including classical positive implication, with a generalised form of *reductio ad absurdum* involving some form of self-application in the labels. This chapter is used as the basis for section 2.1.

For a philosophical account of the generality of the construction-theoretic interpretation see e.g. Tait's 'Against Intuitionism: Constructive Mathematics is part of Classical Mathematics' [Tait (1983)] (p. 182):

> "I believe that, with certain modifications, this idea [propositions as types of their proofs] provides an account of the meaning of mathematical propositions which is adequate, not only for constructive mathematics, but for classical mathematics as well. In particular, the pseudo Platonism implicit in the truth functional account of classical mathematics is, on this view, eliminated. The distinction between constructive and classical rests solely on what principles are admitted for constructing an object of a given type."

- It is *resource aware*: disciplines of abstraction on label-variables reflect the disciplines of assumption withdrawing peculiar to the logic being considered.
- It is not limited to logics with/without Gentzen's *structural rules*, such as *contraction*, *exchange* and *weakening*, though these are indeed reflected naturally in the disciplines of abstraction. (Here we could think of structures (*constellations of labels*) other than sets, multisets.)
- It is *natural-language-friendly* in the sense that it provides a convenient way of modelling natural language phenomena (such as anaphora, 'universal' indefinites, etc. Cf. Gabbay and Kempson's [Gabbay and Kempson (1991, 1992)] work on relevance, labelled abduction and wh-construal) by the underlying device of *keeping track of proof steps*, thus accounting for dependencies. (Here it may be worth mentioning the

potential connections with H. Kamp's *Discourse Representation Theory* [Kamp and Reyle (1993)] and with K. Fine's [Fine (1985)] account of *Reasoning with Arbitrary Objects*.)

- It provides a 'natural' environment whereby the connections between model-theoretic (Skolem, Herbrand) and proof-theoretic (Gentzen, Prawitz) accounts of the theory of provability are more 'visible'. The division of tasks into two distinct (yet harmonious) dimensions, namely label-formula (i.e. functional-logical), allows the handling of 'second order' objects such as function-names to be done via the functional calculus with abstractors, thus keeping the 'logical' calculus first-order. Cf. Gabbay and de Queiroz' [de Queiroz and Gabbay (1995)] 'The functional interpretation of the existential quantifier', presented at *Logic Colloquium '91*, Uppsala.

- It opens the way to a closer connection between Lambek and Scott's [Lambek and Scott (1986)] equational interpretation of deductive systems as categories, and proof-theoretic accounts of proof equivalences. (Here we would draw attention to the potential for spelling out the connections between the unicity conditions of mappings in pullbacks, pushouts and equalisers, and the inductive rôle of η-equality for '$\wedge$', '$\vee$' and '$\rightarrow$', respectively.)

- It offers a *deductive* (as opposed to *model-theoretic*) account of the connections between modal logics and its propositional counterparts when world-variables are introduced in the functional calculus on the labels (i.e. when a little of the semantics is *brought to the syntax*, so to speak). E.g.:

$$\begin{array}{cc}
\Box\text{-}introduction & \Box\text{-}elimination \\
[\mathbb{W} : \mathcal{U}] & \\
\dfrac{F(\mathbb{W}) : A(\mathbb{W})}{\Lambda\mathbb{W}.F(\mathbb{W}) : \Box A} & \dfrac{\mathbb{T} : \mathcal{U} \qquad l : \Box A}{\mathcal{EXTR}(l, \mathbb{T}) : A(\mathbb{T})}
\end{array}$$

where '$\mathcal{U}$' would be a collection of 'worlds' (where a world can be taken to be, e.g., structured collections (lists, bags, trees, etc.) of labelled formulas) and '$F(\mathbb{W})$' is an expression which may depend on the world-variable '$\mathbb{W}$'. The conditions on $\Lambda\mathbb{W}$-abstraction will distinguish different $\Box$'s, in a way that is parallel to the distinction of various implications by conditions on λx-abstraction in:

$$\begin{array}{cc}
\rightarrow\text{-}introduction & \rightarrow\text{-}elimination \\
[x : A] & \\
\dfrac{b(x) : B}{\lambda x.b(x) : A \rightarrow B} & \dfrac{a : A \qquad f : A \rightarrow B}{APP(f, a) : B}
\end{array}.$$

- It offers a convenient framework where various notions of equality can be studied (including the λ-calculus-like β-, η- and ξ- equalities), and whose applications to the formalisation of propositional equality and the definite article, as well as a proof theory for descriptions, are of general interest.
- By incorporating means of manipulating referents and dependency functions (the objects from the 'functional' side) it provides an adequate (and logic-free) middle ground between procedural and declarative approaches to logic, where it makes sense to ask both 'what is the proof theory of model theory?' and 'what is the model theory of proof theory?'
- It offers a natural deduction based explanation for the disjunction-conjunction ambiguity, which may appear in some ordinary language interpretations of logic. The most illustrious example is Girard's [Girard (1989)] defence of a 'disjunctive conjunction' as finding its counterpart in ordinary language when 'proofs are interpreted as actions' (see example later in the section on 'resource handling').

Labels and 'constructivity as explicitation'

In a paper on a sequent calculus for classical linear logic, Girard rightly points out the intimate connections between constructivity and explicitation:

> "Constructivity should not be confused with its ideological variant 'constructiv*ism*' which tries to build a kind of *countermathematics* by an *a priori* limitation of the methods of proofs; it should not either be characterized by a list of technical properties: e.g. disjunction and existence properties. Constructivity is the possibility of extracting the information *implicit* in proofs, i.e. constructivity is about *explicitation*." [Girard (1991)] (p. 255)

Now, one of the aims of inserting a label alongside formulas (accounting for the steps made to arrive at each particular point in the deduction) is exactly that of making *explicit* the use of formulas (and instances of formulas and individuals) throughout a deduction. At this stage it may be relevant to ask how one can be more explicit than this: the functional aspect (related to names of individuals, instances of formulas, names of contexts, etc.) is handled by devices which are of a different nature and origin from the ones which handle the logical aspect, namely, connectives and quantifiers. By using labels/terms alongside formulas, we can

(1) keep track of proof steps (giving local control),

(2) handle 'arbitrary' names (via variable abstraction operators)

and our labelled natural deduction system gives us at least three advantages over the usual plain natural deduction systems:

(1) It benefits from the harmony between

- the functional calculus on terms and
- the logical calculus on formulas.

(2) It takes care of 'contexts' and 'scopes' in a more explicit fashion.
(3) Normalisation theorems may be proved via techniques from term rewriting.

As an example of how explicitation is indeed at the heart of a labelled system, let us look at how the inference rules for quantifiers are formulated:

$\forall$-*introduction*

$$\frac{\begin{array}{c}[x:D]\\ f(x):F(x)\end{array}}{\Lambda x.f(x):\forall x^D.F(x)}$$

$\forall$-*elimination*

$$\frac{a:D \qquad c:\forall x^D.F(x)}{EXTR(c,a):F(a)}$$

$\exists$-*introduction*

$$\frac{a:D \qquad f(a):F(a)}{\varepsilon x.(f(x),a):\exists x^D.F(x)}$$

$\exists$-*elimination*

$$\frac{e:\exists x^D.F(x) \qquad \dfrac{[t:D,g(t):F(t)]}{d(g,t):C}}{INST(e,\acute{g}\grave{t}d(g,t)):C}$$

Note that the individuals are explicitly introduced as labels (new variables) along-side the domain of quantification, the latter being explicitly introduced as a formula: e.g. '$a:D$', 'a' being an individual from domain 'D'.

Some of the difficulties of other systems of natural deduction can be easily overcome. For example, the handling of *inclusive* logics, cf. [Fine (1985)] (Chapter 21, page 205):

> "An *inclusive* logic is one that is meant to be correct for both empty and non-empty domains. There are certain standard difficulties in formulating a system of inclusive logic. If, for example, we have the usual rules of UI [$\forall$-*elim*], EG [$\exists$-*intr*] and conditional proof [$\rightarrow$-*intr*], then the following derivation of the theorem $\forall x F x \supset \exists x F x$ goes through (...) But the formula $\forall x F x \supset \exists x F x$ is not valid in the empty domain; the antecedent is true, while the consequent is false."

Here the difficulty of formulating a system of inclusive logic does not exist simply because the individuals are taken to be part of the calculus: recall that the labelled natural deduction presentation system is made of a functional calculus on the terms, and a logical calculus of deductions on the formulas. It requires that the names of individuals be introduced in the functional part in order for the quantifiers to be introduced and eliminated. This is not the case for plain natural deduction systems: there is no direct way to handle either terms or function symbols in a deduction without the labels. E.g., in

$$\frac{\dfrac{[\forall x.F(x)]}{F(t)}}{\dfrac{\exists x.F(x)}{\forall x.F(x) \rightarrow \exists x.F(x)}}$$

the term t is not explicitly introduced as an extra assumption, as it would be the case in the informal reading of the above deduction ('let t be an arbitrary element from the domain').

Using the functional interpretation, where the presence of terms and of the domains of quantification make the framework a much richer instrument for deduction calculi, we have

$$\frac{\dfrac{[t:D] \qquad [z:\forall x^D.F(x)]}{EXTR(z,t):F(t)}}{\dfrac{\varepsilon x.(EXTR(z,x),t):\exists x^D.F(x)}{\lambda z.\varepsilon x.(EXTR(z,x),\boxed{t}):\forall x^D.F(x) \rightarrow \exists x^D.F(x)}}$$

Here the presence of a free variable (namely 't') indicates that the assumption '[t : D]' remains to be discharged. By making the domain of quantification explicit, one does not have the antecedent (vacuously) true and the consequent trivially false in the case of empty domain: the proof of the proposition is still depending on the assumption 'let t be an element from D', i.e. that the type 'D' is presumably non-empty. To be categorical, the above proof would still have to proceed one step, as in

$$\underbrace{\frac{\dfrac{\dfrac{[t:D] \qquad [z:\forall x^D.F(x)]}{EXTR(z,t):F(t)}}{\dfrac{\varepsilon x.(EXTR(z,x),t):\exists x^D.F(x)}{\lambda z.\varepsilon x.(EXTR(z,x),\boxed{t}):\forall x^D.F(x) \rightarrow \exists x^D.F(x)}}}{\lambda t.\lambda z.\varepsilon x.(EXTR(z,x),t):D \rightarrow (\forall x^D.F(x) \rightarrow \exists x^D.F(x))}}_{\text{no free variable}}$$

Now we look at the proof-construction ('$\lambda t.\lambda z.\varepsilon x.(EXTR(z,x),t)$'). We can see no free variables, thus the corresponding proof is categorical, i.e. does not rely on any assumption.

An alternative to the explicitation of the first-order variables and their domains via labels and formulas is given in Lambek and Scott's [Lambek and Scott (1986)] definition of an intuitionistic type theory with equality. The idea is to define the consequence relation with the set of variables to be used in a derivation made explicit as a subscript to the '⊢':

"We write $\vdash$ for $\vdash_\varnothing$, that is, for $\vdash_X$ when X is the empty set. The reason for the subscript X on the entailment symbol becomes apparent when we look at the following 'proof tree':

$$\frac{\dfrac{\forall_{x \in A}\varphi(x) \vdash \forall_{x \in A}\varphi(x)}{\forall_{x \in A}\varphi(x) \vdash_x \varphi(x)} \qquad \dfrac{\exists_{x \in A}\varphi(x) \vdash \exists_{x \in A}\varphi(x)}{\varphi(x) \vdash_x \exists_{x \in A}\varphi(x)}}{\dfrac{\forall_{x \in A}\varphi(x) \vdash_x \exists_{x \in A}\varphi(x)}{\forall_{x \in A}\varphi(x) \vdash \exists_{x \in A}\varphi(x)}}$$

where the last step is justified by replacing every free occurrence of the variable x (there are none) by the closed term a of type A, *provided* there is such a closed term. Had we not insisted on the subscripts, we could have deduced this in any case, even when A is an empty type, that is, when there are no closed terms of type A." (p. 131)

Note that the introduction of a subscript to the symbol of consequence relation is rather *ad hoc* device, whereas in our case the pattern will fit within the general framework of labels and formulas, being also applicable to identify the phenomenon of *inclusiveness* in modal logics, i.e. the case of *serial* modal logics where there is always an accessible world from the current world (axiom schema '$\Box A \to \Diamond A$').

Labels, connectives, consequence relation and structures

Theorems of a certain logic are well-formed formulas which can be demonstrated to be true regardless of other formulas being true. That is, a complete proof of a theorem relies on no assumptions. So, whenever we construct proofs in natural deduction, we need to look at the rules which discharge assumptions. This is only natural, because when starting from hypotheses and arriving at a certain thesis we need to say that the hypotheses imply the thesis. So, we need to look at the rules of inference which allow us to 'discharge' assumptions (hypotheses) without introducing further assumptions. It so happens that the *introduction* rules for the conditionals (namely, implication, universal quantifier, necessity) do possess this useful feature. They allow us to 'get rid of hypotheses' by making a step from 'given the hypotheses, and arriving at premise' to 'hypotheses imply thesis'. Let us look at the introduction rules for implication in the plain natural deduction style:

$$\to\text{-}introduction$$

$$\frac{\begin{array}{c}[A]\\ B\end{array}}{A \to B}$$

Note that the hypothesis 'A' was *discharged*, and by the introduction of the implication, the conclusion '$A \to B$' (hypothesis 'A' implies thesis 'B') was reached.

Now, if we introduce labels alongside formulas this 'discharge' of hypotheses
will be reflected on the label of the intended conclusion by a device which makes
the arbitrary name introduced as the label of the corresponding assumption 'lose its
identity', so to speak. It is the device of 'abstracting' a variable from a term containing one or more 'free' occurrences of that variable. So, let us look at how the rule
given above looks like when augmented by inserting labels alongside formulas:

$$\frac{\begin{array}{c}[x : A]\\ b(x) : B\end{array}}{\lambda x.b(x) : A \to B}$$

Notice that when we reach the conclusion the arbitrary name 'x' loses its identity
simply because the abstractor 'λ' binds its free occurrences in the term '$b(x)$'
(which in its turn may have none, one or many free occurrence(s) of 'x').[3] Just
think of the more usual variable binding mechanism on the formulas: being simply
a place-marker, the 'x' has no identity whatsoever in x-quantified formulas such
as $\forall x.P(x)$ and $\exists x.P(x)$.

As we can see from the rule $\to$-*introduction* (and generally from the *introduction* rule of any conditional) the so-called 'improper' inference rules, to use a
terminology from Prawitz's [Prawitz (1965)] *Natural Deduction*, leaving *room for
manoeuvre* as to how a particular logic can be handled just by adding conditions
on the discharge of assumptions that would correspond to the particular logical
discipline one is adopting (linear, relevant, ticket entailment, intuitionistic, classical, etc.). The side conditions can be 'naturally' imposed, given that a degree of
'vagueness' is introduced by the form of those improper inference rules, such as
the rule of $\to$-*introduction*:

$$\frac{\begin{array}{c}[x : A]\\ b(x) : B\end{array}}{\lambda x.b(x) : A \to B}$$

Note that one might (as some authors do) insert an explicit sign between the assumption '$[x : A]$' and the premise of the rule, namely '$b(x) : B$', such as e.g. the
three vertical dots, making the rule look like:

$$\frac{\begin{array}{c}[x : A]\\ \vdots\\ b(x) : B\end{array}}{\lambda x.b(x) : A \to B}$$

---

[3]N.B. The notation '$b(x)$' indicates that '$b(x)$' is a functional term which depends on 'x', and not
the application of 'b' to 'x'.

to indicate the element of vagueness. There is no place, however, for the introduction of side conditions on those rules which do not allow for such a 'room for manoeuvre', namely those rules which are not improper inference rules. In his account of linear logic via a (plain) natural deduction system, Avron [Avron (1988)] introduces what we feel rather 'unnatural' side conditions in association with an inference rule which is not improper, namely the rule of $\wedge$-*introduction*.[4]

Now, while in our labelled natural deduction we explore the nuances in the interpretation of conditionals simply by a careful analysis of the properties of the connective itself, there may be cases where the discipline of assumption withdrawal is dictated from the 'outside', so to speak. In the logics which deal with exceptions, priorities, defaults, revisions, etc., there is the need to impose a structure on the collection of assumptions (hypotheses) in such a way that it disturbs as little as possible the basic properties of the connectives. It is for these logics that we need to define structured *constellations* of labelled formulas, and an appropriate proof theory for them. It is as if one needs to study the (meta-)logical sign used for consequence relation, i.e. the turnstile '$\vdash$', *as a connective*. Thus, we shall need to define the notion (and its formal counterpart) of *structural cut*, via the use of explicit data type operations (*à la* Guttag [Guttag (1977)]) over the structured collection of formulas.

The proof theory for the (meta-)logical connective '$\vdash$' which relates *structured*

---

[4]See p. 163:

$$\frac{A \qquad B}{A \wedge B}(*)$$

and condition (3) on p. 164: "For $\wedge$Int we have side condition that A and B should depend on exactly the same multiset of assumptions (condition $(*)$). Moreover, the shared hypotheses are considered as appearing *once*, although they seem to occur *twice*".

Nevertheless, in the framework of our labelled natural deduction we are still able to handle different notions of conjunction and disjunction in case we need to handle explicitly the *contexts* (structured collections of formulas). This is done by introducing names (variables) for the contexts as an extra parameter in all our *introduction* and *elimination* rules for the logical connectives, and considering those names as identifiers for structured collections (sets, multisets, lists, trees, etc.) of labelled formulas: e.g. '$A(\mathbb{S})$' would read 'an occurrence of the formula A is stored in the structure $\mathbb{S}$'. So, in the case of conjunction, for example, we would have:

$\wedge$-*introduction*

$$\frac{a_1 : A_1(\mathbb{S}) \qquad a_2 : A_2(\mathbb{T})}{\langle a_1, a_2 \rangle : (A_1 \wedge A_2)(\mathbb{S} \odot \mathbb{T})} \quad \text{(in sequent calculus: } \frac{\mathbb{S} \vdash a_1 : A_1 \qquad \mathbb{T} \vdash a_2 : A_2}{\mathbb{S} \odot \mathbb{T} \vdash \langle a_1, a_2 \rangle : A_1 \wedge A_2})$$

where the '$\odot$' operator would be compatible with the data structures represented by $\mathbb{S}$ and $\mathbb{T}$. For example, if $\mathbb{S}$ and $\mathbb{T}$ are both taken to be sets, and '$\odot$' is taken to be set union, then we have a situation which is similar to the rule of classical Gentzen's sequent calculus (augmented with labels alongside formulas). If, on the other hand, we take the structures to be multisets, and '$\odot$' multiset union, we would have Girard's linear logic and the corresponding conjunctions ($\otimes/\&$) depending on whether '$\mathbb{S}$' is distinct or identical to '$\mathbb{T}$'.

constellation of labelled formulas will be pursued in detail in a paper on a 'labelled sequent calculus'.

Labels and non-normal modal logics

The use of labels alongside formulas allows for the use of analogies to be made which may be useful in understanding certain concepts in logic where there is less declarative content, such as *relevance* and *non-normality*. Having accounted for relevance in the proof theory of relevant implication, we shall use the analogy between implication and necessity made clear when both are characterised by labelled proof calculus, in order to carry over the reasoning to obtain a proof theory of non-normal necessity.

It is common to define *non-normal* modal logics as systems of modal logics where the *necessitation* rule, i.e.:

$$\frac{\vdash A}{\vdash \Box A}$$

is not a valid inference rule. The class of *regular* modal logics is defined as non-normal modal logics where the rule of *regularity*, i.e.:

$$\frac{\vdash A \to B}{\vdash \Box A \to \Box B}$$

replaces the necessitation rule, somehow making the *necessity* weaker. In order to use modal connectives to formalise concepts like belief and knowledge, one would like to avoid pathologies arising out of the way the inference rule of necessitation, namely, that all tautologies are believed (resp. known); that from 'believed A' and 'believed $A \to B$' one infers 'believed B' (i.e. *omniscience*).

Now, notice that the expression 'provable A' (i.e. '$\vdash A$') is *vague*, but if we know *how* A was proved, we might wish to be more careful in inferring that A is *necessarily true*. One way of knowing how A was proved is by looking at the whole deduction of A. Another way is by *keeping track of proof steps* via a labelling mechanism. After all, one of the main motivations for developing a *labelled* proof theory is to bring some objects from the meta-level (i.e. names of individuals, function symbols, 'resource'-related information, names of collections of formulas, etc.) back into the proof calculus itself. It is as if we wish to make proof theory more *semantics-driven*, yet retaining the good features of Gentzen's [Gentzen (1935)] analysis of deduction.

Labelling: A new paradigm for the functional interpretation

The functional interpretation of logical connectives, the so-called Curry–Howard interpretation [Curry (1934); Howard (1980)], provides an adequate framework for the establishment of various calculi of logical inference. Being an 'enriched' system of natural deduction, in the sense that terms representing proof-constructions (the *labels*) are carried alongside formulas, it constitutes an apparatus of great usefulness in the formulation of logical calculi in an operational manner. By uncovering a certain harmony between a functional calculus on the labels and a logical calculus on the formulas, it proves to be instrumental in giving mathematical foundations for systems of logic presentation designed to handle meta-level features at the object-level via a labelling mechanism, such as, e.g. D. Gabbay's [Gabbay (1994)] *Labelled Deductive Systems*.

Here we demonstrate that the introduction of 'labels' is of great usefulness not only for the understanding of the proof-calculus itself, but also for the clarification of its connections with model-theoretic interpretations. For example, the Skolem-type procedures of introducing 'new' names (in order to eliminate the existential quantifier, for example) and 'discharging' them at the end of a deduction (i.e. concluding that the deduction could have been made without those new names previously introduced) are directly reflected in the *elimination* rules for connectives like $\lor$ and $\exists$. From this perspective one realises that for plain systems it may be true that "natural deduction is only satisfactory for $\to$, $\forall$, $\land$; the connectives $\lor$ and $\exists$ receive a very *ad hoc* treatment: the elimination rule for "$\lor$" [and "$\exists$"] is with the presence of an extraneous formula C," [Girard (1989)] (p. 34) although the situation becomes very different in the presence of labels alongside formulas. In a natural deduction system with labels alongside formulas we realise that there is nothing 'extraneous' about the arbitrary formula 'C'. For example, the rule of $\exists$-*elimination* (shown below) states that

(1) if an arbitrary formula C can be obtained from a deduction which uses both the individual's name newly introduced ('let it be t') and the assumption that it does have the required property ('let $g(t)$ be the evidence for $P(t)$', g being a newly introduced name for the way in which the evidence for $P(t)$ depends on t), then

(2) this formula C can be obtained *without* using the new names at all.

Condition (1) is reflected in the label expression alongside the formula C in the premise being required to show a free occurrence of the new names (i.e. t and g), and requirement (2) is shown to be met when the new names become bound (by the ´-abstractor) in the label expression alongside the same formula C in the conclusion.

$$\vee\text{-}elimination \qquad \frac{p : A_1 \vee A_2 \quad \begin{array}{c}[s_1 : A_1]\\ d(s_1) : C\end{array} \quad \begin{array}{c}[s_2 : A_2]\\ e(s_2) : C\end{array}}{CASE(p, \acute{s_1}d(s_1), \acute{s_2}e(s_2)) : C}$$

$$\exists\text{-}elimination \qquad \frac{e : \exists x^D.P(x) \qquad \begin{array}{c}[t : D, g(t) : P(t)]\\ d(g,t) : C\end{array}}{INST(e, \acute{g}\acute{t}d(g,t)) : C}$$

In our labelled natural deduction, the λ-abstractor shall be reserved for the constructor of label expressions associated with the connective of implication, i.e. '$\rightarrow$'. Although all abstractors (λ, ´, ε, etc.) originate in Frege's value-range technique, there is an important difference: whereas the λ-abstractor accompanies an introduction of an implication in the logical calculus, here our ´-abstractor reflect the discharge of assumptions by binding the corresponding variables, but are not accompanied by the introduction of a conditional connective on the logical side. In other words, we shall need a more general view of the device of abstraction, which may not necessarily be connected with the implication connective, but rather with the discharge of assumptions in the logical calculus.

In passing, note that the abstractors are all reminiscent of Frege's device of transforming functions into objects by forming value-range terms with the help of the notation '$\acute{\varepsilon}f(\varepsilon)$', where the sign ('´', which is the smooth breathing for Greek vowels, according to Dummett [Dummett (1991)]) plays the rôle of an abstractor [Frege (1891)].[5] (We shall come back to this point later on when we describe the rôle of the labels in our framework.)

By means of a suitable harmony between, on the one hand, a functional calculus with abstractors, and, on the other hand, a logical calculus with connectives and quantifiers, we want to show that the labelling device is more than just an extra syntactical ornament. It is a useful device to 'put back into the object language' the (lost) capability of handling names (and eventual dependencies), e.g. names of instances of formulas, names of individuals, names of collections of formulas, etc., which is kept at the meta-level in plain logic presentation systems. The lack of this capability makes it difficult (or at least 'unnatural') for the latter systems to handle a class of logics called 'resource' logics where non-declarative features such as 'how many times a formula was used to obtain another formula', 'which order certain formulas were used in order to obtain another formula as a conclusion', etc., do have a crucial rôle to play. The lack of a naming device such as the labelling mechanism, we shall contend, also obscures the connections with some fundamental theorems given by the model-theoretic interpretations.

[5] See, e.g., the partial English translation of *Grundgesetze I* [Furth (1964)]. See also Peano's [Peano (1889)] original device for functional notation.

Moreover, by using labels one can keep track of all proof steps. This feature of labelled natural deduction systems helps to recover the 'local' control virtually lost in plain natural deduction systems. The 'global' character of the so-called *improper* inference rules (i.e. those rules who do not only manipulate premises and conclusions, but also involve assumptions, such as, e.g., →-*introduction*, ∨-*elimination*, ∃-*elimination*, etc.) is made 'local' by turning the 'discharge functions' into appropriate disciplines for variable-binding via the device of 'abstractors'. As a matter of fact, the use of abstractors to make arbitrary names lose their identity is not a new device: it was already used by Frege in his *Grundgesetze I* [Frege (1893)]. (Think of the step from 'assuming $x : A$ and arriving at $b(x) : B$' to 'assert $\lambda x.b(x) : A \to B$' as making the arbitrary name 'x' lose the identity it had in $b(x)$ as a name. The binding transforms the name into a mere place-marker for arguments in substitutions.) From this perspective, critical remarks of the sort "Natural Deduction uses global rules (...) which apply to whole deductions, in contrast to rules like elimination rule for →, which apply to formulas" [Girard (1989)] (page 35) will have much less impact than they have w.r.t. 'plain' natural deduction systems.

In what follows we demonstrate how labelled natural deduction presentation systems based on our 'Fregean' functional interpretation (resulting from a generalisation of the Curry–Howard interpretation) can help us provide a more intuitive and more general alternative to most logic presentation systems in the Gentzen tradition. We shall discuss how some 'problematic' aspects of plain natural deduction can be overcome by the use of labels alongside formulas.

In particular, we shall be dealing with the following aspects of natural deduction:

- the loss of 'local' control caused by the structure of the so-called *improper* inference rules (i.e. rules which discharge assumptions);
- the lack of an appropriate treatment of η-type normalisation (an *elimination* followed by an *introduction*) for connectives other than implication; this shall point the way to providing an answer to the problem related to non-confluence of Natural Deduction systems with disjunction and η-normalisation[6]; furthermore, we shall be looking at the connections between the rôle of η-equality for ∧, ∨ and → in guaranteeing unicity in the forms of the proofs, and the unicity condition involved in the category-theoretic devices of *pullbacks*, *pushouts* and *equalisers*, respectively;
- the absence of a formal account of the rôle of *permutative* reductions with respect to the functional calculus on the labels, as well as, to some extent,

---

[6]Recently answered in [de Oliveira and de Queiroz (2005)].

with respect to the logical calculus on the formulas; we shall define these reductions as ζ-reductions;

- the lack of a generalised form of the (classical) *reductio ad absurdum*; we shall demonstrate what form the rule of *reductio ad absurdum* should take if one does not have a distinguished propositional constant (namely '$\mathcal{F}$', the *falsum*);

- the absence of a clear account of first-order quantification and the rôle of labels in predicate formulas;

- the lack of an appropriate link with classical results in proof theory obtained via model-theoretic means, such as Skolem's and Herbrand's resolution theorems;[7]

- the absence of appropriate devices for directly handling referents, dependency functions, and equality, thus making it difficult to see how natural deduction techniques can be used to handle descriptions, or even offer a reasonable account of Herbrand's decision procedure for predicate logic, as well as of unification;

- the lack of a 'natural' substitute for the semantic notion of accessibility relation, thus making it difficult to provide a generalised proof theory for modal logics.

---

[7]E.g., to the best of our knowledge, there is no *deduction*-based analysis of Leisenring's studies of the connections between ε-like calculi [Leisenring (1969)], choice principles and the resolution theorems of Skolem and Herbrand. The two 'worlds' (proof theory and model theory) are usually seen as so disjoint as to disencourage a more integrating approach. With an analysis of deduction via our labelled system, we would wish to bridge some of the gaps between these two worlds.

Chapter 1

Labelled Natural Deduction

1.1 The rôle of the labels

Unlike axiomatic systems, natural deduction proofs need not start from axioms. More interested in the *structure* of proofs, Gentzen conceived his natural deduction as a logical system which would make explicit the *structural properties of connectives*. Inference rules would be made of not simply premises and conclusions (as inference rules in Hilbert-style axiomatic systems happened to be, e.g. *modus ponens*), but also of *assumptions*. They would also be framed into a pattern of *introduction* and *elimination* rules, with a logical *principle of inversion* underlying the harmony between those two kinds of rules. Connectives which behaved like conditionals (most notably, implication, but also the universal quantifier) would have *introduction* rules doing the job of *withdrawing assumptions* in favour of the corresponding conditional statement.

As defined in the logic literature, a *valid* statement would be one which would rely on no assumptions. Now, in order to test for the validity of statements using Gentzen's method of natural deduction, one would need to make sure that, by the time a deduction of the statement was achieved, *all* assumptions had been withdrawn. Thus, whenever we were to construct proofs in natural deduction, we would need to look at the rules which could withdraw assumptions. This would only be too natural, because when starting from hypotheses and arriving at a certain thesis we need to say that the hypotheses imply the thesis. So, we need to look at the rules of inference which allow us to 'discharge' assumptions (hypotheses) without introducing further assumptions. As we mentioned above, it so happens that the *introduction* rules for the conditionals (namely, implication, universal quantifier, necessity) do indeed perform the task for us. They allow us to 'get rid of the hypotheses' by making a step to 'hypotheses imply thesis'. Let us look at the introduction rules for implication in the plain natural deduction style:

$$\rightarrow \text{-}introduction$$

$$\frac{\begin{array}{c}[A]\\ B\end{array}}{A \rightarrow B}$$

Note that the hypothesis 'A' is discharged, and by the introduction of the implication, the conclusion 'A $\rightarrow$ B' (i.e. hypothesis 'A' implies thesis 'B') is reached.

Now, if we introduce labels alongside formulas this 'discharge' of hypotheses will be reflected on the label of the intended conclusion by a device which makes the arbitrary name introduced as the label of the corresponding assumption 'lose its identity', so to speak. It is the device of 'abstracting' a variable from a term containing one or more 'free' occurrences of that variable. So, let us look at how the rule given above looks like when augmented by putting labels alongside formulas:

$$\frac{\begin{array}{c}[x:A]\\ b(x):B\end{array}}{\lambda x.b(x):A \rightarrow B}$$

Notice that now by the time we reach the conclusion the arbitrary name 'x' loses its identity because the abstractor 'λ' binds its free occurrences in the term 'b', which in its turn may have one or more free occurrences of 'x' (the notation '$b(x)$' indicates that 'b' is a functional term which depends on 'x').

The moral of the story here is that the last inference rule of any complete proof must be the introduction rule of a conditional, simply because those are the rules which do the job we want: discharging assumptions already made, without introducing any further assumptions. We are now speaking in more general terms ('conditional', rather than 'implication') because the introduction rules for the universal quantifier and the necessitation connectives, namely:

$$\forall\text{-}introduction \qquad \square\text{-}introduction$$

$$\frac{\begin{array}{c}[x:D]\\ f(x):P(x)\end{array}}{\Lambda x.f(x):\forall x^D.P(x)} \qquad \frac{\begin{array}{c}[\mathbb{W}:\mathcal{U}]\\ F(\mathbb{W}):A(\mathbb{W})\end{array}}{\Lambda \mathbb{W}.F(\mathbb{W}):\square A}$$

also discharge old assumptions without introducing new ones. (The treatment of 'necessity' — '$\square$' — by means of the functional interpretation will be sketched in Chapter 7, and is given in more detail in [de Queiroz and Gabbay (1997)].)

Our motto here is that all labelled assumptions must be discharged by the end of a deduction, and we should be able to check this very easily just by looking at the label of the intended conclusion and check if all 'arbitrary' names (labels) of hypotheses are bound by any of the available abstractors. (As we will see later on, abstractors other than 'λ' and 'Λ' will be used. We have already mentioned '', and 'ε'.) So, in a sense our proofs will be *categorical* proofs, to use a terminology from [Anderson and Belnap Jr. (1975)].[1] The connection with the realisability interpretation will be made in the following sense: *e* realises P iff *e* is a *complete* object (no free-variables). For stronger logics (e.g. classical positive logic) there will be additional ways of binding free variables of the label expression which may establish an extended harmony between the functional calculus on the labels and the logical calculus on the formulas. (Here we have in mind the *generalised reductio ad absurdum* defined in [Gabbay and de Queiroz (1992)], where a λ-abstraction binds a variable occurring as function and as argument in the label expression, and there is no introduction of an implication.)

The device of variable-binding, and the idea of having terms representing incomplete 'objects' whenever they contain free variables, were both introduced in a systematic way by Frege in his *Grundgesetze*. As early as 1893 Frege developed in his *Grundgesetze I* what can be seen as the early origins of the notions of *abstraction* and *application*, when showing techniques for transforming functions (expressions with free variables) into value-range terms (expressions with no free variables) by means of an 'introductory' operator of abstraction producing the *Werthverlauf* expression,[2] e.g. '$\acute{\varepsilon}f(\varepsilon)$', and the effect of its corresponding 'eliminatory' operator '$\cap$' on a *value-range* expression.[3]

The idea of forming *value-range* function-terms by abstracting from the corresponding free variable is in fact very useful in representing the handling of

---

[1] "A proof is *categorical* if all hypotheses in the proof have been discharged by use of $\rightarrow$-I, otherwise *hypothetical*; and *A* is a *theorem* if *A* is the last step of a categorical proof." [Anderson and Belnap Jr. (1975)] (p. 9).

[2] [Frege (1893)] (§3, p. 7), translated as *course-of-values* in [Furth (1964)] (p. 36), and *value-range* in most other translations of Frege's writings, including the translation of [Frege (1891)] (published in [McGuinness (1984)]) where the term first appeared.

[3] Cf. [Frege (1893)] (§34, p. 52ff), (translated in [Furth (1964)] (p. 92)):

"(...) it is a matter only of designating the value of the function $\Phi(\xi)$ for the argument Δ, i.e. $\Phi(\Delta)$, by means of 'Δ' and '$\acute{\varepsilon}\Phi(\varepsilon)$'. I do so in this way:

$$'\Delta \cap \acute{\varepsilon}\Phi(\varepsilon)'$$

which is to mean the same as '$\Phi(\Delta)$'."
(Note the similarity to the rule of functional *application*, where 'Δ' is the argument, '$\acute{\varepsilon}\Phi(\varepsilon)$' is the function, and '$\cap$' is the application operator 'APP'.)
Expressing how important he considered the introduction of a variable-binding device for the functional calculus (recall that the variable-binding device for the logical calculus had been introduced earlier in *Begriffsschrift*), Frege says:
"The introduction of a notation for courses-of-values [value-ranges] seems to me to be one of the most important supplementations that I have made of my *Begriffsschrift* since my first publication on this subject." [*Grundgesetze I*, §9, p. 15f.]

assumptions within a natural deduction style calculus. In particular, when the natural deduction presentation system is based on a 'labelling' mechanism the binding of free variables in the labels corresponds to the discharge of respective assumptions. In the sequel we shall be using 'abstractors' (such as 'λ' in '$\lambda x.f(x)$') to bind free-variables and discharge the assumption labelled by the corresponding variable.

We consider a formula to be a theorem of the logical system if it can be derived in a way that its corresponding 'label' contains no free variable, i.e. the deduction rests on no assumption. In other words, a formula is a theorem if it admits a categorical proof to be constructed.

1.1.1 *Dividing the tasks: A functional calculus on the labels, a logical calculus on the formula*

We have seen that the origins of variable binding mechanisms, both on the formulas of logic (the propositions) and on the expressions of the functional calculus (the terms), go back at least as far as Frege's early investigations on a 'language of concept writing'. Although the investigations concerned essentially the establishment of the basic laws of logic, for Frege the functional calculus would have the important rôle of demonstrating that arithmetic could be formalised simply by defining its most basic laws in terms of rules of the 'calculus of concept writing'. Obviously, the calculus defined in *Begriffsschrift* [Frege (1879)], in spite of its functional style, was primarily concerned with the 'logical' side, so to speak. The novel device of binding free variables, namely the universal quantifier, was applicable to propositional functions. Thus, *Grundgesetze* [Frege (1893, 1903)] was written with the intention of fulfilling the ambitious project of designing a language of concept-writing which could be useful to formalise mathematics. Additional mechanisms to handle the functional aspects of arithmetic (e.g. equality between number-expressions, functions over number-expressions, etc.) had to be incorporated. The outcome of Frege's second stage of investigations also brought pioneering techniques of formal logic, this time with respect to the handling of functions, singular terms, definite descriptions, etc. An additional mechanism of variable binding was introduced, this time to bind variables of functional expressions, i.e. expressions denoting individuals, not truth-values.

Summarising, we can see the pioneering work of Frege in its full significance if we look at the two sides of formal logic he managed to formulate a calculus for

(1) the 'logical' calculus on formulas (*Begriffsschrift*)
(2) the 'functional' calculus on terms (*Grundgesetze*)

As a pioneer in any scientific activity one is prone to leave gaps and loopholes to be later filled by others. It happened with Frege that a big loophole was discovered earlier than he would himself have expected: Russell's discovery of the antinomies of his logical notion of set was a serious challenge. There may be several ways of explaining why the resulting calculus was so much susceptible to that sort of challenge. We feel particularly inclined to think that the use of devices which were designed to handle the so-called 'objects', i.e. expressions of the functional calculus, ought to have been kept apart from, and yet harmonised with, the logical calculus on the formulas. Thus, here we may start wondering what might have been the outcome had Frege kept the two sides separate and yet harmonious.

Let us for a moment think of a connection to another system of language analysis which would seem to have some similarity in the underlying ontological assumption, with respect to the idea of dividing the logical calculus into two dimensions, i.e. functional versus logical. The semantical framework defined in Montague's [Montague (1970)] intensional logic makes use of a distinction among the semantic *types* of the objects handled by the framework, namely e, t and s, in words: *entities*, *truth-values*, and *senses*. The idea was that logic (language) was supposed to deal with objects of three kinds: names of entities, formulas denoting truth-values, and possible-worlds/contexts of use. Now, here when we say that we wish to have the bi-dimensional calculus, we are saying that the entities which are *namable* (i.e. individuals, possible-worlds, etc.) will be dealt with separately from (yet harmoniously with) the logical calculus on the formulas, by a calculus of functional expressions. Whereas the variables for individuals are handled 'naturally' in the interpretation of first-order logic with our labelled natural deduction, the introduction of variables to denote *contexts*, or *possible-worlds* (structured collection of labelled formulas), as in [de Queiroz and Gabbay (1997)], is meant to account for Montague's *senses*.

1.1.2 *Reassessing Frege's two-dimensional calculus*

In our attempt to reassess the benefits of having those two sides working together, we would like to insist on the two sides being treated separately. Thus, instead of binding variables on the formulas with the device of forming 'value-range' expressions as Frege does,[4] we shall have a clear separation of functional versus

---

[4]Cf. the following opening lines of *Grundgesetze I*, §10:

"Although we laid it down that the combination of signs '$\acute{\varepsilon}\Phi(\varepsilon) = \acute{\alpha}\Psi(\alpha)$' has the same denotation as '$\overset{a}{\smile}\!\!\!-\ \Phi(a) = \Psi(a)$', this by no means fixes completely the denotation of a name like '$\acute{\varepsilon}\Phi(\varepsilon)$'."

Note that both the abstractor ' ' and the universal quantifier '$\smile$' are used for binding free variables of formulas of the logical calculus such as 'Φ' and 'Ψ'. In our labelled natural deduction we shall take the separation 'functional versus logical' more strictly than Frege himself did. While the

logical devices. We still want to have the device of forming propositional functions, so we still need to have the names of variables of the functional calculus being carried over to take part in the formulas of the logical side. That will be dealt with accordingly when we describe what it means to have predicate formulas in a labelled system. Nevertheless, abstractors shall only bind variables occurring in expressions of the functional calculus, and quantifiers shall bind variables occurring in formulas of the logical calculus. For example, in

$$\forall\text{-}introduction \qquad\qquad \exists\text{-}introduction$$
$$[x : D]$$
$$\frac{f(x) : P(x)}{\Lambda x.f(x) : \forall x^D.P(x)} \qquad \frac{a : D \qquad f(a) : P(a)}{\varepsilon x.(f(x), a) : \exists x^D.P(x)}$$

whilst the abstractors 'Λ' and 'ε' bind variables of the functional calculus, the quantifiers '$\forall$' and '$\exists$' bind variables of the logical calculus, even if the same variable name happens to be occurring in the functional expression as well as in the logical formula.

Notice that although we are dealing with the two sides independently, the harmony seems to be maintained: to each discharge of assumption in the logical calculus there will correspond an abstraction in the functional calculus. In the case of our quantifier rules, we observe that the *introduction* of the universal quantifier is made with the arbitrary name x being bound in both sides (functional and logical) at the same time. In the existential case the 'witness' a is kept unbound in the functional calculus, whilst in the formula the binding is performed.

This is not really the place to discuss the paradoxes of Frege's formalised set theory, but it might be helpful to single out one particularly relevant facet of his 'mistake'. First, let us recall that the development of mechanisms for handling both sides of a calculus of concept writing, namely the logical and the functional, would perhaps recommend special care in the harmonising of these two sides. We all know today (thanks to the intervention of the likes of M. Furth [Furth (1964)], P. Aczel [Aczel (1980)], M. Dummett [Dummett (1973, 1991a)], H. Sluga [Sluga (1980)], and others) that one of the fundamental flaws of Frege's attempt to put the two sides together was the so-called 'Law V' of *Grundgesetze*, which did exactly what we shall avoid here in our 'functional interpretation' of logics: using functions where one should be using propositions, and vice versa. The 'Law V' was stated as follows:

$$\vdash (\acute{\varepsilon}f(\varepsilon) = \acute{\alpha}g(\alpha)) = (\overset{a}{-\!\!\cup\!\!-} f(a) = g(a))$$

---

abstractors will be used to bind variables in the functional calculus, the quantifiers will be used to bind variables in the logical calculus. Obviously, variables may be occurring in both 'sides', but in each side the appropriate mechanism will be used accordingly.

Here we have equality between terms — i.e. $\acute{\varepsilon}f(\varepsilon) = \acute{\alpha}g(\alpha)$ and $f(a) = g(a)$ — on par with equality between truth-values — i.e. the middle equality sign.

In his thorough analysis of Frege's system, Aczel makes the necessary distinction by introducing the sign for propositional equality [Aczel (1980)]:

$$(\lambda x.f(x) \doteq \lambda x.g(x)) \leftrightarrow \forall x.(f(x) \doteq g(x)) \text{ is true}$$

where '$\doteq$' stands for propositional equality, and '$\leftrightarrow$' is to mean logical equivalence (i.e. 'if and only if').[5]

Despite the challenges to his theories of formal logic, Frege's tradition has remained very strong in mathematical logic. Indeed, there is a tendency among the formalisms of mathematical logic to take the same step of 'blurring' the distinction between the functional and the logical side of formal logic. As we have already mentioned, Frege introduced in the *Grundgesetze* the device of binding variables in the functional calculus,[6] in addition to the variable-binding device presented in *Begriffsschrift*, but allowed variables occurring in the formulas to be bound not only by the quantifier(s), but also by a device of the functional calculus, namely the 'abstractors'. One testimony to the strength of Frege's legacy which is particularly relevant to our purposes here is the formalism described in Hilbert and Bernays' [Hilbert and Bernays (1934, 1939)] book where various calculi of singular terms are established. One of these calculi was called the ε-calculus, and consisted of an extension of first-order logic by adding the following axiom schema:

$$(\varepsilon_1) \qquad A(a) \rightarrow A(\varepsilon_x A(x)) \qquad \text{(for any term '}a\text{')}$$
$$(\varepsilon_2) \qquad \forall x.(A(x) \leftrightarrow B(x)) \rightarrow (\varepsilon_x A(x) = \varepsilon_x B(x))$$

where any term of the form '$\varepsilon_x A(x)$' is supposed to denote a term 't' with the property that '$A(t)$' is true, if there is one such term.

Now, observe that the addition of these new axioms has to be proven 'harmless' to the previous calculus, namely the first-order calculus with bound variables, in the sense that no formulas involving only the symbols of the language of the old calculus which was not previously a theorem, is a theorem of the new calculus.

---

[5]Later in this book we shall be dealing with the problem of handling equality on the 'logical side', so to speak: we demonstrate how to provide an analysis of deduction (Gentzen style, i.e. via rules of *introduction* and *elimination* with appropriate labelling discipline) for a proposition saying that two expressions of the functional calculus denote the same object. In order to explain the properties of this new propositional connective we will be discussing the issues of 'extensional versus intensional' approaches to equality. An analysis of propositional equality via our labelled natural deduction may serve as the basis for a proof theory for descriptions.

[6]In fact, Frege had already introduced the device which he called *Werthverlauf* in his article on 'Function and Concept' [Frege (1891)], which, in its turn, may have been inspired by Peano's functional notation [Peano (1889)].

For that one has to prove the fundamental theorems stating that the new calculus is only a 'conservative extension' the old calculus (First and Second ε-Theorems).

The picture becomes slightly different when we follow somewhat more strictly the idea of dividing, as sharply as we can, the two tasks: let all that has to do with entities to be handled by the functional calculus on the labels, and leave only what is 'strictly logical' to the logical calculus on the formulas. So, in the case of ε-terms, we shall not simply replace an existentially quantified variable in a formula (e.g. 'x' in '$\exists x.A(x)$') by an ε-term involving a formula (e.g. '$A(\varepsilon_x A(x))$'). Instead, we shall use 'ε' as an abstractor binding variables of the functional calculus, as we have seen from our rule of $\exists$-*introduction* shown previously. In other words, we do not have the axioms (or rules of inference) for the existential quantifier plus other axiom(s) for the ε-symbol. We shall be presenting the existential quantifier with its usual logical calculus on the formulas, alongside our ε-terms taking care of the 'functional' side. Therefore

$\exists$-*introduction* $\exists$-*elimination*

$$\frac{a:D \qquad f(a):P(a)}{\varepsilon x.(f(x),a) : \exists x^D.P(x)} \qquad \frac{e:\exists x^D.P(x) \qquad \begin{array}{c}[t:D, g(t):P(t)]\\ d(g,t):C\end{array}}{INST(e, \acute{g}\acute{t}d(g,t)) : C}$$

(side condition: the variables t and g must occur free in the term $d(g,t)$ labelling C).

Notice that here our concern with the 'conservative extension' shall be significantly different from the one Hilbert and Bernays had [Hilbert and Bernays (1934, 1939)]. We have the ε-symbol appearing on the label (the functional side, so to speak), and it is only introduced alongside the corresponding existential formula. (More details of our treatment of the peculiarities of the existential quantifier are given in Chapter 3, as well as in [de Queiroz and Gabbay (1995)].)

1.2   Canonical proofs and normalisation

Since Heyting's definition of each (intuitionistic) logical connective in terms of proof conditions (as opposed to the then usual truth-valuation technique), there emerged a whole tradition within mathematical logic of replacing the declarative concept of truth-functions by its procedural counterpart proof-conditions. By providing a 'language-based' (as opposed to Brouwer's language*less*) explanation of intuitionistic mathematics, Heyting put forward a serious alternative approach to the usual truth-tables-based definitions of logical connectives, which was adequate for a certain tradition in the philosophy of language and philosophy of mathematics, namely the so-called anti-realist tradition.

With the advent of Gentzen's 'mathematical theory of proofs', its corresponding classification of 'natural deduction' inference rules into *introduction* and *elimination*, and the principle (advocated by Gentzen himself) saying that the conditions under which one can assert a logical proposition (formalised by the *introduction* rules) define the meaning of its major connective,[7] an intuitionistic proof-theoretic approach to semantics was given a (meta-)mathematical status. Later, the philosophical basis of this particular approach to intuitionism via proof theory is found in Dummett and Prawitz, its main advocates. In his book on the foundations of (language-based) intuitionistic mathematics, Dummett advocates that "the meaning of each [logical] constant is to be given by specifying, for any sentence in which that constant is the main operator, what is to count as a proof of that sentence, it being assumed that we already know what is to count as a proof of any of the constituents." [Dummett (1977)] (p. 12)

A further refinement of the notion of meaning as being determined by the proof-conditions is given by P. Martin-Lof when he elegantly points out the crucial and often neglected distinction between *propositions* and *judgements*, and introduces the notion of *canonical* (or *direct*) proof. By making an attempt to formalise the basic principles of a particular strand of intuitionism called *constructive* mathematics, as practiced by, e.g. E. Bishop in *Foundations of Constructive Analysis* [Bishop (1967)], his explanations of meaning in terms of *canonical* proofs further advocates the replacement of truth-valuation-based accounts of meaning by a canonical-proof-based one.[8]

---

[7]When commenting on the rôle of the natural deduction rules of *introduction* and *elimination*, Gentzen says:

"The introductions represent, as it were, the 'definitions' of the symbols concerned, and the eliminations are no more, in the final analysis, than the consequences of these definitions." ([Gentzen (1935)], p. 80 of the English translation.)

[8]In a series of lectures entitled 'On the meanings of the logical constants and the justifications of the logical laws', Martin-Lof presents philosophical explanations concerning the distinction between propositions and judgements, as well as the connections between, on the one hand, Heyting's explanation of propositions in terms of proofs rather than truth-values, and, on the other hand, the principle that "the meaning of a proposition is determined by what it is to verify it, or what counts as a verification of it" [Martin-Lof (1985)] (p. 43). Although in those lectures the emphasis appears to be on a sort of phenomenological interpretation of the notions of proposition and judgement, in what concern the formalisation of the concepts, the explanations suggest the important rôle of the definition of direct/canonical proofs for the proof-based account of meaning. In a subsequent paper the interpretation is carried a step further, and the connections with Gentzen's claim are spelled out in a more explicit fashion:

"The intuitionists explain the notion of proposition, not by saying that a proposition is the expression of its truth conditions, but rather by saying, in Heyting's words, that a proposition expresses an expectation or an intention, and you may ask, An expectation or an intention of what? The answer is that it is an expectation or an intention of a proof of that proposition. And Kolmogorov phrased essentially the same explanation by saying that a proposition expresses a problem or task (Ger. *Aufgabe*). Soon

It does not seem unreasonable, however, to say that the use of natural deduction and the semantics of assertability conditions advocated by the intuitionists is not the only way to provide a language-based account of the meaning of logical symbols. If for nothing else, the guiding principle that assertability conditions constitute the main semantical device is clearly not in tune with the proclaimed source of inspiration for the critique of the truth-valuation approach and its replacement by explanations of proof conditions.[9] So, there appears to be room for

---

afterwards, there appeared yet another explanation, namely, the one given by Gentzen, who suggested that the introduction rules for the logical constants ought to be considered as so to say the definitions of the constants in question, that is, as what gives the constants in question their meaning. What I would like to make clear is that these four seemingly different explanations actually all amount to the same, that is, they are not only compatible with each other but they are just different ways of phrasing one and the same explanation." (.,.) "If you interpret truth conditions in this way, you see that they are identical with the introduction rules for the logical constants as formulated by Gentzen. So I have now explained why, suitably interpreted, the explanation of a proposition as the expression of its truth conditions is no different from Gentzen's explanation to the effect that the meaning of a proposition is determined by its introduction rules."

[Martin-Lof (1987)] (pp. 410, 411)

Cf. also: "The introduction rules say what are the canonical elements (and equal canonical elements) of the set, thus giving its meaning." [Martin-Lof (1984)] (p. 24)

Observe that the principle of 'meaning is determined by the assertability (proof) conditions' is unequivocally advocated. The introduction rules of Gentzen-type natural deduction are said to constitute definitions (as Gentzen himself had advocated earlier). Here one might start to wonder what class of definition (abbreviation, presentation, etc.) these introduction rules would fall into. And indeed, as it stands, the principle of 'introduction rules as definitions' does not seem to find a place in virtually any classification of definitions used in mathematical logic, even less so in, e.g. Frege's [Frege (1914)] classification of definitions into *constructive* and *analytic* [de Queiroz (1987)].

Perhaps it should be remarked here that in spite of the 'unequivocal' position expressed by P. Martin-Lof, there still seems to be room for interpretation. Cf., e.g. the following observation in M. Dummett's *The Logical Basis of Metaphysics*:

"Intuitively, Gentzen's suggestion that the introduction rules be viewed as fixing the meanings of the logical constants has no more force than the converse suggestion, that they are fixed by the elimination rules; intuitive plausibility oscillates between these opposing suggestions as we move from one logical constant to another. Per Martin-Lof has, indeed, constructed an entire meaning-theory for the language of mathematics on the basis of the assumption that it is the elimination rules that determine meaning." [Dummett (1991b)] (p. 280)

[9]Cf., e.g.:

"*We no longer explain the sense of a statement by stipulating its truth-value in terms of the truth-values of its constituents, but by stipulating when it may be asserted in terms of the conditions under which its constituents may be asserted.* The justification for this change is *how we in fact learn to use these statements*: furthermore, the notions of truth and falsity cannot be satisfactorily explained so as to form a basis for an account of meaning once we leave the realm of effectively decidable statements." [Dummett (1959)] (p. 161)

(The italics is ours.)

Cf. also:

"As pointed out by Dummett, this whole way of arguing with its stress on communication and the role of the language of mathematics is inspired by ideas of Wittgenstein and is very different from Brouwer's rather solipsistic view of mathematics as a languageless activity. Nevertheless, as it seems,

the adoption of a different one, such as, for example, instead of advocating the replacement of the truth-valuation systems by explanations of proof-conditions, one can propose to have the explanation of how the *elimination* inferences act on the result of *introduction* steps, i.e. the β-normalisation procedure, as the main semantical device. The normalisation procedure would be looked at, not merely as a meta-mathematical device introduced to prove consistency (as it is usually seen by proof-theorists), but it would be seen as the formal explanation of the 'functionality' of the corresponding logical sign. Although some other principle might be found to be more appropriate, this is, in fact, what we adopt here.

1.2.1 *Canonical proofs*

Instead of Heyting's explanation of the logical constants solely in terms of proofs (or, 'canonical' proofs as in [Martin-Lof (1984)]), the explanations given by the approach to the functional interpretation taken here involve *both* the notion of *canonical proofs* and that of *normalisation of noncanonical proofs*.

The *canonical* proofs are explained as:

a proof of the proposition:	has the canonical form of:
$A_1 \wedge A_2$	$\langle a_1, a_2 \rangle$ where a_1 is a proof of A_1 and a_2 is a proof of A_2
$A_1 \vee A_2$	$inl(a_1)$ where a_1 is a proof of A_1 or $inr(a_2)$ where a_2 is a proof of A_2 ('*inl*' and '*inr*' abbreviate 'into the left disjunct' and 'into the right disjunct', respectively)
$A \rightarrow B$	$\lambda x.b(x)$ where $b(a)$ is a proof of B provided a is a proof of A

---

it constitutes the best possible argument for some of Brouwer's conclusions. (...)

I have furthermore argued that the rejection of the platonistic theory of meaning depends, in order to be conclusive, on the development of an adequate theory of meaning along the lines suggested in the above discussion of the principles concerning meaning and use. Even if such a wittgensteinian theory did not lead to the rejection of classical logic, it would be of great interest in itself." [Prawitz (1977)] (p. 18)

We have previously endeavoured to demonstrate that the so-called semantics of use advocated by Wittgenstein did not involve simply assertability conditions, but it also accounted for the explanation of (immediate) consequences [de Queiroz (1989)]. So, as it seems, it would be unreasonable to call a theory of meaning based on *assertability conditions* a 'Wittgensteinian theory'.

$\forall x^D.P(x)$ $\Lambda x.f(x)$ where $f(a)$ is a proof of $P(a)$
                                     provided a is an arbitrary individual
                                     chosen from the domain D

$\exists x^D.P(x)$ $\varepsilon x.(f(x), a)$ where a is an individual
                                     (witness) from the domain D,
                                     and $f(a)$ is a proof of $P(a)$

As the reader can easily notice, the explanation of the logical connectives in terms of canonical proofs only cover the rules of *introduction*.[10] They constitute an explanation of the conditions under which one can form a canonical (direct) proof of the corresponding proposition. Its counterpart in Gentzen's natural deduction is the *introduction* rule, now enriched with the 'witnessing' construction which shall be handled by the functional calculus on the labels which we have mentioned above.

Thus, the corresponding formal presentations *a la* natural deduction are, e.g.:

∧-*introduction*

$$\frac{a_1 : A_1 \qquad a_2 : A_2}{\langle a_1, a_2 \rangle : A_1 \wedge A_2}$$

∨-*introduction*

$$\frac{a_1 : A_1}{inl(a_1) : A_1 \vee A_2} \qquad \frac{a_2 : A_2}{inr(a_2) : A_1 \vee A_2}$$

→-*introduction*

$$\frac{\begin{array}{c}[x : A]\\ b(x) : B\end{array}}{\lambda x.b(x) : A \to B}$$

∀-*introduction*[11]

$$\frac{\begin{array}{c}[x : D]\\ f(x) : P(x)\end{array}}{\Lambda x.f(x) : \forall x^D.P(x)}$$

---

[10]When looking at some of those informal explanations of intuitionistic connectives, such as the one via canonical proofs (or, indeed, the one via a realisability predicate given in [Kleene (1945)]), one is often tempted to see the explanations given for '∧', '→' and '∀' as covering the procedures corresponding to *elimination* rules. For the sake of the argument, however, let us stick to the usual intuitionistic account of Heyting's semantics.

[11]Note that in our formulation, where the domain over which one is quantifying is explicitly stated, the *introduction* of the universal quantifier *does* require the discharge of an assumption, namely the assumption which indicates a choice of an arbitrary individual from the domain. This account of the Universal Generalisation does not run into the difficulties related to the classification of ∀-*introduction* either as proper or as improper inference rule, because, similarly to the case of →-*introduction*, this

∃-*introduction*

$$\frac{a : D \qquad f(a) : P(a)}{\varepsilon x.(f(x), a) : \exists x^{D}.P(x)}$$

When constructing an 'ε'-term, we make an 'inverse' substitution: we replace all the occurrences of 'a' in '$f(a)$' by a new variable 'x' which is bound by the 'εx'-constructor.

1.2.2 *Normalisation*

Within the functional interpretation the operators forming the canonical proofs are usually referred to as the 'constructors' (e.g. 'λ', '$\langle,\rangle$', '*inl/inr*', 'Λ', 'ε'), whereas the eliminatory operators which form the noncanonical proofs are referred to as the 'DESTRUCTORS' [de Queiroz and Maibaum (1990)].

To recall our previous discussion, we know that, according to the intuitionistic semantics of canonical proofs, a proposition is characterised by the explanation of the conditions under which one can prove it. The procedure of exhibiting the canonical elements of a type (the canonical proofs of a proposition), which gets formalised by the *introduction* rules, is the key semantic device for the intuitionistic account of meaning. The functional interpretation, however, which accounts for the match between functional calculus on the labels and the logical calculus on the formulas, does not have to abide by the Heyting-like account of meaning. One can, for example, take the explanation of the convertibility (normalisation) relation as the key semantical device, such as it is done in Tait's intensional interpretation of Gödel's T, and the result is an account of the meaning of logical signs which does not rely on intuitionistic principles.[12]

rule requires the discharge of an assumption and would therefore have to be classified as an improper inference rule if one were to use Prawitz's [Prawitz (1965)] terminology. Cf. Fine's remark on such difficulties concerning the classification of ∀-*introduction*:

"Some of the rules require the discharge of suppositions and so have to be classified as improper. Others are so obviously proper that it seems absurd to classify them in any other way. The only real choice concerns universal generalisation (∀); this requires no discharge of suppositions and might, intuitively, be classified as either proper or improper. In considering any proposed account of validity therefore, it must be decided what status this rule is to have." [Fine (1985)] (p. 72).

[12] When explaining the doctrine that a proposition is the type of its proofs, and suggesting that this might have been implicit in Brouwer's writings, Tait says:

"Now, although I shall not have time to explain my views, I believe that, with certain modifications, this idea [propositions are types of their proofs] provides an account of the meaning of mathematical propositions which is adequate, not only for constructive mathematics, but for classical mathematics as well." [Tait (1983)] (p. 182)

This is clearly a departure from Heyting's strictly intuitionistic principles, and, as it seems, it comes as no surprise, given that Tait's semantical instrument is *convertibility* (*normalisation*), rather than canonical proofs.

An important step in the characterisation of the logical connectives which is not covered by the (language-based) intuitionistic account of meaning based on proofs is the explanation of the functionality of the logical sign in the calculus. In the formal apparatus, this means that the explanation as to how the DESTRUCTORS operate on terms built up by the constructors, i.e. the explanation of what (immediate) consequences one can draw from the corresponding proposition, does not play a major semantical rôle for the intuitionists.[13] Within the general (not just intuitionistic) functional interpretation, which we believe might have been implicit in Tait's semantics of convertibility [Tait (1965, 1967)], this aspect is given by the so-called β-normalisation rules. They have the rôle of spelling out the effect of an *elimination* inference on the result of *introduction* steps.

1.2.2.1 *β-type reductions*

The explanation of the normalisation of noncanonical proofs, i.e. those which contain 'redundant' steps identified by an *introduction* inference immediately followed by an *elimination* inference, are framed in the following way (where '$\triangleright_\beta$' represents '*β-converts/normalises to*'):

∧-reduction

$$\cfrac{\cfrac{a_1 : A_1 \qquad a_2 : A_2}{\langle a_1, a_2 \rangle : A_1 \wedge A_2} \wedge\text{-}intr}{FST(\langle a_1, a_2 \rangle) : A_1} \wedge\text{-}elim \qquad \triangleright_\beta \qquad a_1 : A_1$$

$$\cfrac{\cfrac{a_1 : A_1 \qquad a_2 : A_2}{\langle a_1, a_2 \rangle : A_1 \wedge A_2} \wedge\text{-}intr}{SND(\langle a_1, a_2 \rangle) : A_2} \wedge\text{-}elim \qquad \triangleright_\beta \qquad a_2 : A_2$$

∨-reduction

$$\cfrac{\cfrac{a_1 : A_1}{inl(a_1) : A_1 \vee A_2} \vee\text{-}intr \quad \cfrac{[x : A_1] \quad [y : A_2]}{d(x) : C \quad e(y) : C}}{CASE(inl(a_1), \acute{x}d(x), \acute{y}e(y)) : C} \vee\text{-}elim \ \triangleright_\beta \ \cfrac{a_1 : A_1}{d(a_1/x) : C}$$

$$\cfrac{\cfrac{a_2 : A_2}{inr(a_2) : A_1 \vee A_2} \vee\text{-}intr \quad \cfrac{[x : A_1] \quad [y : A_2]}{d(x) : C \quad e(y) : C}}{CASE(inr(a_2), \acute{x}d(x), \acute{y}e(y)) : C} \vee\text{-}elim \ \triangleright_\beta \ \cfrac{a_2 : A_2}{e(a_2/y) : C},$$

where '⁀' is an abstractor which forms value-range terms such as '$\acute{x}d(x)$' where 'x' is bound, discharging the corresponding assumption labelled by x.

---

[13] In an analysis of the relevant aspects of proof-theoretic semantics [de Queiroz (1991)], we have suggested that the *introduction* rules only cover one aspect, namely the grammatical (formational) aspect: they only say how to construct a proof, leaving untouched the aspect as to how to de-construct (challenge) this same proof.

→*-reduction*

$$\frac{a:A \quad \dfrac{\begin{array}{c}[x:A]\\ b(x):B\end{array}}{\lambda x.b(x):A\to B}\to\text{-}intr}{APP(\lambda x.b(x),a):B}\to\text{-}elim \quad \triangleright_\beta \quad \begin{array}{c}a:A\\ b(a/x):B\end{array}$$

∀*-reduction*

$$\frac{a:D \quad \dfrac{\begin{array}{c}[x:D]\\ f(x):P(x)\end{array}}{\Lambda x.f(x):\forall x^D.P(x)}\forall\text{-}intr}{EXTR(\Lambda x.f(x),a):P(a)}\forall\text{-}elim \quad \triangleright_\beta \quad \begin{array}{c}a:D\\ f(a/x):P(a)\end{array}$$

∃*-reduction*

$$\frac{\dfrac{a:D \quad f(a):P(a)}{\varepsilon x.(f(x),a):\exists x^D.P(x)}\exists\text{-}intr \quad \dfrac{[t:D,g(t):P(t)]}{d(g,t):C}}{INST(\varepsilon x.(f(x),a),\acute{g}\acute{t}d(g,t)):C}\exists\text{-}elim \quad \triangleright_\beta$$

$$\begin{array}{c}a:D, f(a):P(a)\\ d(f/g,a/t):C,\end{array}$$

where '´' is an abstractor which binds the free variables of the label, discharging the corresponding assumptions made in eliminating the existential quantifier, namely the 'Skolem'-type assumptions '$[t:D]$' and '$[g(t):P(t)]$', forming the value-range term '$\acute{g}\acute{t}d(g,t)$' where both the Skolem-constant 't', and the Skolem-function 'g', are bound. In the ∃-*elimination* the variables 't' and 'g' must occur free at least once in the term alongside the formula 'C' in the premise, and will be bound alongside the same formula in the conclusion of the rule.

It is useful to compare our definition of the β-normalisation rules with the original definitions given by Prawitz for plain natural deduction systems [Prawitz (1965, 1971)]. Whereas in the latter there was the need to refer to whole branches of deductions (which in the Prawitz terminology were referred to as $\Pi_1(a)$, $\Sigma(t)$, $\Sigma_2(a)$, etc.), here we only need to refer to assumptions, premises and conclusions. The relevant information on the dependency of premises from names (variables, constants, etc.) occurring in the assumptions are to be recorded in the label alongside the formula in the respective premise by whatever proof step(s) eventually made from assumptions to premises. It would seem fair to say that this constitutes an improvement on the formal presentation of proof reductions, reflecting the (re-)gain of local control by the use of labels.[14]

---

[14]Obviously, the first steps towards such improvement was already made by Martin-Lof in the definition of an intuitionistic theory of types [Martin-Lof (1971, 1975b)], but here we want to see it applicable to a wide range of logics.

1.2.2.2 β-equality

By using equality to represent the β-convertibility ('$\rhd_\beta$') relation between terms we can present the *reductions* in the following way[15]:

$\wedge$-*β-equality*

$$FST(\langle a_1, a_2 \rangle) =_\beta a_1 \qquad\qquad SND(\langle a_1, a_2 \rangle) =_\beta a_2$$

$\vee$-*β-equality*

$$CASE(inl(a_1), \acute{x}d(x), \acute{y}e(s_2)) =_\beta d(a_1/x)$$
$$CASE(inr(a_2), \acute{x}d(x), \acute{y}e(s_2)) =_\beta e(a_2/y)$$

$\rightarrow$-*β-equality*

$$APP(\lambda x.b(x), a) =_\beta b(a/x)$$

$\forall$-*β-equality*

$$EXTR(\Lambda x.f(x), a) =_\beta f(a/x)$$

$\exists$-*β-equality*

$$INST(\varepsilon x:(f(x), a), \acute{g}\acute{t}d(g, t)) =_\beta d(f/g, a/t).$$

Remark. Here it is useful to think in terms of 'DESTRUCTORS acting on constructors', especially in connection with the fact that a proof containing an *introduction* inference followed by an *elimination* step is only β-normalisable at that point if the *elimination* has as *major* premise the formula produced by the previous *introduction* step. For example, as remarked by Girard *et al.* [Girard *et al.*

---

[15]The reader may find it unusual that we are here indexing the (definitional) equality with its kind (β, η, ξ, ζ, etc.). But we shall demonstrate that it makes sense in the context of the functional interpretation to classify (and name) the equalities: one has distinct equalities according to the distinct logical equivalences on the deductions. For example, in the presentation of a set of proof rules for a certain logical connective, the second *introduction* rule is meant to show when two canonical proofs are to be taken as equal, so it is concerned with ξ-equality. The *reduction* rule shows how non-canonical expressions can be brought to normal form, so it is concerned with β-equality. Finally, the *induction* rule shows that by performing an introduction step right after an elimination inference, one gets back to the original proof (and corresponding term), thus it concerns η-equality.

As it will be pointed out later on, it is important to identify the kind of definitional equality, as well as to have a logical connective of 'propositional equality' in order to be able to reason about the functional objects (those to the left hand side of the ':' sign). The connective will have an 'existential' flavour: two referents are verified to be equal if there exists a reason (composition of rewrites) for asserting it. For example, one might wish to prove that for any two functional objects of $\rightarrow$-type, if they are equal then their application to all objects of the domain type must result in equal objects of the codomain type.

(1989)],[16] despite involving an →-*introduction* immediately followed by an →-*elimination*, the following proof fragment is not β-normalisable:

$$\frac{\dfrac{[x : A]}{\dfrac{b(x) : B}{\lambda x.b(x) : A \to B}} \to \text{-intr} \qquad c : (A \to B) \to D}{APP(c, \lambda x.b(x)) : D} \to \text{-elim.}$$

Here the major premise of the *elimination* step is not the same formula as the one produced by the *introduction* inference. Moreover, it is clear that the DESTRUCTOR '*APP*' is not acting on the term built up with the constructor '$\lambda x.b(x)$' in the previous step, but it is operating on an unanalysed term 'c'.

1.2.2.3 *η-type reductions*

In a natural deduction proof system there is another way of making 'redundant' steps that one can make, apart from the above '*introduction* followed by *elimination*'.[17] It is the exact inverse of this previous way of introducing redundancies: an *elimination* step is followed by an *introduction* step. As it turns out, the convertibility relation will be revealing another aspect of the 'propositions-are-types' paradigm, namely that there are redundant steps which from the point of view of the definition/presentation of propositions/types are saying that given any arbitrary proof/element from the proposition/type, it must be of the form given by the *introduction* rules. In other words, it must satisfy the '*introduction* followed by an *elimination*' convertibility relation. In the typed λ-calculus literature, this 'inductive' convertibility relation has been referred to as 'η'-convertibility.[18]

---

[16] Chapter Sums in Natural Deduction, Section Standard Conversions.

[17] Some standard texts in proof theory, such as Prawitz' classic survey [Prawitz (1971)], have referred to this proof transformation as 'expansions'. Here we are referring to those proof transformations as *reductions*, given that our main measuring instrument is the label, and indeed the label is reduced.

[18] The classification of those η-conversion rules as inductive rules was introduced by the methodology of defining types used in our reformulated Type Theory described in [de Queiroz and Maibaum (1990)], and first presented publicly in [de Queiroz and Smyth (1989)]. It seems to have helped to give a 'logical' status which they were given previously in the literature.

In a discussion about the Curry–Howard isomorphism and its denotational significance, Girard *et al.* say:

"Denotationally, we have the following (*primary*) equations

$$\pi^1\langle u, v \rangle = u \qquad \pi^2\langle u, v \rangle = v \qquad (\lambda x^U.v)u = v[u/x]$$

together with the *secondary* equations

$$\langle \pi^1 t, \pi^2 t \rangle = t \qquad \lambda x^U.tx = t \quad (x \text{ not free in } t)$$

which have never been given adequate status."   [Girard *et al.* (1989)] (p. 16)
Cf. also:

The '$\triangleright_\eta$'-convertibility relation then defines the *induction* rules:

∧-*induction*

$$\dfrac{\dfrac{c:A_1 \wedge A_2}{FST(c):A_1}\wedge\text{-}elim \qquad \dfrac{c:A_1 \wedge A_2}{SND(c):A_2}\wedge\text{-}elim}{\langle FST(c),SND(c)\rangle : A_1 \wedge A_2}\wedge\text{-}intr \qquad \triangleright_\eta \qquad c:A_1 \wedge A_2$$

∨-*induction*

$$\dfrac{c:A_1 \vee A_2 \quad \dfrac{[x:A_1]}{inl(x):A_1 \vee A_2}\vee\text{-}intr \quad \dfrac{[y:A_2]}{inr(y):A_1 \vee A_2}\vee\text{-}intr}{CASE(c,\hat{x}inl(x),\hat{y}inr(y)):A_1 \vee A_2}\vee\text{-}elim\,\triangleright_\eta$$

$$c:A_1 \vee A_2$$

→-*induction*

$$\dfrac{\dfrac{[x:A] \quad c:A\rightarrow B}{APP(c,x):B}\rightarrow\text{-}elim}{\lambda x.APP(c,x):A\rightarrow B}\rightarrow\text{-}intr \qquad \triangleright_\eta \qquad c:A\rightarrow B$$

where c does not depend on x.

∀-*induction*

$$\dfrac{\dfrac{[t:D] \qquad c:\forall x^D.P(x)}{EXTR(c,t):P(t)}\forall\text{-}elim}{\Lambda t.EXTR(c,t):\forall t^D.P(t)}\forall\text{-}intr \qquad \triangleright_\eta \qquad c:\forall x^D.P(x)$$

where x does not occur free in c.

∃-*induction*

$$\dfrac{c:\exists x^D.P(x) \quad \dfrac{[t:D] \qquad [g(t):P(t)]}{\varepsilon y.(g(y),t):\exists y^D.P(y)}\exists\text{-}intr}{INST(c,\hat{g}\hat{t}\varepsilon y.(g(y),t)):\exists y^D.P(y)}\exists\text{-}elim \quad \triangleright_\eta \quad c:\exists x^D.P(x)$$

"Let us note for the record the analogues of $\langle \pi^1 t, \pi^2 t\rangle \triangleright t$ and $\lambda x.tx \triangleright t$:

$$\varepsilon_{\mathsf{Emp}} t \triangleright t \qquad\qquad \delta x.(\iota^1 x)y.(\iota^2 y)t \triangleright t$$

Clearly the terms on both sides of the '$\triangleright$' are denotationally equal."   (Ibid., p. 81.)

Here 'δ' is used instead of '$CASE$', and 'ι^1'/'ι^2' are used instead of 'inl'/'inr' respectively.

Later on, when discussing the coherence semantics of the lifted sum, a reference is made to a rule which we here interpret as the induction rule for ∨-types, no mention being made of the rôle such 'equation' is to play in the proof calculus:

"Even if we are unsure how to use it, the equation

$$\delta x.(\iota^1 x)y.(\iota^2 y)t = t$$

plays a part in the implicit symmetries of the disjunction."   (Ibid., p. 97.)

By demonstrating that these kind of conversion rules have the rôle of guaranteeing minimality for the non-inductive types such as the logical connectives (not just →, ∧, ∨, but also ∀, ∃) characterised by types, we believe we have given them adequate status. (That is to say: the rules of η-reduction state that any proof of $A \rightarrow B$, $A \wedge B$, $A \vee B$, will have in each case a unique form, namely $\lambda x.y$, $\langle a, b\rangle$, $inl(a)/inr(b)$, resp.)

In the terminology of [Prawitz (1971)], these rules (with the conversion going from right to left) are called *immediate expansions*. Notice, however, that whilst in the latter the purpose was to bring a derivation in full normal form to expanded normal form where all the minima formulas are atomic, here we are still speaking in terms of *reductions*: the large terms alongside the formulas resulting from the derivation on the left are reduced to the smaller terms alongside the formula on the right. Moreover, the benefit of this change of emphasis is worth pointing out here: whereas in the Prawitz plain natural deduction the principal measure is the degree of formulas (i.e. minimal formulas, etc.) here the labels (or proof constructions) take over the main rôle of measuring instrument. The immediate consequence of this shift of emphasis is the replacement of the notion of *subformula* by that of *subdeduction*, which not only avoids the complications of proving the subformula property for logics with 'Skolem-type' connectives (i.e. those connectives whose *elimination* rules may violate the subformula property of a deduction, such as $\vee$, $\exists$, $\doteq$), but it also seems to retake Gentzen's analysis of deduction in its more general sense. That is to say, the emphasis is put back into the deductive properties of the logical connectives, rather than on the truth of the constituent formulas.

Remark. Notice that the mere condition of '*elimination* followed by *introduction*' is not sufficient to allow us to perform an η-conversion. We still need to take into consideration what subdeductions we are dealing with. For example, in:

$$\frac{c : A \vee A \quad \dfrac{[x : A]}{inl(x) : A \vee B} \vee\text{-intr} \quad \dfrac{[y : A]}{inl(y) : A \vee B} \vee\text{-intr}}{CASE(c, \hat{x}inl(x), \hat{y}inl(y)) : A \vee B} \vee\text{-elim} \not\triangleright_\eta$$

$$c : A \vee A$$

we have a case where an $\vee$-*elimination* is immediately followed by an $\vee$-*introduction*, and yet we are not prepared to accept the proof transformation under η-conversion. Now, if we analyse the subdeductions (via the labels), we observe that

$$CASE(c, \hat{x}inl(x), \hat{y}inl(y)) \neq_\eta c$$

therefore, if the harmony between the functional calculus on the labels and the logical calculus on the formulas is to be maintained, we have good enough reasons to reject the unwanted proof transformation.

1.2.2.4 η-equality

In terms of rewriting systems where '=' is used to represent the 'reduces to' relation, indexed by its kind, i.e. β-, η-, ξ-, ζ-, etc., conversion, the above *induction* rules become:

$\wedge$-η-*equality*

$$\langle FST(c), SND(c) \rangle =_\eta c$$

$\vee$-η-*equality*

$$CASE(c, \hat{x} inl(x), \hat{y} inr(y)) =_\eta c$$

$\rightarrow$-η-*equality*

$$\lambda x. APP(c, x) =_\eta c$$

provided x does not occur free in c.

$\forall$-η-*equality*

$$\Lambda t. EXTR(c, t) =_\eta c$$

provided c has no free occurrences of x.

$\exists$-η-*equality*

$$INST(c, \hat{g} \hat{t} \varepsilon y.(g(y), t)) =_\eta c.$$

The presentation taken by each of the rules above does indeed reveal an 'inductive' character: they all seem to be saying that if any arbitrary element 'c' is in the type then it must be reducible to itself via an *elimination* step with the DESTRUCTOR(s) followed by an *introduction* step with the constructor(s).

1.2.2.5 ζ-type reductions: The permutative reductions turned unidirectional

For the connectives that make use of 'Skolem'-type procedures of opening lo-cal branches with new assumptions, locally introducing new names and making them 'disappear' (or lose their identity via an abstraction) just before coming out of the local context or scope, there is another way of transforming proofs, which goes hand-in-hand with the properties of 'value-range' terms resulting from abstractions.

In the literature these proof transformations are called 'permutative' reduc-tions because there is no *preferred* deduction: either one is considered to be as near to the normal form as the other. Nevertheless, if we impose an order between the two, and say that the more the 'extraneous' formulas of the rule of $\vee$-*elimination* are pushed upwards the better is the deduction with respect to approaching the normal form. Thus:

∨-*(permutative) reduction*

$$
\frac{\begin{array}{ccc} & [s_1 : A_1] & [s_2 : A_2] \\ p : A_1 \vee A_2 & d(s_1) : C & e(s_2) : C \\ \hline CASE(p, \acute{s_1}d(s_1), \acute{s_2}e(s_2)) : C \end{array}}{w(CASE(p, \acute{s_1}d(s_1), \acute{s_2}e(s_2))) : W} \quad \triangleright\varsigma
$$

$$
\frac{\begin{array}{ccc} & [s_1 : A_1] & [s_2 : A_2] \\ & d(s_1) : C & e(s_2) : C \\ p : A_1 \vee A_2 & w(d(s_1)) : W & w(e(s_2)) : W \\ \hline CASE(p, \acute{s_1}w(d(s_1)), \acute{s_2}w(e(s_2))) : W \end{array}}{}
$$

∃-*(permutative) reduction*

$$
\frac{\begin{array}{cc} & [t : D, g(t) : P(t)] \\ e : \exists x^D.P(x) & d(g,t) : C \\ \hline INST(e, \acute{g}\acute{t}d(g,t)) : C \end{array}}{w(INST(e, \acute{g}\acute{t}d(g,t))) : W} \quad \triangleright\varsigma
$$

$$
\frac{\begin{array}{cc} & [t : D, g(t) : P(t)] \\ & d(g,t) : C \\ e : \exists x^D.P(x) & w(d(g,t)) : W \\ \hline INST(e, \acute{g}\acute{t}w(d(g,t))) : W \end{array}}{}
$$

All this means that the deduction on the left is further away from its normal form than the one on the right of the '▷' sign.

Missing conversions. Of course this creates a problem for the uniqueness of normal forms, as A. de Oliveira has shown [de Oliveira (1995)].

As we have demonstrated [de Oliveira and de Queiroz (1995)], by proving the *termination* and *confluence* properties for the *term rewriting system* associated to the *LND* system (*TRS-LND*) [de Oliveira and de Queiroz (2005)], we have in fact proved the normalization and strong normalization theorems for the *LND* system, respectively. The *termination* property guarantees the existence of a normal form of the *LND*-terms, while the *confluence* property is its uniqueness. Thus, because of the Curry–Howard isomorphism, we have that every *LND* derivation converts to a normal form and it is unique.

The significance of applying this technique in the proof of the normalization theorems lies in the presentation of a simple and computational method, which allowed the discovery of a new basic set of transformations between proofs, which we baptized as "ι (iota)-reductions" [de Oliveira (1995)]. With this result, we obtained a confluent system which contains the η-reductions. Traditionally, the η-reductions have not been given an adequate status, as rightly pointed out by Girard in [Girard *et al.* (1989)] (p. 16), when he defines the *primary* equations, which correspond to the β-equations and the *secondary* equations, which are the η-equations. Girard says that the system given by these equations is consistent and decidable, however he notes the following:

> "Although this result holds for the whole set of equations, one only ever considers the first three. It is a consequence of the *Church–Rosser property* and the *normalization theorem* (...)" [Girard *et al.* (1989)]

The first three equations, referred to by Girard, are the *primary* ones, i.e. β-equations.

Applying the so-called *completion procedure*, proposed by Knuth and Bendix in [Knuth and Bendix (1970)], to *TRS-LND*, the following term, which causes a non-confluence in the system, is produced (i.e. a divergent critical pair is generated):

$$w(CASE(c, \hat{x}inl(x), \hat{y}inr(y))).$$

This term can be rewritten in two different ways[19]:

1. $\triangleright_\eta w(c)$ 2. $\triangleright_\zeta CASE(c, \hat{x}w(inl(x)), \hat{y}w(inr(y)))$

The method of Knuth and Bendix says that when a terminating system is not confluent it is possible to add rules in such a way that the resulting system becomes confluent. Thus applying this procedure to *TRS-LND*, a new rule is added to the system:

$$CASE(c, \hat{x}w(inl(x)), \hat{y}w(inr(y))) \triangleright_\iota w(c).$$

Since terms represent proof-constructions in the *LND* system, this rule defines a new transformation between proofs:

ι-reduction-∨

$$\cfrac{c : A_1 \vee A_2 \quad \cfrac{\cfrac{[x : A_1]}{inl(x) : A_1 \vee A_2} \vee \text{-}intr}{w(inl(x)) : W}r \quad \cfrac{\cfrac{[y : A_2]}{inr(y) : A_1 \vee A_2} \vee \text{-}intr}{w(inr(y)) : W}r}{CASE(c, \hat{x}w(inl(x)), \hat{y}w(inr(y))) : W} \vee \text{-}elim \; \triangleright_\iota \quad \cfrac{c : A_1 \vee A_2}{w(c) : W}r.$$

---

[19]The η-reduction for the ∨ connective is framed as follows:

$$\cfrac{c : A_1 \vee A_2 \quad \cfrac{[x : A_1]}{inl(x) : A_1 \vee A_2} \vee \text{-}intr \quad \cfrac{[y : A_2]}{inr(y) : A_1 \vee A_2} \vee \text{-}intr}{CASE(c, \hat{x}inl(x), \hat{y}inr(y)) : A_1 \vee A_2} \vee \text{-}elim \; \triangleright_\eta \; c : A_1 \vee A_2$$

('ˆ' is an abstractor, similarly to 'λ')
and the ζ-reduction is defined as follows:

$$\cfrac{\cfrac{p : A_1 \vee A_2 \quad [s_1 : A_1] \quad [s_2 : A_2]}{CASE(p, \hat{s_1}d(s_1), \hat{s_2}e(s_2)) : C}}{w(CASE(p, \hat{s_1}d(s_1), \hat{s_2}e(s_2))) : W}r \; \triangleright_\zeta \; \cfrac{p : A_1 \vee A_2 \quad \cfrac{[s_1 : A_1]}{\cfrac{d(s_1) : C}{w(d(s_1)) : W}r} \quad \cfrac{[s_2 : A_2]}{\cfrac{e(s_2) : C}{w(e(s_2)) : W}r}}{CASE(p, \hat{s_1}w(d(s_1)), \hat{s_2}w(e(s_2))) : W}.$$

Whilst 'r' is usually restricted to an *elimination* rule, we have relaxed this condition: it is only required that 'r' does not discharge any assumptions from the other (independent) branch, i.e. that the auxiliary branches do not interfere with the main branch.

Similarly, the *iota* reduction for the existential quantifier is defined [de Oliveira (1995)], since, similarly to $\vee$, the quantifier $\exists$ is a "Skolem type" connective (i.e. in the *elimination* inference for this type of connective is necessary to open local assumptions):

ι-reduction-$\exists$

$$\dfrac{c:\exists x^D.P(x) \qquad \dfrac{[t:D] \quad [g(t):P(t)]}{\dfrac{\varepsilon y.(g(y),t):\exists y^D P(y)}{w(\varepsilon y.(g(y),t)):W}}\exists\text{-intr}}{INST(c,\acute{g}tw(\varepsilon y.(g(y),t))):W}\exists\text{-elim} \quad \triangleright_\iota \quad \dfrac{c:\exists x^D P(x)}{w(c):W}\mathsf{r}$$

(where 'ε' is an abstractor).

With this result, we believe that we have answered the question as to why the η-reductions are not considered in the proofs of the normalization theorems (*confluence* requires ι-reductions). However, by applying a computational and well-defined method, the completion procedure, it seems that this problem of the non-confluence caused by η-reductions are solved.

Theorem 1.1 ([de Oliveira (1995)]). *Every proof in the* LND *system has a unique normal form.*[20]

1.2.2.6 ζ-type equality

Now, if the functional calculus on the labels is to match the logical calculus on the formulas, we must have the following ζ-equality (read 'zeta'-equality) between terms:

$$w(CASE(p,\acute{s}_1 d(s_1),\acute{s}_2 e(s_2)),u) =_\zeta$$
$$CASE(p,\acute{s}_1 w(d(s_1),u),\acute{s}_2 w(e(s_2),u))^{[21]}$$

---

[20] In fact, when we have disjunction and η-rules, uniqueness is not guaranteed. It is necessary to add new transformation rules baptised in [de Oliveira and de Queiroz (2005); de Oliveira (1995)] as 'ι' ('iota') rules.

[21] When defining 'Linearised sum', Girard *et al.* [Girard *et al.* (1989)] give the following equation as the term-equality counterpart to the permutative reduction:

"Finally, the commuting conversions are of the form

$$\mathsf{E}(\delta\, x.u\, y.v\, t) \triangleright \delta\, x.(\mathsf{E}u)\, y.(\mathsf{E}v)\, t$$

where E is an elimination."   [Girard *et al.* (1989)] (p. 103)

Note the restriction on the step corresponding to the operator 'E' (which corresponds to our 'w'): it has to be an elimination.

In our ζ-equality the operator 'w' does not have to be an eliminatory operator, but it only needs to be such that it preserves the dependencies of the term coming from the main branch, namely the step must preserve the free variables on which our 'p' depends. In other words, w cannot be an abstraction over free variables of p.

Our generalised ζ-equality also finds parallels in the literature on equational counterparts to commutative diagrams of category theory. For example, in the definition of *binary sums* given by A. Poigne

for disjunction, and

$$w(INST(e, \acute{g}\acute{t}d(g,t)), u) =_\zeta INST(e, \acute{g}\acute{t}w(d(g,t),u))$$

for the existential quantifier.

Note that both in the case of '∨' and '∃' the operator 'w' could be 'pushed inside' the value-range abstraction terms. In the case of disjunction, the operator could be pushed inside the '´'-abstraction terms, and in the ∃-case, the 'w' could be pushed inside the '´'-abstraction term.

In terms of the proof theory, these reductions imply that the newly opened branches must be independent from the main branch. And, indeed, notice that in the proof-trees above, the step coming after the *elimination* of the connective concerned (∨-, ∃-*elimination*) is taken to be as general as possible, provided that it does not affect the dependencies on the main branch (i.e. '$p : A_1 \vee A_2$', '$e : \exists x^D P(x)$', respectively). (E.g. any deduction step involving discharge of assumptions may disturb the dependencies.) Those reductions will then uncover β-type redundancies which may be hidden by an ∨-, ∃- *elimination* rule. Perhaps for this reason, in the literature it is common to restrict that particular step to a deduction to an *elimination* rule where the formula 'C' is to be its major premise.[22]

The restriction to the case when the step is an *elimination* rule seems to be connected with the idea that the *permutative* conversions are brought in to help recover the so-called *subformula property*.[23] We would prefer to see the rôle of those

---

[Poigné (1992)], the counterpart of our ζ-equality for disjunction appears as:

$$h \circ case(f, g) = case(h \circ f, h \circ g)$$

where '∘' is the basic operation of composition. Note that the function 'h' can be pushed inside the '*case*'-term, similarly to our ζ-equality where the 'w' can be pushed inside the '´-abstraction terms of our $CASE$-expression.

[22] When commenting on the requirements of permutative reductions, Prawitz remarks:

"It has been remarked by Martin-Löf that it is only necessary to require in the ∨E- and ∃E-reductions that the lowest occurrence of C is the major premiss of an elimination. A reduction of this kind can then always be carried out and we can sharpen the requirements as to the normal form accordingly." [Prawitz (1971)] (p. 253ff)

And, indeed, for his proof of the strong validity lemma (p. 295) Prawitz needs the condition on the permutative reductions that the step after the ∨-*elimination* (resp. ∃-*elimination*) be also an *elimination* inference.

No restriction to an *elimination* step is mentioned in [Martin-Löf (1975a)]. Rather, it is required that the dependencies be preserved:

"(...) the *permutative* rules for ∨ and ∃, (...) provided the inference from C to D neither binds any free variable nor discharges any assumption in the derivation of $A \vee B$ and $(\exists x)B[x]$, respectively." [Martin-Löf (1975a)] (p. 100f)

Cf. also other standard texts in the literature where the restriction is unnecessarily imposed: Troelstra and van Dalen's [Troelstra and van Dalen (1988)] (p. 534ff) and Girard *et al.*'s [Girard *et al.* (1989)] definitions of *permutative conversions* have the requirement that the step following the ∨- (∃-)*elimination* be an '*E*-rule' (*Elimination* rule).

[23] Cf. [Girard *et al.* (1989); Troelstra and van Dalen (1988)].

rules of proof transformation as that of guaranteeing a 'pact of non-interference' between the main branch and those new branches created by the elimination rules of 'Skolem-type' connectives ($\vee$, $\exists$, $\doteq$).

Thus, in the more general case, it seems as though the restriction (to the case where the formula 'C' is a major premise of an *elimination* inference) is unnecessary. And this is because we can have the following conversion using an *introduction* inference instead:

$$
\cfrac{\cfrac{p : A_1 \vee A_2 \quad \cfrac{[x : A_1]}{d(x) : C} \quad \cfrac{[y : A_2]}{e(y) : C}}{CASE(p, \acute{x}d(x), \acute{y}e(y)) : C}}{inl(CASE(p, \acute{x}d(x), \acute{y}e(y))) : C \vee U} (*) \qquad \rhd_\zeta
$$

$$
\cfrac{p : A_1 \vee A_2 \quad \cfrac{\cfrac{[x : A_1]}{d(x) : C}}{inl(d(x)) : C \vee U} \quad \cfrac{\cfrac{[y : A_2]}{e(y) : C}}{inl(e(y)) : C \vee U}}{CASE(p, \acute{x}inl(d(x)), \acute{y}inl(e(y))) : C \vee U}.
$$

One can readily notice that the $\vee$-*introduction* step marked '$(*)$' does not affect the dependencies (i.e. does not involve any assumption discharge), so the constructor '*inl*' can be pushed inside the ´-abstraction terms. The same holds if, instead of $\vee$-*introduction*, one performs an $\wedge$-*introduction* as in:

$$
\cfrac{\cfrac{p : A_1 \vee A_2 \quad \cfrac{[x : A_1]}{d(x) : C} \quad \cfrac{[y : A_2]}{e(y) : C}}{CASE(p, \acute{x}d(x), \acute{y}e(y)) : C} \quad u : U}{\langle CASE(p, \acute{x}d(x), \acute{y}e(y)), u \rangle : C \wedge U} \qquad \rhd_\zeta
$$

$$
\cfrac{p : A_1 \vee A_2 \quad \cfrac{\cfrac{[x : A_1]}{d(x) : C} \quad u : U}{\langle d(x), u \rangle : C \wedge U} \quad \cfrac{\cfrac{[y : A_2]}{e(y) : C} \quad u : U}{\langle e(y), u \rangle : C \wedge U}}{CASE(p, \acute{x}\langle d(x), u \rangle, \acute{y}\langle e(y), u \rangle) : C \wedge U}
$$

and, clearly:

$$
\langle CASE(p, \acute{x}d(x), \acute{y}e(y)), u \rangle =_\zeta CASE(p, \acute{x}\langle d(x), u \rangle, \acute{y}\langle e(y), u \rangle),
$$

given that the pairing operation can be pushed inside the ´-abstraction terms without disturbing the dependencies. One can readily see that the $\wedge$-*introduction* is harmless with respect to the dependencies. (Note that the same observation applies to ζ-reduction of $\exists$, and, as we shall see later on, to the permutative reduction of $\doteq$.)

Chapter 2

The Functional Interpretation
of Implication

2.1 Introduction

The so-called Curry–Howard[1] interpretation [Curry (1934); Curry and Feys (1958); Howard (1980); Tait (1965)] is known to provide a rather neat

---

[1]We would rather call it the 'Curry–Howard–*Tait*' interpretation, given that it is Tait's intensional semantics based on convertibility which allows flexibility in obtaining a functional interpretation for many other logics including those which do not abide by the tenets of intuitionism. In the opening lines of his seminal paper on a proof of Gentzen's result via Gödel's functional interpretation modified by taking convertibility as the key semantical notion, Tait says:

"Instead of assuming functionals of higher type, we may regard the *definitional schemata* for the p.r. functionals simply as *rules of computation*, i.e. for transforming symbols. On this view, Gödel's result may be interpreted as a consistency proof relative to the quantifier-free theory of p.r. functionals (his system *T*), which in turn must be justified by a proof that all the constant numerical terms of the theory can be transformed by the rules of computation into unique numerals." ([Tait (1965)] (p. 176). Our *emphasis*.)

Note that Tait's intensional interpretation is mainly concerned with rules of computation as definitional schemata, perhaps even before the obvious concern of 'uniqueness of convertibility as consistency'. In other words, in Tait's intensional interpretation the characterisation (definition) of each type is given by the rules of conversion (normalisation), rather than by its canonical terms. It uses a semantical principle which is very different from Heyting-like 'a proposition is defined by laying down what counts as a (canonical) proof of the proposition'. (Cf. [Martin-Löf (1984)] (p. 24): "The introduction rules say what are the canonical elements (and equal canonical elements) of the set [type], thus giving its meaning,") A semantical principle concerned primarily with rules of *normalisation* as rules which say 'how the logical sign functions', as it is advocated in [de Queiroz (1988)], would seem to be more faithful to Tait's intensional interpretation than the principle concerned with what counts as a (canonical) proof of the corresponding proposition. Tait's intensional interpretation has in fact given the so-called functional interpretation a degree of freedom such that it is not restricted to the intuitionistic case, as it seems to be the case for the Curry–Howard interpretation as originally formulated.

Indeed, in another paper, when explaining the idea that a proposition can be seen as the type of its proofs, and suggesting that this might have been implicit in Brouwer's writings, Tait says:

"Now, although I shall not have time to explain my views, I believe that, with certain modifications, this idea [propositions are types] provides an account of the meaning of mathematical propositions which is adequate, not only for constructive mathematics, but for classical mathematics as well." [Tait (1983)] (p. 182)

27

term-functional account of intuitionistic implication. Could one refine the interpretation to obtain an almost as good account of other neighbouring implications,
including the so-called 'resource' implications (e.g. linear, relevant, etc.)?

We answer this question positively by demonstrating that just by working with
side conditions on the rule of *assertability conditions* for the connective representing implication ('$\rightarrow$') one can characterise those 'resource' logics. The idea stems
from the realisation that whereas the *elimination* rule for conditionals (of which
implication is a particular case) remains virtually unchanged no matter what kind
of conditional one has (i.e. linear, relevant, intuitionistic, classical, etc., all have
modus ponens), the corresponding *introduction* rule carries an element of vagueness which can be explored in the characterisation of several sorts of conditionals.
The rule of $\rightarrow$-*introduction* is classified as an 'improper' inference rules, to use a
terminology from [Prawitz (1965)]. Now, the so-called improper rules leave room
for manoeuvre as to how a particular logic can be obtained just by imposing conditions on the discharge of assumptions that would correspond to the particular
logical discipline one is adopting (linear, relevant, ticket entailment, intuitionistic,
classical, etc.). The side conditions can be 'naturally' imposed, given that a degree
of 'vagueness' is introduced by the presentation of those improper inference rules,
such as the rule of $\rightarrow$-*introduction*,

$$\frac{\begin{array}{c}[A]\\B\end{array}}{A \rightarrow B},$$

which states that starting from assumption 'A', and arriving at 'B' via an unspecified number of steps, one can discharge the assumption and conclude that 'A'
implies 'B'.

Note that one might (as some authors do) insert an explicit indication between
the assumption '$[A]$' and the premise of the rule, namely 'B', such as for example
the three vertical dots, making the rule look like:

$$\frac{\begin{array}{c}[A]\\\vdots\\B\end{array}}{A \rightarrow B}$$

drawing attention to the element of vagueness. The more specific we wish to be
about what the three dots ought to mean, the more precise we will be with respect
to what kind of implication we shall be dealing with.

This is clearly a departure from Heyting's strictly intuitionistic principles, and, as it seems, it comes
as no surprise, given that Tait's semantical instrument is *convertibility* (*normalisation*), rather than
assertability conditions.

In the Curry–Howard interpretation, where one of the guiding principles is the matching of the functional calculus on the terms with the logical calculus on the formulas, such a rule involves a λ-*abstraction*. That is, the *introduction* rule displayed above becomes

$$\frac{\begin{array}{c} [x : A] \\ \vdots \\ b(x) : B \end{array}}{\lambda x.b(x) : A \to B}$$

according to which starting from 'A' labelled by 'x' and, after a certain number of steps, arriving at 'B' labelled by a term which may contain 'x' as a subterm, we can conclude that 'A implies B' is labelled by that same term where the occurrences of x are bound.

Here one also finds a considerable amount of vagueness in the abstraction discipline: for example, when abstracting 'x' from 'b' to make '$\lambda x.b$', 'b' could have one, many, or even no free occurrences of 'x' (occurring as function, as argument, as both, etc.). Logic presentation systems based on the functional interpretation are particularly useful in handling a not-quite-declarative feature of resource logics such as linear logic and relevant logic, namely the special requirement saying that in order to allow '$A \to B$' to be derived as a theorem the assumption 'A' must be used in order to obtain 'B'. This is because they are based on the identification of propositions with types and of proofs/constructions with elements, thus allowing the manipulation of proofs/constructions in the object language. In a proof system this means that one can 'keep track of proof steps', so to speak, because the terms/constructions carry the whole history of the deduction.

There is another aspect to the Curry–Howard interpretation which is worth pointing out: it underlies the identification of propositions with types as in Gödel's T [Gödel (1958)], as well as the adoption of conversion (i.e. normalisation) as the key semantical notion, as in Tait's intensional interpretation [Tait (1965, 1967)]. Based on an extension of the interpretation, obtained by introducing modified λ-abstractions, we provide a classification of different systems of propositional implication (concatenation,[2] **W**,[3] ticket entailment, linear, relevant, entailment, strict, minimal,[4] intuitionistic, classical). The classification works by establishing which axioms of the implicational calculus are allowed to be derived by means of the inference rules of '$\to$'-type, subject to side conditions on the rule of $\to$-

---

[2] A case study in Gabbay (1989).

[3] This system of implication is called '**BB′I**' in [Komori (1990)], and '**T$\to$ $-$ W**' (i.e. '**T$\to$**' minus *contraction*) in [Anderson and Belnap Jr. (1975)].

[4] In the sense of [Johansson (1936)].

introduction.[5] As each axiom corresponds to the type-scheme[6] of a stratified pure term of combinatory logic [Curry (1934); Curry and Feys (1958); Hindley and Seldin (1986)], we can classify combinators through systems of implication and vice-versa.[7] For example, I, B, B' and C are linear, whilst S is not linear but is relevant (indeed, the derivation of the axiom corresponding to the type-scheme of S involves a multiple and branching assumption discharge); K is not relevant but is (minimal) intuitionistic (the derivation of the axiom for K involves a non-relevant/vacuous assumption discharge); etc. By treating the proposition '$A \to B$' as an $\to$-*type* of λ-terms, we shall demonstrate how to formalise re-source logics with the framework of the functional interpretation by working on side conditions on the λ-calculus *abstraction* rule ($\to$-*introduction*). We shall exhibit derivations of desired axioms by identifying the combinators (from combinatory logic) to whose type-scheme they correspond, as well as conditions to invalidate derivations of undesired axioms for each particular system of implication weaker than intuitionistic implication.[8] So, by restricting λ-abstractions we

---

[5]The side conditions on the discharge of assumptions will be guided by the distinguishing characteristics of each implication. Our 'raw data', so to speak, is made by the λ-abstraction rule rather than by the axioms. So, e.g. instead of saying that "it seems plausible to consider the following axiomatic system as capturing the notion of relevance ..." [Anderson and Belnap Jr. (1975)] (p. 20), we shall rather say that 'it seems reasonable to adopt non-vacuous assumption discharge (and, therefore, non-vacuous λ-abstractions) for relevant implication'.

[6]For an elegant presentation of the notion of type-schemes (and 'principal type-schemes'), including its relevance to the formulae-as-types interpretation, see Chapter 14 of [Hindley and Seldin (1986)].

[7]We should like to thank Dr Kosta Došen for pointing out that similar work on classifying subsystems of implication was done by Y. Komori [Komori (1983, 1989, 1990)] and H. Ono [Ono (1988); Ono and Komori (1985)], although their framework was not the Curry–Howard interpretation with natural deduction, but Gentzen's sequent calculi. We are grateful to Profs Komori and Ono for having sent us some still unpublished typescripts and manuscripts. Dr Došen has also informed us about the 'Lambek calculus' [Lambek (1958)] (implicational calculus without the so-called 'structural' rules), and van Benthem's interpretations (e.g. [van Benthem (1989)], as well as about his own work Došen [Došen (1988, 1989)].

[8]In his treatise on λ-calculus [Barendregt (1981)], Barendregt refers to a dissertation by G. Helman as an application of restricted λ-abstraction to relevant logic and other non-classical logics:

"The formulae-as-types idea gave rise to several investigations connecting typed λ-calculus, proof theory and some category theory, (...). Another direction is the connection between subsystems of logic and restricted versions of the typed λ-calculus (e.g. relevance logic and the typed λI-calculus), see Helman [1977]."   [Barendregt (1981)] (p. 572)

So far we have not got hold of Helman's work [Helman (1977)], but it looks as though there might be strong connections with part of what we are doing here.

An interesting attempt at classifying subsystems of implication through the formulae-as-types interpretation is also reported in [Hindley and Meredith (1990)]. The emphasis there, however, is on the so-called 'rule of condensed detachment'. The fragments of typed λ-calculus obtained by modified λ-abstractions (BCKW-, BCIW-, BCK-, and BCI-terms, in §3.5) are strictly based on the ones obtained by Curry's 'bracketing algorithm' developed in [Curry and Feys (1958)] (§6A3, p. 190):

Curry & Feys: Hindley & Meredith:

restrict the supply of definable terms thus getting *less* types 'A' demonstrably non-empty. Less non-empty types means less theorems which by its turn means a weaker logic. How about stronger logics, intermediate logics between intuitionistic and classical logic? What mechanism do we have for characterising them? Obviously, we need to increase the stock of definable (existing) terms so that more types can be shown to be non-empty, i.e. more theorems are available. We do not want to just throw in (stipulate) existence of functionals, in an ad hoc manner just to obtain the intermediate logic we want. We should put forward some reasonable principles, the kind natural to λ-calculus functional environment, and show that adopting them yields corresponding logics.

We chose one simple principle, which is that of completing a functional diagram. If 'φ' is a *monotonic* increasing or a *monotonic* decreasing functional on types, then one can ask the question whether we can complete the following diagram:

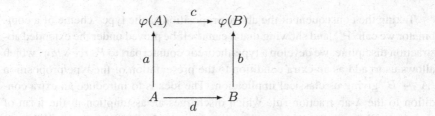

If we assume that we have enough functions to complete the diagram, then we get a logic $L\varphi$ of all formulas (as types) which are non-empty types. If we start with a logic which is weaker than intuitionistic logic (obtained by restricting λ-abstraction) and we add a φ-diagram to be closed, we can get logics which are incomparable with intuitionistic logic and yet have a well defined functional interpretation. The lattice of such logics is yet to be studied. To take an example of a φ which can yield classical implicational logic, let '$\varphi(X) \equiv (A \to B) \to X$'. The '$a$' and '$b$' arrows, which make, respectively, '$A \to ((A \to B) \to A)$' and '$B \to ((A \to B) \to B)$', are provable with intuitionistic implication. The arrow 'c', which corresponds to

$$((A \to B) \to A) \to ((A \to B) \to B)$$

(a) $[x].X \equiv \mathsf{K}X$ $\lambda^*x.Y \equiv \mathsf{K}Y$ *if x does not occur in Y,*

(b) $[x].x \equiv \mathsf{I}$ $\lambda^*x.x \equiv \mathsf{I}$,

(d) $[x].YZ \equiv \mathsf{B}YZ$ $\lambda^*x.UV \equiv \mathsf{B}U(\lambda^*x.V)$ *if x occurs in V but not in U,*

(e) $[x].YZ \equiv \mathsf{C}YZ$ $\lambda^*x.UV \equiv \mathsf{C}(\lambda^*x.U)V$ *if x occurs in U but not in V,*

(f) $[x].YZ \equiv \mathsf{S}YZ$ $\lambda^*x.UV \equiv \mathsf{S}(\lambda^*x.U)(\lambda^*x.V)$ *if x occurs in both U and V.*

(where 'Y' and 'Z' may or may not contain x).

is also provable with intuitionistic implication if we allow '$A \to B$' be used as both ticket and minor premise of the *modus ponens* ($\to$-*elimination*). So, instead of full classical logic where for any 'A' and for any 'B' such that either 'A' is true or '$A \to B$' is true ('B' could be the false, i.e. '$\mathcal{F}$'), thus filling the '*d*' arrow, we have a weakened excluded middle: either '$(A \to B) \to A$' or '$A \to B$'. As a parallel to Curry's [Curry (1950)] proof that

$$(A \vee (A \to B)) \to (((A \to B) \to A) \to A)$$

we show that under a certain extended λ-abstraction discipline, namely that a single abstraction cancels free occurrences of the variable which appear both as higher and as lower type (on the logical calculus it means that an implicational formula was used both as minor and as major premise of the rule of *modus ponens*), one can prove that

$$(((A \to B) \to A) \vee (A \to B)) \to (((A \to B) \to A) \to ((A \to B) \to B))$$

Taking the consequent of the above proposition as the type-scheme of a combinator we call 'P'', and showing that it can itself be proved under the extended abstraction discipline, we develop a type-theoretic counterpart to *Peirce's law*, which allows us to add as an extra condition to the presentation of the type/proposition '$A \to B$' giving us classical implication. The idea is to introduce an extra condition to the λ-abstraction rule which discharges an assumption in the form of '$y : A \to B$' forming a λ-abstraction term as a member of the antecedent (lower) type 'A', under the condition that 'B' is obtained from the assumption '$A \to B$', the latter being used as both minor assumption and major assumption of a modus ponens ($\to$-*elimination*).[9] This extra condition gives us a combinator we here call 'P''.

Another example that we could provide in this context would be Łukasiewicz' k-valued implicational logics, where we would have '$\varphi(X) \equiv (A^{k-1} \to B) \to X$'.[10]

---

[9] In [Seldin (1989)], he deals with Curry's inferential counterpart to Peirce's law, in a way which is very different from the one dealt with here. He attempts to change/extend the normalisation procedure (*à la* [Prawitz (1965, 1971)]) for the connective of implication to cover the newly introduced rule. In our case, the normalisation rules remain fixed, while the *introduction* rule (which involves a λ-abstraction) is changed/extended accordingly.

[10] This is an attempt at further extending the Curry–Howard interpretation to a sound semantical instrument which can be capable of handling logics as rule-based calculi (as opposed to truth-value-based calculi). In other words, we believe that the interpretation should not be restricted to the intuitionistic case, therefore we want the so-called *Peirce's axiom* to be provable. But, obviously, we want to do it on the conditions that an assumption can be discharged where it would not be possible to discharge it just by using the rules of the calculus for the intuitionistic case. (As we shall see below, the classical *reductio ad absurdum* involves the discharge of an assumption of 'higher' order such as '$A \to \mathcal{F}$' into a 'lower' order conclusion such as 'A'. This means going in the opposite direction of

Summarising, what we do is essentially to extend the Curry–Howard functional interpretation, by starting from intuitionistic implication and moving in both directions, namely:

- downwards: restricting λ-abstraction
  - Linear (drop *contraction* and *weakening*)
  - Ticket Entailment (drop *permutation*)
  - Relevant (drop *weakening*)

- upwards: completing diagrams by 'lifting' (e.g. from 'A' to '$(A^n \rightarrow B) \rightarrow A$')
  - Classical 2-valued implication (add *Peirce's axiom*)
  - Łukasiewicz' finitely valued implication (add *Rose's axiom*)

The use of the term 'resource' has its origins in Gabbay's (1989, 1990) investigations of systems of implication through a technique which combines features of the object language and the meta-language: the Metabox technique, based on a proof methodology similar to the ones developed in Jaśkowski [Jaśkowski (1934)] and Fitch [Fitch (1952)]. In order to illustrate the use of the Metabox technique, an algorithmic proof system methodology based on *labelled deductive systems* (LDS) described in Gabbay (1994), a class of logics called *resource* logics is defined as a subclass of the more comprehensive *controlled derivation* (CD)-logics, the latter term covering all logics where it is important to know exactly from

---

the usual assumption discharge of →-*introduction*.) We are obviously moving away from the strictly intuitionistic principles underlying the framework of, e.g. Howard:

"Results following from cut elimination in P($\supset$) (e.g.) the nonderivability of Peirce's Law ($\alpha \supset \beta$. $\supset \alpha$) $\supset \alpha$) seem to be obtainable at least as easily from the normalizability of constructions." [Howard (1980)] (p. 483)

We are trying to do something similar to what Curry did which is to devise systems of implication *including* the classical case. N.B.: classical *implication*, not classical logic, and we shall end up with something like an *implicational* logic (to use a term of van Benthem [van Benthem (1989)]; e.g. we do not want '$A \lor \neg A$' to be provable regardless of A, but we want '$\neg\neg A \rightarrow A$' to be a theorem under the condition of negation being defined as '$\neg A \equiv A \rightarrow \mathcal{F}$'. It is as if we are trying to find principles of construction giving us a classical proof of '$a : A$'. Cf. Tait's [Tait (1983)] (p. 182f):

"The distinction between a constructive and classical [mathematics] rests solely on what principles are admitted for constructing an object of a given type. And since all constructive principles are classical too, constructive mathematics is a part of classical mathematics. A classical proof $a : A$ is constructive just in case it is obtained using only certain principles of construction."

For a consistency proof of our modified framework, it is sufficient to show that one cannot prove a proposition which does not have a conditional (i.e. implication, universal quantifier, necessity, etc.) as its major connective. The notion of *provable* here is similar to [Martin-Löf (1972)] (p. 96): "A formula is *provable* if there is a deduction of it all of whose assumptions have been discharged." (Recall that only the *introduction* rules of conditionals discharge (old) assumptions without introducing new ones.)

Moreover, as we shall see below, such a consistency proof would find a parallel in [Martin-Löf (1972)] *consistency theorem*: "*No atomic formula is provable.*" (Ibid., p. 102).

which assumptions, using which inference rules and at what order a conclusion is drawn. Here, instead of using the Metabox technique, which deals with proofs via a 'labelling' discipline where each assumption/step is given a new atomic label ('*label* : *formula*'), we deal with those *resource* logics via a type-theoretic presentation system based on the Curry–Howard functional interpretation where the form of judgement '*proof* : *proposition*' finds an immediate parallel with the one used in *LDS*.[11]

In Gabbay (1989) the different logical implications are presented in a Hilbert system as:

Linear
$A \to A$ (*reflexivity*)
$(A \to B) \to ((C \to A) \to (C \to B))$ (*left transitivity*)
$(A \to B) \to ((B \to C) \to (A \to C))$ (*right transitivity*)
$(A \to (B \to C)) \to (B \to (A \to C))$ (*permutation*)

Modal T- strict
Add the schema below to linear implication:
$(A \to (A \to B)) \to (A \to B)$ (*contraction*)

Relevant
Add the schema below to linear implication:
$(A \to (B \to C)) \to ((A \to B) \to (A \to C))$ (*distribution*)

---

[11] With respect to the rôle of labels in deductive systems, we have found an interesting remark by Lambek and Scott in their book on *An Introduction to Higher Order Categorical Logic*:

"Logicians should note that a deductive system is concerned not just with unlabelled entailments or sequents $A \to B$ (as in Gentzen's proof theory), but with deductions or proofs of such entailments. In writing $f : A \to B$ we think of f as the 'reason' why A entails B."   [Lambek and Scott (1986)] (p. 47)

In the framework we discuss here, the 'reason' is represented by the witnessing of a closed λ-term (such as, e.g. '$\lambda x.x : A \to A$'). In Gabbay (1989, 1990), where the main data consist of axioms in a Hilbert-style presentation, the label put on a formula is an auxiliary tool which plays a crucial rôle in the description of the proof methodology, similarly to the 'numerical marks' used by Anderson and Belnap [Anderson and Belnap Jr. (1975)] (Chapter I, §3) in their technique of marking the relevance conditions:

"(...) the easiest way of handling the matter [relevance in proofs] is to use classes of numerals to mark the relevance conditions (...): (1) one may introduce a new hypothesis $A_{\{k\}}$, where k should be different from all subscripts on hypotheses of proofs to which the new proof is subordinate; (2) from A_a and $A \to B_b$ we may infer $B_{a \cup b}$; (3) from a proof of B_a from the hypothesis $A_{\{k\}}$, we may infer $A \to B_{a - \{k\}}$, provided k is in a; and (4) reit[eration] and rep[etition] retain subscripts (where a, b, c, range over sets of numerals)."   [Anderson and Belnap Jr. (1975)] (p. 22)

(Minimal) Intuitionistic
Add the schema below to relevant implication:
$A \rightarrow (B \rightarrow A)$ (*weakening*)

(Full) Intuitionistic
Add the schema below to minimal implication:
$\mathcal{F} \rightarrow A$ (*absurdity*)

Classical
Add the schema below to intuitionistic implication:
$((A \rightarrow B) \rightarrow A) \rightarrow A$ (*Peirce's law*)

Now, according to the Curry–Howard interpretation one can treat a proposition of the form '$A \rightarrow B$' as a $\rightarrow$-*type* of λ-terms, and to say that the proposition is true is the same as to say that one can find a pure closed λ-term which is contained in it. That is the main principle underlying the so-called constructive notion of validity, which supports the 'propositions are types' identification, and whose seminal ideas stem from Curry's theory of functionality, Howard's formulae-as-types notion of construction, and Tait's notion of convertibility which establish the connections between cut-elimination and normalisation.[12]

Notational remark. In what follows, the sign '$\square$' denotes the end of a definition, and '$\blacksquare$' indicates the end of a proof.

2.2 Origins

The idea of reading a formula as a type originates with Curry [Curry (1934)][13] and is used to give a λ-calculus interpretation of an intuitionistic theorem. A formula of intuitionistic implicational logic is a theorem if and only if, when read as a type, it can be shown to be non-empty using the rules of term-construction, namely *abstraction* and *application*. By varying the natural abstraction principles available in the λ-calculus, we are able to extend the point of view of formulae-as-types to some weak systems of implication (relevance, linear, etc.) as well as to a

---

[12]"H. Curry (1958) has observed that there is a close correspondence between *axioms* of positive implicational propositional logic, on the one hand, and *basic combinators* on the other hand. (...) The following notion of construction, for positive implicational propositional logic, was motivated by Curry's observation. More precisely, Curry's observation provided *half* the motivation. The other half was provided by W. Tait's discovery of the close correspondence between cut elimination and reduction of λ-terms (W. W. Tait, 1965)." [Howard (1980)] (p. 480)

[13] As noted in [Hindley and Meredith (1990)], perhaps the most explicit sign of Curry's 'awareness' of the parallel between the axiom-schemes of implication and the type-schemes of combinators is recorded in [Curry (1942)](footnote 2, p. 60).

system which is stronger than intuitionistic, namely classical implicational logic. The weaker logics are called resource logics in the framework of *LDS* of Gabbay (1989, 1990). The research reported here can also be understood in the spirit of *LDS*, where the labels are not terms of a certain algebra (as in Gabbay 1989, 1990) but typed λ-terms. As pointed out in Gabbay (1989, 1990), the framework of *LDS* generalises the usual consequence relation '$A_1, \ldots, A_n \vdash A$' between formulas to the more general notion '$t_1 : A_1, \ldots, t_n : A_n \vdash t : A$' where the '$t_i$'s are labels. The logical 'unit' in *LDS* is not a well-formed formula 'A' but a labelled well-formed formula '$t : A$', 't' being a label which conveys some 'meta-level' information about 'A'. In modal logic 't' can be a possible world index, and in the resource logics (which include intuitionistic, linear and relevant logics) the label 't' indicates what assumptions and rules we used to prove 'A'. The Curry–Howard interpretation can be viewed as a labelling scheme for intuitionistic well-formed formulas and here we try to generalise this scheme for other neighbouring logics situated below and above intuitionistic logic. We take Church's λ-calculus and Curry's combinatory logic as the building blocks supporting our framework.

Let us then take a standard definition of the terms and operators needed to obtain a λ-calculus, and let us examine the abstraction rule more closely. In his treatise *The Lambda Calculus* Barendregt defines:

"2.1.1. Definition. (i) *Lambda terms* are words over the following alphabet:

$v_0, v_1, \ldots$	variables,
λ	abstractor,
$(\,,)$	parentheses.

(ii) The set of λ-terms is defined inductively as follows:
  (1) $x \in \Lambda$;
  (2) $M \in \Lambda \Longrightarrow (\lambda x.M) \in \Lambda$;
  (3) $M, N \in \Lambda \Longrightarrow (MN) \in \Lambda$;
where x in (1) or (2) is an arbitrary variable."

[Barendregt (1981)] (p. 22)

Note that, apart from the clear distinction between *argument* and *function* made in (3), namely,

(1) 'M' as the *function* is higher than
(2) 'N' as the *argument*

there is an element of vagueness in case (2) in the definition of λ-terms, in particular λ-abstraction terms. When *abstracting* a variable from a term, one may encounter different situations, namely,

(a) M may have *no* free occurrence of x:

  (a.1) M is an open term, but contains no free occurrence of x;

  (a.2) M is a closed term, thus contains no free variable at all.

(b) M may have *one* free occurrence of x:

  (b.1) M may be of the form '(Tx)' (or '$APP(T, x)$'), i.e. x is occurring as an argument.

  (b.2) M may be of the form '(xT)' (or '$APP(x, T)$'), i.e. x is occurring as a function.

(c) M may have *more than one* free occurrence of x:

  (c.1) the λ-*abstraction* may cancel *exactly one* of the free occurrences of x.

  (c.2) the λ-*abstraction* may cancel *all* free occurrences of x.

  (c.2.1) all free occurrences in the same *order* (*argument/function*) that is;

  (c.2.2) all free occurrences regardless of *order*.

Again, in (3), where *application* is being defined (which can be done by juxtaposition as in '(MN)', or by an explicit operator '$APP(M, N)$' as in the terminology used here in this book), 'M' is assumed to be of 'higher' order than 'N', i.e. 'M' is supposed to be the 'course-of-values' of a function, while 'N' is assumed to be the argument. While (2) leaves some room for varying the λ-abstraction discipline, (3) leaves no room for manoeuvre as to how one can have (stronger or weaker) λ-calculi. This seems to match with the fact that all implications have *modus ponens*, each implication differing only in the way assumptions are discharged (one by one, in strict order, relevantly, etc.).

Now, by working with appropriate refinements of the aforementioned vague elements of the rule of *abstraction* one can use the simple typed λ-calculus together with the Curry–Howard interpretation to formalise a number of systems of implication, as we shall demonstrate below.[14]

---

[14]The classification of a number of systems of implication has been made by various people in the context of the 'Lambek calculus' [Lambek (1958)], giving rise to what is sometimes referred to as the 'categorial hierarchy', and has been used by many of those interested in the connections between the language of category theory, λ-calculus and proof theory, such as, e.g.:

"The general linguistic framework which arises here is that of a *Categorial Hierarchy* of different logical calculi ('categorial engines'). At the lower end lies the standard calculus of Ajdukiewicz [Ajdukiewicz 1935], at the upper end lies the full constructive, or intuitionistic conditional logic, whose derivations correspond to arbitrary lambda/application terms. In between lies a whole spectrum, not necessarily linearly ordered, of calculi with stronger or weaker intermediate rules of inference. For instance, one important principle of classification concerns the number of occurrences of premises which may be withdrawn in one application of conditionalization. Only *one* occurrence at a time was withdrawn in Examples 1 and 3. This particular restriction gives a very natural intermediate logic, which was already studied by Lambek as early as 1958, and is often called after him."   [van

2.3   Types and propositions

As pointed out above, the identification of propositions with types of their proofs/constructions (the latter indicates the distinction from proof-trees), usually referred to as the 'formulae-as-types' notion of construction, goes back at least to Curry's investigations on the functionality and the principal type-schemes of combinators [Curry (1934)]. (To be fair, one should remember that the harmony between the two facets of formal logic, namely, a functional calculus on terms and a logical calculus on formulas, finds its early origins in Frege's *Begriffschrift* (1879) and *Grundgesetze* (1893) respectively.) Now, to see how and within what context this isomorphism was 'discovered' it might be useful to go back to Curry's original paper on a 'theory of functionality' [Curry (1934)].

Curry's theory of functionality. The definitions of the 'building blocks of mathematical logic', as Schönfinkel [Schönfinkel (1924)] called them, were originally given by means of rules of 'conversion'. Taking *application* as the most primitive notion, a combinator was then characterised by how its application to an ordered list of combinatory objects would operate on that list, what resulting object was to be obtained. So, writing '$\to$' to mean 'converts to', variables $(x, y, z, \ldots)$ to denote arbitrary combinatory objects, and '$app(x, y)$' to mean 'the application of x to y', the combinators would be defined as:

$$
\begin{aligned}
\mathsf{I}x &\to x \\
\mathsf{B}xyz &\to app(x, app(y, z)) \\
\mathsf{C}xyz &\to app(app(x, z), y) \\
\mathsf{W}xy &\to app(app(x, y), y) \\
\mathsf{S}xyz &\to app(app(x, z), app(y, z)) \\
\mathsf{K}xy &\to x
\end{aligned}
$$

This was essentially the original characterisation of the combinators given (independently) by Schönfinkel and Curry in the late 1920. In his 1934 paper, however, Curry wanted to have the combinators characterised by their 'functionality' (i.e. given objects of certain types, the combinator would return an object of a certain type). In order to do that he needed to assign 'types' to the combinatory objects. Obviously, the principle of application remained valid, only extended with symbols for the types of objects, i.e.:

$$
\text{(APP)} \qquad \frac{y : A \qquad x : A \to B}{app(x, y) : B}
$$

Benthem (1990)] (p. 90)

   More recently, the Lambek calculus has also been used to provide a 'formulae-as-types' interpretation of a hierarchy of sub-intuitionistic logics [Wansing (1990)].

meaning: if y is of type A and x is of type $A \rightarrow B$ then the application of x to y results in an object of type B.

Thus, by arriving at a conclusion that an object '$app(x, y)$' is of type 'Q' we could follow (APP) backwards and give x a type of the form '$P \rightarrow Q$' for some P, and y the type P itself. Following this procedure one would then be able to find the right typing for the combinators. In order to see the procedure working, let us try to find the type of the combinator 'B', for example. First, we have

$$B xyz \rightarrow app(x, app(y, z))$$

We start by assigning the resulting object, namely '$app(x, app(y, z))$', a type, say Q. So,

$$app(x, app(y, z)) : Q$$

Then, while the type of x must be of the form $P \rightarrow Q$ for some P, $app(y, z)$ must have this same P as its type. So, we have

$$x : P \rightarrow Q$$
$$app(y, z) : P$$

From the latter we can deduce that while y must be of type $R \rightarrow P$ for some R, z must have the same R as its type. We have now identified the type of our arbitrary combinatory objects:

$$x : P \rightarrow Q$$
$$y : R \rightarrow P$$
$$z : R$$

Now, we need to find the type of the combinator B itself, which takes x, y, z (in this order) and produces $app(x, app(y, z))$ of type Q. To do that, let us move backwards from the resulting type towards the beginning of the ordered list of input objects, introducing an '$\rightarrow$' every time we move one position up in the list. So, from:

$$app(x, app(y, z)) : Q$$

we remove the variable on the left-hand side, adding its type to the right-hand side:

$$B xyz : Q$$
$$B xy : (R \rightarrow Q)$$
$$B x : ((R \rightarrow P) \rightarrow (R \rightarrow Q))$$
$$B : (P \rightarrow Q) \rightarrow ((R \rightarrow P) \rightarrow (R \rightarrow Q))$$

and we obtain the functionality (typing) of 'B' as

$$(P \rightarrow Q) \rightarrow ((R \rightarrow P) \rightarrow (R \rightarrow Q))$$

If we repeat the same procedure for the other combinators we find the following type-schemes:

$$I : P \to P$$
$$C : (P \to (Q \to R)) \to (Q \to (P \to R))$$
$$W : (P \to (P \to Q)) \to (P \to Q)$$
$$S : (P \to (Q \to R)) \to ((P \to Q) \to (P \to R))$$
$$K : P \to (Q \to P)$$

which can be read as the axiom schemes of implicational logic if '$\to$' is read as 'implies'.

While Curry's 'discovery' was essentially concerned with implicational logic, the whole framework was given a more precise presentation in Howard's [Howard (1980)] investigations on the isomorphism between natural deduction proofs of theorems of Heyting's arithmetic (including conjunctions, disjunctions and quantifiers) and terms of the λ-calculus. In fact, the identification of proposition with types, also used in Gödel's [Gödel (1958)] functional interpretation of intuitionistic logic via a system T of finite types, has played an important rôle in most developments of some key notions of modern logic such as 'constructive validity' (see, e.g. Läuchli [Läuchli (1965)],[15] 1970), [Scott (1970)]) and 'theory of constructions' (see, e.g. [Goodman (1970)]), as well as in some attempts at reconciling category theory with constructive logics (such as, e.g. [Lambek and Scott (1986)]). In most of the literature, the interpretation has been looked at as only applicable to intuitionistic reasoning. Here we demonstrate that there is no reason to restrict the applicability of the functional interpretation to the intuitionistic case. Its true origins are in fact to be found in Frege's two-faceted calculus: a functional calculus on the terms — *Grundgesetze* (1893) — and a logical calculus on the formulas — *Begriffschrift* (1879).

In order to make 'logical' sense of the division of task into a functional calculus on the labels and a logical calculus on the formulas, leading to the identification between propositions and types ('$a : A$' reading: the functional object 'a' is of the type denoted by the logical formula 'A'), we can recall that it was in Frege's (1879) *Begriffschrift* that the symbol '$\vdash$' (then meaning 'is true') first appeared. It was motivated by the need for characterising a proposition, which Frege called 'thought' (Frege's horizontal stroke, i.e. '$—A$': 'A is a thought') and distinguishing it from a judgement (Frege's vertical stroke, i.e. '$\vdash A$': 'A is a true thought'). In other words, to say that 'A' is true we must have already established that 'A' is

---

[15]"*Theorem: A is a tautology of intuitionistic propositional calculus if and only if t(A) is definably non-empty.*"

a proposition. So, when saying that:

we are saying on one level that 'A' is a *proposition*, and on another level that 'A is true' is a *judgement*. A *judgement* of the form 'A is true' can only be made on the basis of the existence of a proof of the proposition 'A'. Contrary to the classical view, a proposition is not the same as a truth value. And in contrast to the traditional proof-theoretic account of propositions and inference rules, a logical inference is to be made from *judgement(s)* to *judgement*, and not from *proposition(s)* to *proposition*. Both premises and conclusions of inference rules are not *propositions* as in the usual case even in traditional natural deduction presentations of logics, but *judgements*. This seems to be a highly relevant refinement of the usual formalisation of mathematical procedures into rules of inference, such as for example, natural deduction style *à la* Gentzen [Gentzen (1935)]. The difference between usual natural deduction presentation rules which have *propositions* as premises and conclusions, and *intuitionistic type theory* where *judgements* are the objects on which the rules of inference operate, is explained briefly in Martin-Löf's [Martin-Löf (1985)] illuminating account of the often neglected distinction between the two logical concepts of *proposition* and *judgement*.[16]

Now, the identification of propositions with types gives us instruments to deal with *judgements* which include its justification: in '$a : A$' we are basically saying that 'A is true because of a'. (e.g. in '$\lambda x.x : A \to A$' we say that '$A \to A$' is true because we have a closed term '$\lambda x.x$' which inhabits it.)

We shall be using here what has been named *'Meaning-As-USE' type theory* [de Queiroz and Maibaum (1990)], a reformulation of *intuitionistic type theory*

---

[16] It is essentially the written account of a series of lectures entitled 'On the Meanings of the Logical Constants and the Justifications of the Logical Laws' given in Siena, Italy, in April 1983. There he says:

"We must remember that, even if a logical inference, for instance, a conjunction introduction, is written

$$\frac{A \quad B}{A \,\&\, B}$$

which is the way in which we could normally write it, it does not take us from the propositions A and B to the proposition A & B. Rather, it takes us from the affirmation of A and the affirmation of B to the affirmation of A & B, which we may make explicit, using Frege's notation, by writing it

$$\frac{\vdash A \quad \vdash B}{\vdash A \,\&\, B}$$

instead. It is always made explicit in this way by Frege in his writings, and in Principia, for instance. Thus we have two kinds of entities here: we have the entities that the logical operations operate on, which we call propositions, and we have those that we prove and that appear as premises and conclusion of a logical inference, which we call assertions." [Martin-Löf (1985)] (pp. 204–205)

[Martin-Löf (1975, 1982, 1984)], where instead of Martin-Löf's rules for the definition of types/propositions,

formation
introduction
elimination
equality,

we have the following rules, with corresponding purpose:

formation: to show how to form the type-expression as well as when two type-expressions are equal.

introduction: to show how to form the canonical value-expressions via the constructor(s), as well as when two canonical value-expressions are equal.

reduction: to show how to normalise non-canonical value-expressions, by demonstrating the effect of DESTRUCTOR(S) on the terms built up by constructor(s).

induction: minimality rule.

The →-type of λ-terms is presented as[17]:

→-formation

$$\frac{A\ type \quad B\ type}{A \to B\ type}$$

→-introduction

$$\frac{\begin{array}{c}[x:A]\\b(x):B\end{array}}{\lambda x.b(x):A\to B} \qquad \frac{\begin{array}{c}[x:A]\\b(x)=d(x):B\end{array}}{\lambda x.b(x)=\lambda x.d(x):A\to B}$$

→-reduction

$$\frac{a:A \qquad \begin{array}{c}[x:A]\\b(x):B\end{array}}{APP(\lambda x.b(x),a)=b(a/x):B}$$

→-induction

$$\frac{c:A\to B}{\lambda x.APP(c,x)=c:A\to B}$$

□

---

[17]An attempt at finding useful connections between the presentation of an →-type as a theory of λ-terms, and the axiomatic presentation of a λ-theory (such as the one presented in [Barendregt (1981)]) is given in [de Queiroz and Maibaum (1991)].

We shall be concerned here mainly with the first $\rightarrow$-*introduction*. In the actual proof-trees, we shall be making use of $\rightarrow$-*elimination*:

$$\frac{a : A \qquad c : A \rightarrow B}{APP(c, a) : B}$$

Observe that unlike Barendregt's set of *Lambda terms*, which included variables and *application*-terms, our $\rightarrow$-type only contains λ-abstraction terms (the rule of $\rightarrow$-*induction*, which is the counterpart to λ-calculus η-rule, is a kind of formal counterpart to that[18]). Here those Barendregt's terms which are not λ-*abstractions* can only be *subterms*. (Nonetheless, we shall need to refer to the whole of Λ-terms when defining the notion of *saturation gap* of a functional term.) Thus, in order to work with the hidden assumptions of the *abstraction* rule discussed in the previous section, we need to look at the '$b(x)$' of our first $\rightarrow$-*introduction*, which can have the form of any of Barendregt's *Lambda terms*: a variable, an *abstraction*-term, or an *application*-term.

2.4 λ-abstraction and implication

Now, we have to show that one can construct derivations of the Hilbert style axioms given for linear, relevant and intuitionistic implication in Gabbay (1989, 1990), and how one can draw the appropriate distinctions for each implication. At least since Curry's theory of functionality [Curry (1934)] it is well known

---

[18] The form of induction (minimality) rules for non-inductive types such as arrow ($\rightarrow$) type and the product ($\times$) type is discussed in [de Queiroz and Smyth (1989)]. A full account of the 'minimality' rôle of η-convertibility for the other logical connectives interpreted as types is presented in de Queiroz and Gabbay (1991b).

Concerning equality in general, we are here being rather 'loose' on our use of the equality sign. Strictly speaking, we should be indexing the equality with its kind. For example, the second *introduction* rule shows when two canonical elements are equal, so it is concerned with ξ-equality. The *reduction* rule shows how non-canonical expressions can be brought to normal form, so it is concerned with β-equality. Finally, the *induction* rule shows that by performing an introduction step right after an elimination one gets back to the original proof (and corresponding term), thus it concerns η-equality. With the indexes for equality those rules would look like:

$$\frac{\overset{[x:A]}{b(x) =_s d(x):B}}{\lambda x.b(x) =_{s \cdot \xi} \lambda x.d(x):A \rightarrow B} \qquad \frac{\overset{[x:A]}{a:A \quad b(x):B}}{APP(\lambda x.b(x), a) =_\beta b(a/x):B} \qquad \frac{c:A \rightarrow B}{\lambda x.APP(c, x) =_\eta c:A \rightarrow B}$$

where in the first case 's' could be any (sequence of) equality and the '$\cdot$' stands for the concatenation of 's' with 'ξ' (and each equality step is recorded, in line with the 'keeping track of proof steps' motto of the functional interpretation).

Later in the book we shall be discussing the need to identify the kind of definitional equality, as well as the need to have a type of 'propositional equality' in order to be able to reason about the functional objects (those to the left hand side of the ':' sign). For example, one might wish to prove that for any two functional objects of $\rightarrow$-type, if they are equal then their application to all objects of the domain type must result in equal objects of the codomain type.

that there is a correspondence between the type-schemes of combinators and the axioms of intuitionistic implication. Within the propositions-are-types paradigm there is a correspondence between axioms of implication and $\rightarrow$-types which contain λ-terms as elements or proofs/constructions of the corresponding axioms. So, combinators are mathematical objects which correspond to λ-terms, which in their turn are elements/proofs/constructions which belong to an $\rightarrow$-type.[19]

Moreover, just to make clear our own proof methodology, we should say that we read the first rule of $\rightarrow$-*introduction*, namely,

$$\frac{\begin{array}{c} [x : A] \\ b(x) : B \end{array}}{\lambda x.b(x) : A \rightarrow B}$$

as follows: having made the assumption '$x : A$', and arriving at the conclusion '$b(x) : B$' by means of one of the rules available, then we can discharge the assumption by making a λ-abstraction of the assumption-term ('x') over the conclusion-term ('$b(x)$'). In other words, when constructing a proof-tree one can discharge an assumption if there is at least one proof step between the assumption and the conclusion where the assumption is discharged. So, in the construction of a proof of '$\lambda x.x : A \rightarrow A$', as we shall see below, we need at least *reflexivity* in order to arrive at a conclusion of the form '$x : A$' from the assumption '$[x : A]$'. Such a methodological requirement makes our framework slightly different from the one used in Chapter 15 of [Hindley and Seldin (1986)], where a proof of the latter is constructed as follows:

"EXAMPLE 15.4. In any system containing ($\rightarrow$e) and ($\rightarrow$i),

$$\vdash \mathsf{I} \, \varepsilon \, \alpha \rightarrow \alpha.$$

Proof.

$$\frac{\overset{1}{[x \, \varepsilon \, \alpha]}}{\lambda x.x \, \varepsilon \, \alpha \rightarrow \alpha}(\rightarrow \text{i-1}).$$

$\square$"

[Hindley and Seldin (1986)] (p. 208)

---

[19] When approaching a presentation of the Curry–Howard interpretation one has to be warned to specific terminological diversions from conventional logical frameworks. In the framework of the interpretation 'proofs' refer to constructions, and not to the actual proof-trees. Without such a warning on terminology, misconceptions may arise. e.g. in Lambek's:

"The association of entities with proofs becomes even more striking when we compare the free typed Schönfinkel algebra (generated by a set of letters) with pure intuitionistic implicational logic. Then

combinators = proofs."

[Lambek (1980)] (p. 385)

'proofs' should be understood as constructions (terms).

(Here '($\to$ i-1)' indicates that at that particular step the assumption numbered '1' was being discharged by the introduction of the '$\to$'.[20])

Proofs and corresponding conditions for invalidation:

1. $A \to A$ (*reflexivity*)

$$\frac{\dfrac{\dfrac{[x : A]}{x = x : A}}{x : A}}{\lambda x.x : A \to A}$$

and we have the 'identity' construction, which corresponds to combinator 'I' $\equiv$ $\lambda x.x$ [Curry and Feys (1958)] (p. 152); [Hindley and Seldin (1986)] (p. 191).[21] ∎

---

[20] It must be noted, however, that Hindley and Seldin seem to adopt the same reading, in spite of the divergence in the example just mentioned. Cf.:
"It [the rule of $\to$-introduction] is usually written thus:
($\to$i)

$$\frac{\begin{array}{c}[x \ \varepsilon \ \alpha]\\ M \ \varepsilon \ \beta\end{array}}{\lambda x.M \ \varepsilon \ \alpha \to \beta}.$$

(...)
In such a system, rule ($\to$i) is read as "If x $\notin$ FV($L_1 \ldots L_n$), and M ε β is the *conclusion* of a deduction whose not-yet-discharged assumptions are x ε α, L_1 ε δ_1, ..., L_n ε δ_n, then you may deduce

$$(\lambda x.M) \ \varepsilon \ (\alpha \to \beta),$$

and whenever the assumption x ε α occurs undischarged at a branch-top above M ε β, you must enclose it in brackets to show that it has now been discharged." [Hindley and Seldin (1986)] (p. 206) (Our *emphasis*. Note that in the 'I' example, as well as in the 'K' example below (footnote 28), there is no *conclusion* of the form 'M ε β'. The introduction of the '$\to$', and corresponding assumption-discharge, is made straight from the assumption.)

[21] Here we have used one of the general rules of equality available in our type-theoretic framework, namely the *reflexivity* rule:

$$\frac{x : A}{x = x : A}$$

followed by either one of the *equality left* or *right*:

$$\frac{a = b : A}{a : A} \qquad \frac{a = b : A}{b : A}$$

Notice that here we have no structural rules, not even the usual 'repeat' inference step — namely, from the assumption 'A', deduce 'A' itself — taken for granted in most of the Natural Deduction literature. In the frameworks based on the sequent calculus, this step is taken to be an 'axiom', but here we try to avoid introducing axioms, and even *reflexivity* (of the logical calculus, that is) stays away from the primitives of the proof system. It becomes provable or not depending on whether we want reflexivity

2. $(A \rightarrow B) \rightarrow ((C \rightarrow A) \rightarrow (C \rightarrow B))$ *(left transitivity)*

$$\cfrac{\cfrac{\cfrac{\cfrac{\cfrac{[z:C] \qquad [y:C \rightarrow A]}{APP(y,z):A} \qquad [x:A \rightarrow B]}{APP(x,APP(y,z)):B}}{\lambda z.APP(x,APP(y,z)):C \rightarrow B}}{\lambda y.\lambda z.APP(x,APP(y,z)):(C \rightarrow A) \rightarrow (C \rightarrow B)}}{\lambda x.\lambda y.\lambda z.APP(x,APP(y,z)):(A \rightarrow B) \rightarrow ((C \rightarrow A) \rightarrow (C \rightarrow B))},$$

whose resulting closed term corresponds to combinator 'B' $\equiv \lambda x.\lambda y.\lambda z.APP$ $(x, APP(y, z))$ [Curry and Feys (1958)] (p. 152); [Hindley and Seldin (1986)] (p. 191). In terms of a calculus of functions, 'B' would correspond to the functor for the *(left) composition* of two functions.[22] Observe that the λ-abstraction discipline needed for the derivation of this proposition is very 'well-behaved' in the sense that the order (from lower to higher types) has not been violated. Thus a rule of the form:

$$\frac{\vdash A \rightarrow B}{\vdash (B \rightarrow C) \rightarrow (A \rightarrow C)}$$

would still be valid, as in concatenation logic discussed in Gabbay (1989, 1990), although the axiom below *(right transitivity)* would not be derivable. ∎

3. $(A \rightarrow B) \rightarrow ((B \rightarrow C) \rightarrow (A \rightarrow C))$ *(right transitivity)*

$$\cfrac{\cfrac{\cfrac{\cfrac{\cfrac{[z:A] \qquad [x:A \rightarrow B]}{APP(x,z):B} \qquad [y:B \rightarrow C]}{APP(y,APP(x,z)):C}}{\lambda z.APP(y,APP(x,z)):A \rightarrow C}}{\lambda y.\lambda z.APP(y,APP(x,z)):(B \rightarrow C) \rightarrow (A \rightarrow C)}}{\lambda x.\lambda y.\lambda z.APP(y,APP(x,z)):(A \rightarrow B) \rightarrow ((B \rightarrow C) \rightarrow (A \rightarrow C))},$$

on the functional calculus to be a primitive. This way we could have a logic where the reflexivity axiom '$A \rightarrow A$' is not valid for atomic formulas, such as, e.g. Komori's (1990) B-logic where the only axiom is the axiom scheme corresponding to the type-scheme of combinator 'B'.

In other words, by dropping the rule of equality corresponding to *reflexivity* on terms (of the functional calculus) we can 'invalidate' the derivation of the axiom corresponding to the type-scheme of 'I'. This means that we have an explicit condition to 'stop' I-like axioms from being derivable. In [Hindley and Meredith (1990)] (§4.12, p. 98) there is no such explicit condition, resulting in that a BCK-system which is supposed to have type-assignments to formulas of the kind 'B', 'C' and 'K' only, also assigns a type to formulas of the 'I'-kind. The only condition that is imposed when making a $\lambda x.$-abstraction over a term Y in the type-assignment for BCK-formulas is that x occurs in Y at most once. Now, under this condition one can indeed assign a type to $\lambda x.x$. Thus, strictly speaking, Hindley and Meredith's BCK type-assignment should instead be called BCIK type-assignment.

[22] It also guarantees, together with the previous combinator 'I', that there is a left identity function such that, for all $f : A \rightarrow B$, '$f1_A = f$', as in the definition of a *category* as a deductive system in [Lambek and Scott (1986)] (p. 52).

whose resulting closed term corresponds to a combinator which results from applying combinator 'C' to combinator 'B', or what Curry has called combinator 'B'' in [Curry and Feys (1958)] (p. 379), and in [Curry (1963)] (p. 118): 'B'' $\equiv \lambda x.\lambda y.\lambda z.APP(y, APP(x, z))$ ('CB' in [Hindley and Seldin (1986)] (p. 191)). In terms of a calculus of functions, 'B'' would correspond to the *(right) composition* of two functions.[23]

In order to invalidate the derivation above one would have to impose the restriction on the λ-*abstraction* rule such that the abstractions have to occur in the order: 'first increasing functional order on the variables occurring as argument, then decreasing entry order on the variables occurring as function'. ∎

4. $(A \to (B \to C)) \to (B \to (A \to C))$ *(permutation)*

$$\cfrac{\cfrac{[y:B] \qquad \cfrac{[z:A] \qquad [x:A \to (B \to C)]}{APP(x,z):B \to C}}{\cfrac{APP(APP(x,z),y):C}{\cfrac{\lambda z.APP(APP(x,z),y):A \to C}{\cfrac{\lambda y.\lambda z.APP(APP(x,z),y):B \to (A \to C)}{\lambda x.\lambda y.\lambda z.APP(APP(x,z),y):(A \to (B \to C)) \to (B \to (A \to C))}}}}}{}$$

whose resulting closed term corresponds to combinator 'C' $\equiv \lambda x.\lambda y.\lambda z.APP$ $(APP(x, z), y)$ [Curry and Feys (1958)] (p. 152); [Hindley and Seldin (1986)] (p. 191). It is the counterpart to the rule of exchange of Gentzen's sequent calculi. In a calculus of functions it would correspond to the associativity of *composition*.

In order to invalidate the derivation above one would have to impose the restriction on the λ-*abstraction* rule such that the abstractions have to occur in the order: 'functional order into entry order'. ∎

5. $(A \to (A \to B)) \to (A \to B)$ *(contraction)*

$$\cfrac{\cfrac{[y:A] \qquad \cfrac{\boxed{[y:A]} \qquad [x:A \to (A \to B)]}{APP(x,y):A \to B}}{\cfrac{APP(APP(x,y),y):B}{\cfrac{\boxed{\lambda y.}\, APP(APP(x,y),y):A \to B}{\lambda x.\lambda y.APP(APP(x,y),y):(A \to (A \to B)) \to (A \to B)}}}}{},$$

whose resulting closed term corresponds to combinator 'W' $\equiv \lambda x.\lambda y.APP$ $(APP(x, y), y)$ [Curry and Feys (1958)] (p. 152); [Hindley and Seldin (1986)]

---

[23]Similarly to the previous case, this also guarantees the identity to the right '$1_B f = f$', Ibid.

(p. 191). It is also the counterpart to the rule of contraction of Gentzen's sequent calculi.

The assumption '$\boxed{[y : A]}$' is used twice and in a nested way. So, the restriction one has to impose here is rather obvious: a λ-abstraction will cancel *one* free occurrence of the variable at a time. ∎

6. $(A \to (B \to C)) \to ((A \to B) \to (A \to C))$ *(distribution)*

$$
\cfrac{
\cfrac{
\cfrac{
\cfrac{\boxed{[z : A]} \qquad [x : A \to (B \to C)]}{APP(x,z) : B \to C} \qquad \cfrac{\boxed{[z : A]} \qquad [y : A \to B]}{APP(y,z) : B}
}{APP(APP(x,z), APP(y,z)) : C}
}{\boxed{\lambda z.}APP(APP(x,z), APP(y,z)) : A \to C}
}{\cfrac{\lambda y.\lambda z.APP(APP(x,z), APP(y,z)) : (A \to B) \to (A \to C)}{\lambda x.\lambda y.\lambda z.APP(APP(x,z), APP(y,z)) : (A \to (B \to C)) \to ((A \to B) \to (A \to C))}}
$$

whose resulting closed term corresponds to combinator 'S' $\equiv \lambda x.\lambda y.\lambda z.APP$ $(APP(x,z), APP(y,z))$ [Curry and Feys (1958)] (p. 153); [Hindley and Seldin (1986)] (p. 191). ∎

Note that the assumption '$\boxed{[z : A]}$' is used twice, and both occurrences are discharged in one single abstraction '$\boxed{\lambda z.}$'. To obtain a linear implication one has to restrict the discharging abstraction to one occurrence of the assumption only. In other words, each discharge affects only one (linear) path in the proof-tree, instead of affecting all branching occurrences like in the proof above.

7. $A \to (B \to A)$ *(weakening[24])*

Now we want to build a derivation of the above axiom and show where it can be invalidated by the appropriate side condition. By imposing the condition that

---

[24]This axiom essentially represents that 'a true proposition is implied by anything', so we might as well have called it *truth*. As we shall see, a derivation of this axiom from the presentation of the $\to$-type involves a non-relevant abstraction under the condition (a.1) of section 2 above, which is when abstraction is made over an open term which contains no free occurrence of the variable. Additionally, by allowing a non-relevant abstraction over closed terms (condition (a.2) above), such as, e.g. in '$\lambda x.\lambda y.y : B \to (A \to A)$', one can see how to relate this axiom to the following axiom of a deductive system defined in [Lambek and Scott (1986)] (p. 48), 'R2. $A \xrightarrow{\bigcirc_A} T$'. As it is remarked there, the distinguished proposition 'T' stands for the categorical notion of *terminal* object and its existence is connected to the existence of a unique arrow $\bigcirc_A : A \to T$ for all objects A. If one has *permutation* (recall the type-scheme of combinator 'C' above), it is easy to see that:

$$(B \to (A \to A)) \to (A \to (B \to A))$$

and vice-versa.

the *abstraction* can only be made when there is indeed at least one free occurrence of the variable being abstracted from the expression ('$b(x)$' in the case below), one can obtain a 'relevant' *abstraction*:

$\rightarrow$*-introduction*

$$\frac{\begin{array}{c}[x : A]\\ b(x) : B\end{array}}{\lambda x.b(x) : A \rightarrow B}$$

The proof-tree is constructed as follows:

$$\frac{\dfrac{\dfrac{\boxed{[y : B]}}{[x : A]}}{\boxed{\lambda y.x} : B \rightarrow A}}{\lambda x.\lambda y.x : A \rightarrow (B \rightarrow A)}$$

whose resulting closed term corresponds to combinator 'K' $\equiv \lambda x.\lambda y.x$ [Curry and Feys (1958)] (p. 153); [Hindley and Seldin (1986)] (p. 191). It is also the counterpart to the structural rule of thining (weakening) of Gentzen's sequent calculi. ∎

Note that the discharge/abstraction of the assumption '$\boxed{[y : B]}$' is made over the expression 'x' in '$\boxed{\lambda y.x}$', which prevents it from being considered 'relevant', given that the expression 'x' does not contain any free occurrence of 'y'. (Such a 'non-relevant' discharge/abstraction is called 'vacuous discharge' in [Hindley and Seldin (1986)].[25]) So, the restricted λ-abstraction to be adopted in order to invalidate the derivation above is exactly the relevant abstraction, i.e. there must be at least one free occurrence of the variable in the term on which the abstraction is operating. In logical terms, what the relevant abstraction does is to allow us to ensure that within the Curry–Howard interpretation we can have a "systematic handle on relevance in the sense of logical dependence" [Anderson and Belnap Jr. (1975)] (p. 31). The idea is that for 'A' to be relevant to 'B' it must be *necessary*

---

[25] "EXAMPLE 15.3. In any system containing ($\rightarrow$e) and ($\rightarrow$i).

$$\vdash K \ \varepsilon \ \alpha \rightarrow \beta \rightarrow \alpha.$$

Proof. Here is a deduction of the required formula. In it, the first application of ($\rightarrow$-i) discharges all assumptions y ε β that occur. But none in fact occur, so nothing is discharged. This is perfectly legitimate; it is called '*vacuous discharge*', and is shown by '($\rightarrow$i-v)'.

$$\frac{\dfrac{\dfrac{1}{[x \ \varepsilon \ \alpha]}}{\lambda y.x \ \varepsilon \ \beta \rightarrow \alpha}(\rightarrow\text{i-v})}{\lambda xy.x \ \varepsilon \ \alpha \rightarrow \beta \rightarrow \alpha}(\rightarrow\text{i-1}).$$

□"

[Hindley and Seldin (1986)] (p. 208)

to use 'A' in the deduction of 'B' from 'A'. At first sight it may look as though the restriction on λ-abstraction is not by itself sufficient, given that a proof of '$(A \to B) \to ((B \to A) \to (A \to B))$' can be constructed using relevant λ-abstractions only, as in e.g.:

$$\cfrac{\cfrac{\cfrac{\cfrac{\cfrac{\cfrac{[z:A] \quad [x:A \to B]}{APP(x,z):B} \quad [y:B \to A]}{APP(y,APP(x,z)):A}(*) \quad [x:A \to B]}{APP(x,APP(y,APP(x,z))):B}(*)}{\lambda z.APP(x,APP(y,APP(x,z))):A \to B}}{\lambda y.\lambda z.APP(x,APP(y,APP(x,z))):(B \to A) \to (A \to B)}}{\lambda x.\lambda y.\lambda z.APP(x,APP(y,APP(x,z))):(A \to B) \to ((B \to A) \to (A \to B))}$$

However, it is easy to show that the steps marked with '$(*)$' were unnecessary and could be eliminated, and the '$\lambda y.$'-abstraction would have to be made over the term obtained prior to those steps, namely '$APP(x,z)$'. A 'minimal' proof of '$(A \to B) \to ((B \to A) \to (A \to B))$' would have to involve at least one non-relevant abstraction, such as e.g. in:

$$\cfrac{\cfrac{\cfrac{\cfrac{[z:A] \quad \cfrac{[y:B \to A]}{[x:A \to B]}}{APP(x,z):B}}{\lambda z.APP(x,z):A \to B}}{\lambda y.\lambda z.APP(x,z):(B \to A) \to (A \to B)}}{\lambda x.\lambda y.\lambda z.APP(x,z):(A \to B) \to ((B \to A) \to (A \to B))}$$

where the $\lambda y.$-abstraction is a non-relevant abstraction.

As pointed out in [Lambek (1989)] (p. 234), in his *The Calculi of Lambda-Conversion* Church already distinguished the relevant from the non-relevant λ-abstraction.[26]

Most modern textbooks, however, still omit such a restriction in the *abstraction* rule (see, e.g. [Barendregt (1981)], and [Hindley and Seldin (1986)].[27]

---

[26]"If M does not contain the variable x (as a free variable), then $(\lambda x M)$ might be used to denote a function whose value is constant and equal to (the thing denoted by) M, and whose range of arguments consists of all things. This usage is contemplated below in connection with the calculi of λ-K-conversion, but is excluded from the calculi of λ-conversion and λ-δ-conversion — for technical reasons which will appear." [Church (1941)] (pp. 6–7)

[27]Barendregt calls Church's calculus of λ-K-conversion simply λ-Calculus, or 'Classical λ-Calculus', whereas the calculus with 'relevant' abstraction rule (Church's calculus of λ-conversion) is referred to as λI-Calculus [Barendregt (1981)] (Chapter 9):

"Remember the definition of the restricted class Λ_I:
$$x \in \Lambda_I,$$
$$M \in \Lambda_I, x \in \mathrm{FV}(M) \Rightarrow (\lambda x.M) \in \Lambda_I,$$
$$M, N \in \Lambda_I \Rightarrow (MN) \in \Lambda_I."$$

[Barendregt (1981)] (p. 185)

8. $\mathcal{F} \to A$ (absurdity)

We assume that there is a distinguished proposition '$\mathcal{F}$' which is taken to be empty (i.e. no term, whether open or closed, is a member of it), and a distinguished closed term '$\lambda\bot.$' such that

$$\lambda\bot. : \mathcal{F} \to A$$

for any 'A'. We say that '$\mathcal{F} \to A$', for any 'A', is the type-scheme for a combinator we would call '$\mathsf{I}_{\mathcal{F}}$'.

With this axiom, and taking '$\neg A \equiv A \to \mathcal{F}$', we can prove Heyting's axioms involving (intuitionistic) negation,[28] namely,
(i) $\neg A \to (A \to B)$ and
(ii) $((A \to B) \land (A \to \neg B)) \to \neg A$.

(i) $(A \to \mathcal{F}) \to (A \to B)$:

$$\frac{\dfrac{\dfrac{[y:A] \qquad [x:A \to \mathcal{F}]}{APP(x,y) : \mathcal{F}} \qquad \lambda\bot. : \mathcal{F} \to B}{APP(\lambda\bot., APP(x,y)) : B}}{\dfrac{\lambda y.APP(\lambda\bot., APP(x,y)) : A \to B}{\lambda x.\lambda y.APP(\lambda\bot., APP(x,y)) : (A \to \mathcal{F}) \to (A \to B)}}$$

(ii) $((A \to B) \land (A \to (B \to \mathcal{F}))) \to (A \to \mathcal{F})$:

$$\frac{\dfrac{[y:A] \quad \dfrac{[x:(A \to B) \land (A \to (B \to \mathcal{F}))]}{FST(x) : A \to B}}{APP(FST(x),y) : B} \qquad \dfrac{[y:A] \quad \dfrac{[x:(A \to B) \land (A \to (B \to \mathcal{F}))]}{SND(x) : A \to (B \to \mathcal{F})}}{APP(SND(x),y) : B \to \mathcal{F}}}{\dfrac{APP(APP(SND(x),y), APP(FST(x),y)) : \mathcal{F}}{\dfrac{\lambda y.APP(APP(SND(x),y), APP(FST(x),y)) : A \to \mathcal{F}}{\lambda x.\lambda y.APP(APP(SND(x),y), APP(FST(x),y)):((A \to B) \land (A \to (B \to \mathcal{F}))) \to (A \to \mathcal{F})}}}$$

Moreover, to prove two of Ackermann's axioms for negation (quoted in Gabbay 1988), namely:
$(A \to \neg B) \to (B \to \neg A)$
$(A \to \neg A) \to \neg A$,
we put '$\neg A \equiv A \to \mathcal{F}$' and they become, respectively:
$(A \to (B \to \mathcal{F})) \to (B \to (A \to \mathcal{F}))$
$(A \to (A \to \mathcal{F})) \to (A \to \mathcal{F})$

---

[28] See the two axioms below:
X. $\vdash \neg p \to (p \to q)$.
XI. $\vdash ((p \to q) \land (p \to \neg q)) \to \neg p$.   [Heyting (1956)] (p. 101)

which are instances of the type-schemes of combinators 'C' (*permutation*), and 'W' (*contraction*), respectively.

We can also prove intuitionistically that '$\neg\neg(A \vee \neg A)$':

$$\cfrac{\cfrac{\cfrac{\cfrac{\cfrac{[y:A]}{inl(y):A \vee (A \to \mathcal{F})} \quad [x:(A \vee (A \to \mathcal{F})) \to \mathcal{F}]}{APP(x,inl(y)):\mathcal{F}}}{\lambda y.APP(x,inl(y)):A \to \mathcal{F}}}{inr(\lambda y.APP(x,inl(y))):A \vee (A \to \mathcal{F}) \quad [x:(A \vee (A \to \mathcal{F})) \to \mathcal{F}]}}{APP(x,inr(\lambda y.APP(x,inl(y)))):\mathcal{F}}}{\lambda x.APP(x,inr(\lambda y.APP(x,inl(y)))):((A \vee (A \to \mathcal{F})) \to \mathcal{F}) \to \mathcal{F}}$$

Note that we started with '$y : A$', literally 'y is a proof of A', as an assumption. One can also prove the same theorem by starting with '$y : A \to \mathcal{F}$', literally 'y is a proof of $A \to \mathcal{F}$ (i.e. $\neg A$)', but in this case one would need classical implication as we shall see below.

Similarly to the proof of '$((A \vee (A \to \mathcal{F})) \to \mathcal{F}) \to \mathcal{F}$' above, we can prove intuitionistically that '$(\neg A \vee B) \to (A \to B)$':

$((A \to \mathcal{F}) \vee B) \to (A \to B)$:

$$\cfrac{\cfrac{\cfrac{[x:(A \to \mathcal{F}) \vee B] \quad \cfrac{\cfrac{[z:A] \quad [y:A \to \mathcal{F}]}{APP(y,z):\mathcal{F}} \quad \lambda\bot.:\mathcal{F} \to B}{APP(\lambda\bot.,APP(y,z)):B} \quad \cfrac{\cfrac{[t:B]}{t=t:B}}{t:B}}{CASE(x,\acute{y}APP(\lambda\bot.,APP(y,z)),\acute{t}t):B}}{\lambda z.CASE(x,\acute{y}APP(\lambda\bot.,APP(y,z)),\acute{t}t):A \to B}}{\lambda x.\lambda z.CASE(x,\acute{y}APP(\lambda\bot.,APP(y,z)),\acute{t}t):((A \to \mathcal{F}) \vee B) \to (A \to B)}$$

It is proper to remark here that in such an interpretation of *absurdity* we have just given, it is not the case that there are open terms of type '$\mathcal{F}$'. What we are saying here is that there are *closed* terms of the form '$\lambda\bot.$' of type '$\mathcal{F} \to A$' for any 'A'.[29]

----

[29]Howard's observation that by introducing an *absurdity* type such as '$\mathcal{F}$' one introduces open terms, does not seem to be applicable to our case:

"(i) For $\neg$: add a new prime formula f to P($\supset$). Then, for each formula α, introduce a term $A^{f \supset \alpha}$. (...) There *are* open terms of type f; for example, the variable X^f–which is a construction of f$\to$f." [Howard (1980)] (p. 483)

In the present framework, a construction of '$\mathcal{F} \to \mathcal{F}$' is also of the form '$\lambda\bot.$', therefore a closed term. After all, in our interpretation *absurdity* implies *anything*, including *absurdity* itself. But that does not imply that in our framework there is an open term (such as Howard's 'X^f') as a member of a closed type (such as Howard's 'f$\to$f'). (Briefly: similarly to Howard's case, any judgement of the form '$a : \mathcal{F}$' will not be a closed judgement, i.e. it will contain at least one free variable. But, unlike Howard's 'X^f' being a construction of 'f $\to$ f', an open term (such as 'a') could not be a construction of '$\mathcal{F} \to \mathcal{F}$'.) This leaves us to justify the equivalence of a term like '$\lambda x.x$' to the term '$\lambda\bot.$', given that both are terms of type '$\mathcal{F} \to \mathcal{F}$', but we need not worry too much about it.

9. $((A \to B) \to A) \to A$ *(Peirce's law)*[30]

In his *Foundations of Mathematical Logic* [Curry (1963)] Curry presents an 'inferential' counterpart to the axiomatic form of the *Peirce's law* as:

$$\frac{[A \to B]}{\frac{A}{A}} \quad \text{(with labels alongside formulas:} \quad \frac{[x : A \to B]}{\frac{b(x) : A}{\lambda x.b(x) : A}})$$

in [Curry (1963)] (p. 182).

One of the usual presentations of the rule for *reductio ad absurdum* such as:

$$\frac{[\neg A]}{\frac{A}{A}}$$

can be seen as a particular case of *Peirce's law* in its deductive (non-axiomatic) presentation, when negation is introduced (taking $\neg A \equiv A \to \mathcal{F}$):

$$\frac{[A \to \mathcal{F}]}{\frac{A}{A}}$$

where 'B' is instantiated with '$\mathcal{F}$'.[31]

---

[30] As it is well known, this axiom does not find a straight counterpart in the type-schemes of Curry's combinators. Nonetheless, it seems unlikely that Curry intended his theory of functionality to be applicable only to intuitionistic implication. Rather, he appeared to be more interested in defining families of *calculi* of implication, which would also include a calculus of classical implication, such as his LC- (HC-, TC-) systems as classical counterparts to intuitionistic LA- (HA-, TA-) systems, 'A' standing for 'absolute':

"4. The classical positive propositional algebra. In Sec. 4C5 we saw that the scheme

$$(A \supset B) \supset A \leq A \tag{15}$$

was not an elementary theorem scheme of an absolute implicative lattice, and in Sec. 4D1 a classical implicative lattice was defined, in effect, as an implicative lattice for which (15) holds. This classical implicative lattice is here called the *system EC*.

Acting by analogy with the absolute system, we can define classical positive propositional systems HC and TC by adjoining to HA and TA, respectively, postulates in agreement with (15). The postulate for HC is the scheme

Pc $\vdash A \supset B. \supset A :\supset A\ddagger$

which is commonly known as 'Peirce's law'; that for TC is the rule

$$\frac{[A \supset B]}{\frac{A}{A}}\text{,"}$$

Pk

[Curry (1963)] (p. 182)

(The rule 'Pk' was already presented as early as in Curry [Curry (1950)], where due reference to a paper by K. Popper is made.)

So, pursuing what we believe to have been Curry's 'methodology', which was to direct the chief concern at the establishment of calculi (intuitionistic, classical, etc.) rather than at the interpretations, we want to obtain a calculus of classical implication within the 'propositions-are-types' interpretation

Now, if we use the formulation given by Curry to construct the proof-tree for the axiom scheme '$((A \to B) \to A) \to A$' similarly to the one given above, we get:

$$\frac{\dfrac{\boxed{[y : A \to B]} \qquad [x : (A \to B) \to A]}{APP(x,y) : A}}{\dfrac{\boxed{\lambda y.} APP(x,y) : A}{\lambda x.\lambda y.APP(x,y) : ((A \to B) \to A) \to A}} (*)$$

where the step '$(*)$' is justified by Curry's inferential presentation of *Peirce's law* above. It allows the rewriting of '$(A \to B) \to A$' to 'A', in this direction. In the opposite direction the rewriting can be made with a weaker implication such as strict implication, as shown by the following proof-tree:

$$\frac{\dfrac{\boxed{[y : A \to B]}}{\dfrac{[x : A]}{\boxed{\lambda y.} x : (A \to B) \to A}}}{\lambda x.\lambda y.x : A \to ((A \to B) \to A)}$$

(Note that the abstraction '$\boxed{\lambda y.}$' is not a relevant abstraction, but it is an intuitionistic one.)   ■

---

by extending the conditions for closing a term (therefore binding a free variable and discharging an assumption), so that one can obtain a closed 'pure' (?) term for '$((A \to B) \to A) \to A$'.

[31] Note that the axiom corresponding to *Peirce's law* has nothing to do with the axiom which introduces negation (or better, *absurdity*). It only requires that the rules for assumption discharge with λ-*abstraction* be changed to cover the full power of classical implication. There are some slightly different views on this particular point, such as, e.g. Lambek's:

"The negationless formula $A \Leftarrow (A \Leftarrow (B \Leftarrow A))$ is a theorem classically but not in the system without negation."

[Lambek (1980)] (p. 384)

The system without negation referred to by Lambek corresponds to the system we had before introducing the *absurdity* judgement. *Peirce's axiom* cannot be expected to be a theorem of that system, given that the extra conditions of assumption-discharge were not present in it. But those extra conditions are not introduced specifically to allow the handling of negation.

In the presentation of propositional calculus as a deductive system, Lambek and Scott also insist on the fact that it is by introducing (double) negation that one obtains classical implication:

"If we want *classical* propositional logic, we must also require

R7. $\perp \Leftarrow (\perp \Leftarrow A) \to A$."   [Lambek and Scott (1986)] (p. 50)

The point here is that the double negation above (R7) follows from the more general characteristic of classical implication which is captured by *Peirce's law* and which is not intrinsically bound to the introduction of (explicit) negation. In other words, one can have a negationless classical implication by dropping the *absurdity* axiom (or $\lambda \perp$.-*abstraction*) from full intuitionistic implication and adding *Peirce's law*. (Cf. Curry's system TC of *classical positive propositional logic*, discussed in [Curry (1963)].)

As the reader can check, it is straightforward to prove the classical double negation '$((A \to \mathcal{F}) \to \mathcal{F}) \to A$' by using Curry's inferential counterpart to *Peirce's law*.

Moreover, as we have mentioned above, we can prove classically '$\neg\neg(A \vee \neg A)$' starting from the assumption that '$y : \neg A$' (i.e. '$y : A \to \mathcal{F}$').

Indeed, using Curry's $\dfrac{\begin{array}{c}[A \to B]\\ A\end{array}}{A}$, we have:

$$\cfrac{\cfrac{\cfrac{\cfrac{\cfrac{\cfrac{[y : A \to \mathcal{F}]}{inr(y) : A \vee (A \to \mathcal{F})} \quad [x : (A \vee (A \to \mathcal{F})) \to \mathcal{F}]}{APP(x, inr(y)) : \mathcal{F}} \quad \lambda\bot. : \mathcal{F} \to A}{APP(\lambda\bot., APP(x, inr(y))) : A}}{\lambda y.APP(\lambda\bot., APP(x, inr(y))) : A}}{inl(\lambda y.APP(\lambda\bot., APP(x, inr(y))) : A \vee (A \to \mathcal{F}) \quad [x : (A \vee (A \to \mathcal{F})) \to \mathcal{F}]}}{APP(x, inl(\lambda y.APP(\lambda\bot., APP(x, inr(y))) : \mathcal{F}}}{\lambda x.APP(x, inl(\lambda y.APP(\lambda\bot., APP(x, inr(y))) : ((A \vee (A \to \mathcal{F})) \to \mathcal{F}) \to \mathcal{F}}.$$

Despite working well in most cases, Curry's inferential counterpart to Peirce's law does not seem to be sufficient to prove the following theorem of classical implication (taking $\neg A \equiv A \to \mathcal{F}$):

$$((A \to \neg A) \to B) \to ((A \to B) \to B)$$

whose proof is left as an exercise by Curry in his *Foundations of Mathematical Logic* [Curry (1963)] (p. 279). We have been able to prove it using a reformulation of the inferential counterpart to Peirce's axiom, which seems to be fit more easily within the functional interpretation,[32] and is framed as follows:

$$\frac{\begin{array}{c}[A \to B]\\ B\end{array}}{A} \qquad \text{(provided `}A \to B\text{' is used as both minor and ticket)}$$

In the framework of the functional interpretation it would be framed as:

$$\frac{\begin{array}{c}[x : A \to B]\\ b(x, \ldots, x) : B\end{array}}{\lambda x.b(x, \ldots, x) : A} \quad \boxed{\begin{array}{c}\text{`}A \to B\text{'}\\ \text{as minor \&}\\ \text{as ticket}\end{array}}$$

meaning that if from the assumption that a term 'x' belongs to a type of the form '$A \to B$' one obtains a term '$b(x)$' belonging to the consequent 'B' where

---

[32]Notice that with Curry's *reductio ad absurdum* there is no particular configuration in the labels which did not show up in the proof of axiom schemas for weaker logics. We shall see that in our alternative to Curry's rule an interesting configuration in the labels, namely at least an instance of a sort of restricted self-application, will appear.

'x' appears both as a 'higher' and a 'lower' subterm of '$b(x)$', then we can apply a $\lambda x.$-abstraction over the '$b(x,\ldots,x)$' term and obtain a term of the form '$\lambda x.b(x,\ldots,x)$' belonging to the antecedent 'A', discharging the assumption '$x : A \to B$'. With that special proviso we can prove *Peirce's axiom* in the following way:

$$\cfrac{\cfrac{\cfrac{[y : A \to B] \qquad [x : (A \to B) \to A]}{APP(x,y) : A}}{\cfrac{APP(y, APP(x,y)) : B}{\lambda y.APP(y, APP(x,y)) : A} \qquad [y : A \to B]}(*)}{\lambda x.\lambda y.APP(y, APP(x,y)) : ((A \to B) \to A) \to A},$$

where in the step marked with '$(*)$' we have applied our alternative to Curry's *generalised reductio ad absurdum*. The resulting term '$\lambda x.\lambda y.APP(y, APP(x,y))$' which intuitionistically would have a type-scheme of the form '$((A \to B) \to A) \to ((A \to B) \to B)$', we would call '$P'$'. In fact, P' follows from a weakened version of the excluded middle, namely,

$$((A \to B) \to A) \vee (A \to B)$$

in the system of intuitionistic implication extended by allowing two assumptions of the form '$A \to B$' (one used as minor premise and the other used as major premise of a modus ponens) to be discharged by one single $\to$-*introduction*. Notice the (restricted) self-application which is being uncovered in the resulting term: 'y' is being applied to the result of an application of 'x' to 'y' itself.

Furthermore, the resulting alternative presentation of the inferential counterpart to *Peirce's axiom* finds a special case in another one of the standard presentations of the proof-theoretic *reductio ad absurdum*, namely,

$$\cfrac{\begin{array}{c}[\neg A]\\ \mathcal{F}\end{array}}{A} \qquad \text{which can also be presented as} \qquad \cfrac{\begin{array}{c}[A \to \mathcal{F}]\\ \mathcal{F}\end{array}}{A}$$

By using such an alternative to Curry's formulation we can also prove the classical double negation:

$$\cfrac{\cfrac{\cfrac{\cfrac{[y : A \to \mathcal{F}] \qquad [x : (A \to \mathcal{F}) \to \mathcal{F}]}{APP(x,y) : \mathcal{F}} \qquad \lambda\bot. : \mathcal{F} \to A}{APP(\lambda\bot., APP(x,y)) : A}}{APP(y, APP(\lambda\bot., APP(x,y))) : \mathcal{F}} \qquad [y : A \to \mathcal{F}]}{\cfrac{\lambda y.APP(y, APP(\lambda\bot., APP(x,y))) : A}{\lambda x.\lambda y.APP(y, APP(\lambda\bot., APP(x,y))) : ((A \to \mathcal{F}) \to \mathcal{F}) \to A}}.$$

It is possible to extend this *generalised reductio ad absurdum* to the many-valued implicational logics of Łukasiewicz. For the present framework, however, we shall restrict ourselves to 2-valued classical implicational logic with the *Peirce's law*.[33]

---

[33] We are also looking for a proof discipline (and corresponding λ-abstraction discipline) for Łukasiewicz' $\aleph_0$- and m-valued logics which would perhaps gives us an insight as to which minimal discipline to adopt for classical (2-valued) implication. Curry's inferential counterpart of *Peirce's axiom* appears to be a strong candidate, giving that it is general enough to prove all Łukasiewicz-Tarski (1930) axioms for $\aleph_0$-valued logics, namely,

L1. $A \to (B \to A)$
L2. $(A \to B) \to ((B \to C) \to (A \to C))$
L3. $((A \to B) \to B) \to ((B \to A) \to A)$
L4. $((\neg A \to \neg B) \to (B \to A)$
L5. $((A \to B) \to (B \to A)) \to (B \to A)$.

We only need to worry about L3 and L5 (although it is well known that the latter follows from L4), given that L1 and L2 each has a counterpart in the type-scheme of a combinator (K and B', respectively). To be able to deal with m-valued logics within the functional interpretation, we also have to find a λ-abstraction counterpart to a proof discipline for *Rose's axiom* [Rose (1956)] (axiom A5), namely '$((A^{m-1} \to B) \to A) \to A$', which is a kind of generalised form of *Peirce's axiom*. (For $m = 2$ it becomes '$((A \to B) \to A) \to A$', and for $m = 3$ it takes the form of '$((A \to (A \to B)) \to A) \to A$'.)

In fact, Curry's rule, as it stands, will collapse all m-valued logics into classical 2-valued logics by failing to distinguish *Rose's axiom* from *Peirce's axiom*. However, both Curry's inferential counterpart to *Peirce's axiom* and our own alternative formulation mentioned above appear to be capable of dealing with *Rose's axiom* as a general form of *Peirce's axiom*. They would only need to be put in a generalised form taking into account the number of values 'm'. So, for m-valued logics, we would have assumption-discharge rules (i.e. λ-abstraction rules) of the following form:

$$\text{Curry's:} \quad \frac{\begin{array}{c}[y : A^{k-1} \to B]\\ b(y) : A\end{array}}{\lambda y.b(y) : A} \qquad \text{Ours:} \quad \frac{\begin{array}{c}[y : A^{k-1} \to B]\\ b(y, \ldots, y) : B\end{array}}{\lambda y.b(y, \ldots, y) : A} \left|\begin{array}{c}\text{'}A^{k-1} \to B\text{'}\\ \text{as minor \&}\\ \text{as ticket}\end{array}\right.,$$

where $k \geqslant m$, and in our own formulation the proviso is that '$b(y, \ldots, y)$' contains 1 occurrence of 'y' as a 'higher' subterm (i.e. '$A^{k-1} \to B$' used once as ticket) and at least $k - 1$ occurrences as a 'lower' subterm (i.e. '$A^{k-1} \to B$' used $k - 1$ times as minor premiss). As an example, to prove the 3-valued axiom '$((A \to (A \to B)) \to A) \to A$' we have:

$$\frac{\dfrac{\dfrac{[y:A \to (A \to B)] \quad [x:\ldots]}{APP(x,y):A} \quad \dfrac{[y:A \to (A \to B)][x:(A \to (A \to B)) \to A]}{APP(x,y):A} \quad [y:A \to (A \to B)]}{\dfrac{APP(APP(y,APP(x,y)),APP(x,y)):B}{\lambda y.APP(APP(y,APP(x,y)),APP(x,y)):A}}}{\lambda x.\lambda y.APP(APP(y,APP(x,y)),APP(x,y)):((A \to (A \to B)) \to A) \to A} (*)$$

where in the step '$(*)$' the $\lambda y.$-abstraction was made over a term containing 'y' once as a 'higher' subterm and twice as a 'lower' subterm, discharging all three occurrences of the assumption '$y : A \to (A \to B)$'.

Type lifting. We mentioned in the introductory section that the mechanism of completing diagrams is used to obtain the functional interpretation of intermediate logics between intuitionistic and classical implicational logic. One of the principles underlying this mechanism is that of 'type lifting'. Starting from a type 'A', we can lift it up to '$A \rightarrow ((A \rightarrow B) \rightarrow B)$' by a simple deduction procedure, namely,

$$\frac{\dfrac{[x:A] \qquad [y:A \rightarrow B]}{APP(y,x):B}}{\dfrac{\lambda y.APP(y,x):(A \rightarrow B) \rightarrow B}{\lambda x.\lambda y.APP(y,x):A \rightarrow ((A \rightarrow B) \rightarrow B)}},$$

and this shows that we can go from 'A' to '$A \rightarrow ((A \rightarrow B) \rightarrow B)$' for any 'B'.

Now, let us see what happened with our proof of *Peirce's law* in terms of type lifting. We started with

$$((A \rightarrow B) \rightarrow A) \rightarrow A,$$

and ended up with a proof of what would have been (without the use of our *generalised reductio ad absurdum*)

$$((A \rightarrow B) \rightarrow A) \rightarrow ((A \rightarrow B) \rightarrow B).$$

The relationship between the latter and the former, as it can be readily seen, is that a type lifting is made at the outermost consequent, namely 'A'. In other words, while the outermost consequent in the former formula is 'A', the consequent in the latter is '$(A \rightarrow B) \rightarrow B$'. Now, if we proceed by type lifting on the outermost consequent of other super-intuitionistic axiom schema, such as, e.g. Łukasiewicz':

$$((A \rightarrow B) \rightarrow B) \rightarrow ((B \rightarrow A) \rightarrow A)$$
$$((A \rightarrow B) \rightarrow (B \rightarrow A)) \rightarrow (B \rightarrow A),$$

and LC-logic:

$$((A \rightarrow B) \rightarrow C) \rightarrow (((B \rightarrow A) \rightarrow C) \rightarrow C),$$

we end up with

$$((A \rightarrow B) \rightarrow B) \rightarrow ((B \rightarrow A) \rightarrow ((A \rightarrow B) \rightarrow B))$$
$$((A \rightarrow B) \rightarrow (B \rightarrow A)) \rightarrow (B \rightarrow ((A \rightarrow B) \rightarrow B)),$$

and

$$((A \rightarrow B) \rightarrow C) \rightarrow (((B \rightarrow A) \rightarrow C) \rightarrow ((C \rightarrow (B \rightarrow A)) \rightarrow (B \rightarrow A))),$$

which are the formulas provable without applying our *generalised reductio ad absurdum*. Obviously, in the presence of such a (super-intuitionistic) inference rule, and noticing that the proofs of the latter involve the kind of self-application in the labels which we have mentioned above, the former ones will also be provable.

2.5   Consistency proof

Theorems of a certain logic are well-formed formulas which can be demonstrated to be true regardless of other formulas being true, i.e. a complete proof of a theorem relies on no assumptions. So, whenever we construct proofs in natural deduction, we need to look at the rules which discharge assumptions. This is only natural, because when starting from hypotheses and arriving at a certain thesis we need to say that the hypotheses imply the thesis. So, we need to look at the rules of inference which allow us to 'discharge' assumptions (hypotheses) without introducing further assumptions. It so happens that the *introduction* rules for the conditionals (namely, implication, universal quantifier, necessity) do possess this useful feature. They allow us to 'get rid of hypotheses' by making a step from 'given the hypotheses, and arriving at premise' to 'hypotheses imply thesis'. Let us look at the introduction rules for implication in the plain Natural Deduction style:

$\rightarrow$-*introduction*

$$\frac{\begin{array}{c}[A]\\ B\end{array}}{A \to B}.$$

Note that the hypothesis 'A' was discharged, and by the introduction of the implication, the conclusion '$A \to B$' (hypothesis 'A' implies thesis 'B') was reached.

Now, if we introduce labels alongside the formulas this 'discharge' of hypotheses will be reflected on the label of the intended conclusion by a device which makes the arbitrary name introduced as the label of the corresponding assumption 'lose its identity', so to speak. It is the device of 'abstracting' a variable from a term containing zero or more 'free' occurrences of that variable. So, let us look at how the rule given above looks like when augmented by inserting labels alongside the formulas:

$$\frac{\begin{array}{c}[x:A]\\ b(x):B\end{array}}{\lambda x.b(x):A \to B}.$$

Notice that when we reach the conclusion the arbitrary name 'x' loses its identity simply because the abstractor 'λ' binds its free occurrences in the term '$b(x)$' (which in its turn may have none, one or many free occurrence(s) of 'x').[34] Just think of the more usual variable binding mechanism on the formulas: being simply

---

[34]N.B. The notation '$b(x)$' indicates that '$b(x)$' is a functional term which depends on 'x', and not the application of 'b' to 'x'.

a place-marker, the 'x' has no identity whatsoever in x-quantified formulas such
as $\forall x.B(x)$ and $\exists x.B(x)$.

The moral of the story is that the last inference rule of any complete proof
must be the introduction of a conditional which is the rule doing the job we want:
discharging assumptions already made, without introducing any further assumptions. We are now speaking in more general terms now ('conditional', rather than
'implication') because the introduction rules for the universal quantifier and the
necessitation connectives, namely:

$$\begin{array}{cc} \forall - introduction & \Box - introduction \\ [x:D] & [\mathbb{W}:\mathcal{U}] \\ \dfrac{f(x):F(x)}{\Lambda x.f(x):\forall x^D.F(x)} & \dfrac{f(\mathbb{W}):A(\mathbb{W})}{\Lambda \mathbb{W}.f(\mathbb{W}):\Box A}\text{'} \end{array}$$

also discharge old assumptions without introducing new ones.[35]

Our motto here is that all labelled assumptions must be discharged by the end
of the proof, and we should be able to check this very easily just by looking at
the label of the intended conclusion and check if all 'arbitrary' names (labels)
of hypotheses are bound by any of the available abstractors.[36] So, in a sense
our proofs will be *categorical* proofs, to use a terminology from [Anderson and
Belnap Jr. (1975)] *Entailment*.[37] The notion of variable-binding, and the idea
of having terms representing incomplete 'objects' whenever they contain 'free'
variables was introduced in a systematic way by Frege in his *Grundgesetze*. And,

---

[35] The functional interpretation of 'necessity' ('$\Box$') is attempted in Gabbay and de Queiroz (1991a).

[36] The careful reader will have noticed that when proving the axiom '$(\neg A \vee B) \to (A \to B)$' above
we have made use of the abstractor '$'$' in the discharge of assumptions involved in the elimination of
disjunction. In addition to this, in de Queiroz and Gabbay (1991) we have used an 'ε'-abstractor to
construct terms/labels of $\exists$-type, and a '$'$'-abstractor to bind labels of Skolem-type assumptions being
discharged by the rule of $\exists$-*elimination*.

[37] "A proof is *categorical* if all hypotheses in the proof have been discharged by use of $\to$-I, otherwise
hypothetical; and A is a *theorem* if A is the last step of a categorical proof." [Anderson and Belnap Jr.
(1975)] (p. 9)

Only, in our case we also have the other conditionals ($\forall$, $\Box$) to discharge assumptions and therefore
make a proof categorical. For example, the formula

$$\forall y^D.F(y).(\forall x^D.F(x) \to F(y)),$$

given as an example of a first-order theorem in [Quine (1950)] (p. 180), schema (11), can be proved as:

$$\dfrac{\dfrac{\dfrac{[y:D] \qquad [u:\forall x^D.F(x)]}{EXTR(u,y):F(y)}}{\lambda u.EXTR(u,y):\forall x^D.F(x) \to F(y)}}{\Lambda y.\lambda u.EXTR(u,t):\forall t^D.(\forall x^D.F(x) \to F(y))}\text{'}$$

and the last (assumption-discharging) step of the proof was made by a rule of $\forall$-*introduction*, instead
of $\to$-*introduction*.

indeed, as early as 1893 Frege described (in his *Grundegesetze I*) what can be seen as the early origins of the notions of *abstraction* and *application* when showing techniques for transforming functional expressions into completed objects by an 'introductory' operator of abstraction producing the '*Wertverlauf*' expression,[38] e.g. '$\acute{\varepsilon}f(\varepsilon)$', and the effect of its corresponding 'eliminatory' operator '$\cap$' on a *course-of-values* expression.[39]

The idea of forming *course-of-values* expressions by abstracting over the corresponding free variable proves to be very useful in representing the handling of assumptions within a Natural Deduction style calculus. In particular, when the Natural Deduction presentation system is based on a 'labelling' mechanism the binding of free variables in the labels corresponds to the discharge of corresponding assumptions.

We consider a formula to be a theorem of the logical system if it can be derived in a way that its corresponding 'label' contains no free variable, i.e. the deduction rests on no assumption. In other words, a formula is a theorem if it admits a categorical proof to be constructed. To use a term coined by Frege (1891),[40] a proof

---

[38] Frege 1893, §3, p. 7, translated as *course-of-values* in [Furth (1964)] (§3, p. 36)

[39] Frege 1893, §34, p. 52ff, translated in [Furth (1964)] (§34, p. 92):

"(...) it is a matter only of designating the value of the function $\Phi(\xi)$ for the argument Δ, i.e. $\Phi(\Delta)$, by means of "Δ" and '$\acute{\varepsilon}\Phi(\varepsilon)$'. I do so in this way:

$$\text{``}\Delta \cap \acute{\varepsilon}\Phi(\varepsilon)\text{''},$$

which is to mean the same as '$\Phi(\Delta)$'."
(Note the similarity to the rule of functional *application*, where 'Δ' is the argument, '$\acute{\varepsilon}\Phi(\varepsilon)$' is the function, and '$\cap$' is the operator '$APP$'.)

Expressing how pleased he was with the introduction of a variable-binding device for the functional calculus (recall that the variable-binding device for the logical calculus had been introduced earlier in the *Begriffsschrift*), Frege says:

"The introduction of a notation for courses-of-values seems to me to be one of the most important supplementations that I have made of my *Begriffsschrift* since my first publication on this subject." [*Grundgesetze I*, §9, p. 15f.]

It may be appropriate to recall that Peano's early essay on arithmetic (1889), published before Frege's (1891) account of *Wertverlauf*, already contained a device for functional abstraction and application.

[40] When speaking of 'unsaturated' expressions Frege compares the free variables of an expression with holes (empty places):

"(...) we may discern that it is the common element of these expressions that contains the essential peculiarity of a function; i.e. what is present in

$$\text{`}2.x^3 + x\text{'}$$

over and above the letter 'x'. We could write this somewhat as follows:

$$\text{`}2.(\ )^3 + (\ )\text{'}.$$

I am concerned to show that the argument does not belong with a function, but goes together with the function to make up a complete whole; for a function by itself must be called incomplete, in need of supplementation, or 'unsaturated'."

will be categorical if the corresponding proof-construction term is a *saturated* function-term, i.e. there is no variable occurring free in it.

So, following a technique presented in Martin-Löf [Martin-Löf (1972)], we shall prove that our systems of implication are consistent by demonstrating that no judgement of the form '$m : M$', where 'm' is a *saturated* term, for 'M' atomic, can be obtained from the presentation of the →-type. In order to end with a closed term and have all assumptions discharged we need to have proofs ending with at least one →-*introduction*. (Obviously, that is not sufficient, because we must make sure that the proof is normal.)

First of all, we define the notion of *saturation gap*, which shall work as a measure for the saturation of our proof-construction terms (and thus for the *categoricity* of our proofs).

Definition 2.1. Let m be a term built up by variables, λ-abstraction and application. The *saturation gap* (abbr. '*satgap*') of m is defined as the number of occurrences of free variables it contains. It is defined inductively as:

$$satgap(\lambda\bot.) = 0$$
$$satgap(x) = 1 \qquad\qquad\qquad \text{where } x \text{ is a variable}$$
$$satgap(APP(m, n)) = satgap(m) + satgap(n)$$
$$satgap(\lambda x.f(x)) = satgap(f) - nfree(x)$$

where '$nfree(x)$' is the number of free occurrences of x in f which are accessible by the appropriate discipline of assumption discharge (which varies according to the assumption-handling procedure of each logic).   □

Definition 2.2. A proof m of a proposition M (abbr. '$m : M$') is said to be *categorical* if and only if its proof-construction term m has *saturation gap* equal to 0.   □

Later, when speaking of 'value-range' of functions (Ger. '*Wertverlauf*', also translated as 'course-of-values' of functions) Frege gives the reason why the expressions containing no 'empty place' (i.e. free variable) must be taken as completed objects:

"When we have thus admitted objects without restriction as arguments and values of functions, the question arises what it is that we are here calling an object. I regard a regular definition as impossible, since we have here something too simple to admit of logical analysis. It is only possible to indicate what is meant. Here I can only say briefly: An object is anything that is not a function, so that an expression for it does not contain any empty place.

(...) '$\acute{\varepsilon}(\varepsilon^2 - 4\varepsilon)$', is fully complete in itself and thus stands for an object. Value-ranges of functions are objects, whereas functions themselves are not." (Frege 1891, pp. 18f, (p. 147 of the translation).)

Recall that in Frege's terminology, 'functions' are expressions containing 'empty places' (free variables).

Now, recall that the only assumptions allowed in any step of a proof is of the form '$[z : A]$' where 'z' is simply a (new) variable, and 'A' is a (not necessarily atomic) type. If an undischarged occurrence of the assumption 'A' already appears labelled by 'z', then 'z' has to be used for further occurrences of the same undischarged assumption 'A', which means that once a assumption-formula is given a label, it cannot be given a different one from that point onwards. Why adopting such a convention is a question which needs to be answered by looking at the true gist of the Curry–Howard interpretation: the matching between the functional calculus on the labels with the logical calculus on the formulas. In other words, by using labels/terms alongside the formulas, we are able to

(1) keep track of proof steps (gaining local as well as global control) and
(2) handle arbitrary names (via abstractors).

Having these two useful features our functional interpretation framework gives us at least two advantages over the usual plain Natural Deduction systems:

(1) It matches

    (a) the functional calculus on the labels with
    (b) the logical calculus on the formulas.

(2) It takes care of 'contexts' and 'scopes' in a more explicit fashion.[41]

Now, if this matching — functional calculus on the labels, logical calculus on the formulas — is to be taken seriously, we ought to make sure that assumptions have a unique label until they are discharged. If this is not required, our discipline of λ-abstraction on the labels will not be effective in making sure we accept/reject the right theorems. For example, let us take the axiom-schema corresponding to *contraction*, which we know is not a theorem of linear implication:

$$(A \to (A \to B)) \to (A \to B),$$

and let us construct a proof tree without abiding by the convention: different occurrences of an assumption-formula can be labelled by a different variable even

---

[41]As we point out in de Queiroz and Gabbay (1997), this is crucial especially when dealing with the notion of 'contexts' as in modal logic (such as, e.g. the problem of skolemisation across 'worlds'), as well as in the sequent calculus.

before they are discharged. The proof-tree would result in

$$\frac{[y:A] \quad \dfrac{[z:A] \quad [x:A \to (A \to B)]}{APP(x,z):A \to B}}{\dfrac{APP(APP(x,z),y):B}{\dfrac{\lambda z.APP(APP(x,z),y):A \to B}{\dfrac{\lambda y.\lambda z.APP(APP(x,z),y):A \to (A \to B)}{\lambda x.\lambda y.\lambda z.APP(APP(x,z),y):(A \to (A \to B)) \to (A \to (A \to B))}}}},$$

and by allowing the assumption 'A' to be labelled once by 'z' and once by 'y', we did not succeed either in proving the right theorem, or in demonstrating that the initial proposition is not a theorem of linear implication because it violates the discipline of discharging assumptions one by one.

Lemma 2.3. *A proof containing an* introduction *followed by* elimination *such as in*:

$$\frac{m:A \quad \dfrac{\dfrac{[x:A]}{b(x):B}}{\lambda x.b(x):A \to B} \to \text{-}intr}{APP(\lambda x.b(x),m):B} \to \text{-}elim,$$

can be normalised to:

$$\frac{[m:A]}{b(m/x):B}$$

via β*-reduction, and the saturation gap of the label of the end formula (i.e. label* '$b(m)$' *of formula 'B') does not decrease.*

Proof. Standard, except that the binding and substitution of variables will be parameterised by the logic (i.e. assumption-handling discipline). ∎

Lemma 2.4. *Regardless of the logic (i.e. assumption-handling discipline) being used, no atomic proposition has a categorical proof.*

Proof. In the rule of →*-introduction* assumptions are only discharged, so only by introducing an → one can make the saturation gap decrease.

The only odd case is that of the rule of *generalised reductio ad absurdum*, where a λ*-abstraction* term is said to belong to a 'lower' type and the assumption on the higher type is discharged:

$$\frac{[x:A \to B]}{\dfrac{b(x,\ldots,x):B}{\lambda x.b(x,\ldots,x):A}} \boxed{\begin{array}{c} \text{'}A \to B\text{'} \\ \text{as minor \&} \\ \text{as ticket} \end{array}}$$

Here the higher type assumption '$A \to B$' is discharged and the lower type 'A' is asserted to contain a λ-abstraction term '$\lambda x.b(x, \ldots, x)$'. So, it would seem that an atomic formula could have a saturated proof-construction. But to obtain 'A' from the assumption '$A \to B$' one would have to be using assumption '$A \to B$' once as ticket once as minor: the label 'x' appears (at least) once as argument and (at least) once as function, so the restricted self-application. ∎

The point is that a judgement where a λ-term is a member of an atomic 'A' will still have free variables and therefore will not be a closed term.

§5.1 A λ-term can inhabit a type which may not be an $\to$-type

We have seen above that the η-convertibility rule, namely,

$\to$-*induction*

$$\frac{c : A \to B}{\lambda x.APP(c, x) = c : A \to B}$$

states that for any arbitrary term 'c' which inhabits an $\to$-type, it must be susceptible to an application to an argument, giving as a result a term which when abstracted from the argument given for the application, should give back the original λ-abstraction term. Thus, take any arbitrary '$c : A \to B$', then 'c' must be a λ-abstraction term. The η-convertibility rule guarantees that no term other than λ-abstraction terms inhabits $\to$-types.

The guidelines for the definition of logical connectives in the framework of the functional interpretation described here prescribe that one needs to say: (1) what a type contains (*introduction*) by showing how to construct a canonical element, as well as when two canonical elements are equal; (2) how an element of the type is deconstructed (*reduction*); and (3) that any term other than those constructed by the *introduction* rules can inhabit the type. None of the rules (*introduction, reduction, induction*) says anything, however, as to whether a λ-abstraction term can inhabit a type which is not necessarily an $\to$-type. If we remind ourselves that the labelled Natural Deduction rules given for '$\to$' had the type as the focus of attention, we have no difficulties accepting that the rule of *reductio ad absurdum*, namely,

$$\frac{\begin{array}{c}[x : A \to B]\\ b(x, \ldots, x) : B\end{array}}{\lambda x.b(x, \ldots, x) : A} \quad \boxed{\begin{array}{c}\text{'}A \to B\text{'}\\ \text{as minor \&}\\ \text{as ticket}\end{array}}$$

does no harm to the methodology: it only says that a λ-abstraction term can also inhabit a type which may not necessarily be an $\rightarrow$-type. The rule allows us to conclude that the λ-abstraction term '$\lambda x.b(x, \ldots, x)$' inhabits a type of the form 'A', which may or may not be an $\rightarrow$-type.

2.6   Systems of implication and combinators

In Gabbay's (1989) LDS a classification of different systems of implication is given in terms of the axioms chosen from a stock of basic axioms:

1. *Identity*
   $A \rightarrow A$, which corresponds to the type-scheme of combinator 'I'.

2. *Left transitivity*
   $(A \rightarrow B) \rightarrow ((C \rightarrow A) \rightarrow (C \rightarrow B))$, which corresponds to the type-scheme of combinator 'B'.

3. *Right transitivity*
   $(A \rightarrow B) \rightarrow ((B \rightarrow C) \rightarrow (A \rightarrow C))$, which corresponds to the type-scheme of combinator 'B''.

4. *Distribution*
   $(A \rightarrow B) \rightarrow ((A \rightarrow (B \rightarrow C)) \rightarrow (A \rightarrow C))$, which is a variation of the type-scheme of combinator 'S', namely 'SC'.

5m, n. (m, n) *Contraction*
   $(A^m \rightarrow B) \rightarrow (A^n \rightarrow B)$, which is a generalisation of the type-scheme of combinator 'W', namely $(A \rightarrow (A \rightarrow B)) \rightarrow (A \rightarrow B)$.[42]

---

[42]By presenting contraction in its 'generalised' form we can see that in some logics one can still do contraction although not necessarily the well known (2,1)-contraction corresponding to the type-scheme of combinator 'W'. In [Ono and Komori (1985)] it is rightly remarked that Łukasiewicz' n-valued logics, where n is finite and $n > 2$, do not have (2,1)-contraction:

"For each positive integer m, L_m denotes the set of all formulas valid in the $(m + 1)$-*valued model of Łukasiewicz*. We can easily show that the formula $(p \supset p \supset q) \supset (p \supset q)$, which corresponds to the contraction rule, does not belong to L_m for $m \geqq 2$." [Ono and Komori (1985)] (p. 190) Although no mention is made of the fact that it is only a particular kind of contraction that is allowed in Łukasiewicz' L_m, namely $(m + 1, m)$-contraction (cf. axiom 'A4 $(p \rightarrow^k q) \rightarrow (p \rightarrow^{k-1} q)$' of Tuziak's [Tuziak (1988)] L_k.), unlike Girard's linear logic where *no* form of contraction is allowed. In 3-valued Łukasiewicz' L_2, for example, the proposition '$(A \rightarrow (A \rightarrow (A \rightarrow B))) \rightarrow (A \rightarrow (A \rightarrow B))$', which corresponds to (3,2)-contraction, is valid. It is proved in a similar way to (2,1)-contraction

6α. α-weakening

$\alpha \rightarrow (B \rightarrow \alpha)$, which comprises variations of the type-scheme of combinator 'K'. For example, by saying that α has to be in implicational form we get a $K_\rightarrow$ for strict implication.

7γ. γ-permutation

$(A \rightarrow (\gamma \rightarrow C)) \rightarrow (\gamma \rightarrow (A \rightarrow C))$, which corresponds to variations of the type-scheme of combinator 'C'. For entailment implication, one has to impose the condition that γ must be in implicational form (e.g. '$P \rightarrow Q$'), and we here call the corresponding combinator '$C_\rightarrow$'. If we impose that '$A \nvdash \gamma$' in the particular system, we have the variation '$C_{\nvdash}$'.

8. Restart (Peirce's law)

$((A \rightarrow B) \rightarrow A) \rightarrow A$, whose weakened version corresponds to the type-scheme of our combinator 'P''.

Based on the correspondence between the axioms and the type-schemes of combinators, we can do the same classification done in Gabbay (1989), but now in terms of the combinators which would be obtained according to the side conditions on the rule of *assertability conditions* for the logical connective of implication, namely →-*introduction*.

System name:	Combinators:	Abstraction discipline
Concatenation	I, B	one by one, strict order
W	I, B, B'	one by one, order: entry > functional
Linear	I, B, B', C	one by one, full permutation
Modal T-Strict	I, B B', C, W	non-vacuous
Relevant	I, B, B', C, S	non-vacuous
Entailment	I, B, B', S, $C_\rightarrow$	non-vacuous, restricted permutation

that we have proved above, i.e.:

$$\frac{\dfrac{[y:A] \qquad \dfrac{[y:A] \qquad [x:A \rightarrow (A \rightarrow (A \rightarrow B))]}{APP(x,y):A \rightarrow (A \rightarrow B)}}{\dfrac{APP(APP(x,y),y):A \rightarrow B}{\lambda y.APP(APP(x,y),y):A \rightarrow (A \rightarrow B)}}}{\lambda x.\lambda y.APP(APP(x,y),y):(A \rightarrow (A \rightarrow (A \rightarrow B))) \rightarrow (A \rightarrow (A \rightarrow B))}.$$

Observe that the terms/constructions are verbatim those of the proof-tree for the axiom corresponding to the type-scheme of 'W'. This means that in the framework of the formulae-as-types interpretation, to deal with the generalised form of contraction in the same spirit of the present paper we would have to take into account not only of the number of free occurrences of a variable being discharged by the abstraction, but also of the number of occurrences of the proposition 'A'.

Ticket Entailment	I, B, B′, S, SC	non-vacuous, restricted permutation
Strict	I, B, B′, C, K$_\rightarrow$	restricted non-vacuous
Minimal	I, B, B′, C, S, K	vacuous
Intuitionistic	I, B, B′, C, S, K, I$_\mathscr{F}$	vacuous, with *absurdity*
Classical	I, B, B′, C, S, K, P′	intuitionistic plus 'function & argument'

The systems are roughly ordered by proof-theoretic strength, and the system of concatenation logic is considered to be the weakest implicational system for which a reasonable deduction theorem exists.[43] Note that the most primitive combinators are I, B, B′, C and W, instead of I, K and S. These were, in fact the primitive combinators in Curry's earliest results on combinatory logic, unlike Schönfinkel's independent pioneering results using B, C, I, K, and S.[44]

2.7   Finale

With the help of the distinction between a *proposition* ('*A*') and a *judgement* ('*a* : *A*'), together with the merge of a functional calculus on the labels with a logical calculus on the formulas leading to the identification of *propositions* with *types*, and the, one can have proof-objects, which would normally belong to the meta-language, 'coded' into the object language. Such an 'improvement' on the syntactical tools of a proof calculus seems to be particularly helpful in dealing with the so-called resource logics. By making the type-theoretic equivalent of the λ-calculus' *abstraction* rule into a 'resource' *abstraction* where an extra condition is included requiring the existence of at least a free occurrence of the variable being abstracted, one can provide a workable framework to present resource logics via the so-called Curry–Howard interpretation in a simple way.

Concerning the classification of the various logics via the functional interpretation, it is appropriate to point out the fundamental difference between the framework adopted here and the so-called 'structural' logical frameworks, i.e. those based on Gentzen's sequent calculus and the structural properties of sequents, such as Lambek's [Lambek (1958)] calculus, and Girard's [Girard (1987)] linear

---

[43] Here we should mention an attempt by Y. Komori (in a handwritten memo — Komori (1990) — which Prof. Komori kindly sent to us) to answer the question 'What is the weakest *meaningful* logic?' by saying that "[t]he weakest meaningful logic is B Logic," where B is a logic with only one axiom (the one corresponding to the type-scheme of combinator B) and a rule of modus ponens. The B-logic could be Gabbay's 'concatenation' logic without reflexivity.

[44] "The earliest work of Curry (till the fall of 1927), which was done without knowledge of the work of Schönfinkel [Schönfinkel, 1924], used B, C, W, and I as primitive combinators.", [Curry and Feys (1958)] (p. 184).

logic. Whereas in the structural approaches the idea is to amalgamate the structure (structured 'contexts' of formulas) and the logical connectives, giving rise to rather intriguing duplications — e.g. 'left' and 'right' implications (Lambek); additive and multiplicative conjunctions, as well as additive and multiplicative disjunctions (Girard) — , we are here concerned with the properties of the connective itself. By adding a label alongside formulas we get a 'two-dimensional proof calculus', where a functional calculus on the labels and a logical calculus on the formulas work harmoniously in parallel, the label helping to keep track of proof steps and to regain local control (virtually lost when the labels are not there). Take implication, for example. We have seen above that implication can be classified by using only its own properties: the application (modus ponens) defines an order; the abstraction discipline (assumption discharge) may vary, etc. The handling of structural properties of the context of formulas can still be made though at a different level of abstraction.[45]

Concerning other logical connectives, we are currently working on the extension of the classification presented here for the case of implication to first-order quantification, given that in the Curry–Howard interpretation the universal quantifier is dealt with in a similar manner to implication.[46] And indeed, by presenting

---

[45] In Gabbay and de Queiroz (1991b) we consider the introduction of explicit structural operations in order to generalise Gentzen's sequent calculus into a proof calculus where the structured contexts are given names and are handled together with the connectives.

[46] In Howard's account of the formulae-as-types notion of construction he defines constructions as terms built up from prime terms by means of term formation as indicated by *Prime terms*, λ-*abstraction* and *Application*:

"(i) *Type symbols.* The prime type symbols are: 0 and every equation of $H(\supset, \wedge, \forall)$. From these we generate all type symbols by the following two rules:

(a) From α and β get $\alpha \supset \beta$ and $\alpha \wedge \beta$.

(b) From α and a number variable x get $\forall x \alpha$.

(ii) *Prime terms.* These are:

(a) number variables x, y,. . .; constants 0 and 1; function symbols for plus and times,

(b) variables X^α, Y^β,. . .,

(c) certain special terms, mentioned in §8, below, corresponding to axioms and rules of inference of $H(\supset, \wedge, \forall)$.

(iii) λ-*abstraction*:

(a) From F^β get $(\lambda X^\alpha . F^\beta)^{\alpha \supset \beta}$ as in §2.

(b) If x does not occur free in the type symbol of any free variable of F, form $(\lambda x F^\beta)^{\forall x \beta}$.

(iv) *Application*:

(a) From F^α and $G^{\alpha \supset \beta}$ form $(GF)^\beta$ as in §2.

(b) From $G^{\forall x \alpha(x)}$ and t of type 0 form $G(t)^{\alpha(t)}$." [Howard (1980)] (p. 485)

Observe that for both implication ('(i.a)', '(iii.a)' and '(iv.a)') and universal quantification ('(i.b)', '(iii.b)', '(iv.b)') a λ-system is used (with *abstraction* and *application*). Later he gives the following axioms for the reducibility of terms:

"*11. Normalisation of terms*

For the theory of reducibility of terms we postulate the following contraction schemes

(i) $(\lambda X.F(X))^{\alpha \supset \beta} G$ contr $F(G)^{\beta}$

   $(\lambda x.F(x))^{\forall x \alpha(x)} t$ contr $F(t)^{\alpha(t)}$

(...)"

(Ibid., p. 487.)

The equivalent of β-normalisation is postulated to the contraction of terms characterising implication, as well as terms characterising universal quantification.

In the description of his type system F Girard also makes use of the notions of *abstraction* and *application* in the definition of both implication and universal quantification, although in a way which is different from Howard's:

"*Types* are defined starting from *type variables* $X, Y, Z, \ldots$ by means of two operations:

1. if U and V are types, then $U \to V$ is a type.
2. if V is a type, and X a type variable, then $\Pi X.V$ is a type.

There are five schemes for forming *terms*:

1. *variables:* $x^T, y^T, z^T, \ldots$ of type T,

2. *application:* tu of type V, where t is of type $U \to V$ and u is of type U,

3. *λ-abstraction:* $\lambda x^U.v$ of type $U \to V$, where x^U is a variable of type U and v is of type V,

4. *universal abstraction:* if v is a term of type V, then we can form $\Lambda X.v$ of type $\Pi X.V$, so long as the variable X is not free in the type of a free variable of v.

5. *universal application* (sometimes called *extraction*): if t is a term of type $\Pi X.V$ and U is a type, then tU is a term of type $V[U/X]$.

As well as the usual *conversions* for application/λ-abstraction, there is one for the other pair of schemes:

$$(\Lambda X.v)U \rightsquigarrow v[U/X]"$$

[Girard et al. (1989)] (pp. 81–2)

Note that Howard's way is to abstract on *term*-variables ('$(\lambda X^{\alpha}.F^{\beta})^{\alpha \supset \beta}$') for implication and on *number*-variables ('$(\lambda x F^{\beta})^{\forall x \beta}$') for universal quantification, whereas Girard's way is to abstract on *element*-variables ('$\lambda x^U.v$') for implication (clause 3 above) and on *type*-variables ('$\Lambda X.v$') for universal quantification (clause 4 above).

In a review of Girard, Lafont and Taylor's *Proofs and Types*, Howard [Howard (1991)] makes the following comments on the above feature of system F described in the book, which introduces abstraction over type-variables in order to allow the handling of universal quantification within the system:

"The system F contains not only *terms* but also *types* (or, more properly, type symbols) and there is a special abstraction operator Λ for variables X ranging over types. (...) This gives rise to problems of interpretation.

What is a type? From the viewpoint of set theory, a type is a set D. To say that an object x has type D is simply to say that x is a member of D. An interesting variant of this, due to Martin-Löf, is as follows. Recall that a type is denoted by (essentially) a closed formula A. But A denotes a *proposition*. If one regards a proposition as identical with the species of its proofs, one gets an intuitionistic counterpart of the set-theoretic notion of type. None of these interpretations, however, apply to the system **F**: if $\Lambda X.v$ were to denote a function belonging to the type $\Pi X.V$ interpreted as a set, then $\Lambda X.v$ ranges over all types." [Howard (1991)] (p. 761)

Here, similarly to Martin-Löf [Martin-Löf (1984)] we abstract on *element*-variables in both cases, the difference being that for implication we can say that '$\lambda x.b(x) : A \to B$' provided '$b(x) : B$' on the assumption that '$x : A$', whereas for universal quantification '$\Lambda x.b(x) : \forall x^A.B(x)$' provided '$b(x) : B(x)$' (where '$B(x)$' is a type indexed by '$x$') on the assumption that '$x : A$'. But unlike Martin-Löf's unified treatment with a Π-type (and associated definitional equalities distinguishing '$\to$' from '$\forall$', namely, '$A \to B \equiv (\Pi x \in A)B$' and '$(\forall x \in A)B(x) \equiv (\Pi x \in A)B(x)$', [Martin-Löf (1984)] (p. 32)), we follow Howard's and Girard's approach of dealing with implication and universal quantification by using separate type definitions. (Notice that Martin-Löf's 'unified' treatment via a

the universal quantifier in a way such that it shows its parallels with implication as:

$\forall$-*formation*

$$\frac{A\ type \quad B(x)\ type}{\forall x^A.B(x)\ type}$$

$$[x:A]$$

$\forall$-*introduction*

$$[x:A]$$

$$\frac{b(x):B(x)}{\Lambda x.b(x):\forall x^A.B(x)} \qquad \frac{b(x)=d(x):B(x)}{\Lambda x.b(x)=\Lambda x.d(x):\forall x^A.B(x)}$$

$$[x:A]$$

$\forall$-*reduction*

$$[x:A]$$

$$\frac{a:A \quad b(x):B(x)}{EXTR(\Lambda x.b(x),a)=b(a/x):B(a)}$$

$\forall$-*induction*

$$\frac{c:\forall x^A.B(x)}{\Lambda y.EXTR(c,y)=c:\forall x^A.B(x)}$$

$\square$

Π-type would not be capable of distinguishing '$A \to (B \to A)$' from '$A \to \forall x^B.A$', the former being the principal type scheme of 'K', namely 'PK', and the latter the principal type scheme of the vacuous universal quantifier, namely 'ΠK' [Curry (1931)] or 'Π_2' [Curry *et al.* (1972)] (see below).)

Curry's original insight concerning the treatment of universal quantification in a similar way to implication is clear from his early 'The Universal Quantifier in Combinatory Logic' [Curry (1931)]:

"The combinator here defined I have called the formalizing combinator, because by means of it is possible to define the relation of formal implication for functions of one or more variables in terms of ordinary implication. Thus if P is ordinary implication, it follows from Theorem 1 (below) that $(\phi_n P)$ is that function of two functions of n variables, whose value for the given functions $f(x_1, x_2, \ldots, x_n)$ and $g(x_1, x_2, \ldots, x_n)$ is the function $f(x_1, \ldots, x_n) \to g(x_1, \ldots, x_n)$. Thus formal implication for functions of n variables is $(B\Pi_n(\phi_n P))$." [Curry (1931)] (pp. 165–166)

and later in the same paper he gives the axioms:

"(ΠB). (f) $[(x)fx \to (g,x)f(gx)]$.
(ΠC). (f) $[(x,y)f(x,y) \to (x,y)f(y,x)]$.
(ΠW). (f) $[(x,y)f(x,y) \to (x)f(x,x)]$.
(ΠK). (p) $[p \to (x)Kpx]$.
(ΠP). (f,g) $[(x)(fx \to gx) \to ((x)fx \to (x)gx)]$."

(Ibid., p. 170.)

The combined treatment is also made in the second volume of *Combinatory Logic* (with R. Hindley and J. Seldin). In both cases the fundamental rule is a sort of modus ponens (universal instantiation):

"Rule Π. $\Pi X, EU \vdash XU$,
Rule P. $PXY, X \vdash Y$."

[Curry *et al.* (1972)] (p. 427)

one can see the correspondence of type-schemes of combinators, axioms of implication, and axioms of universal quantification:[47]

I:
$$A \to A$$
$$\forall x^A . A(x)$$

C:
$$(A \to (B \to C)) \to (B \to (A \to C))$$
$$\forall x^A . (B \to C(x)) \to (B \to \forall x^A . C(x))^{48}$$

S:
$$(A \to (B \to C)) \to ((A \to B) \to (A \to C))$$
$$\forall x^A . (B(x) \to C(x)) \to (\forall x^A . B(x) \to \forall x^A . C(x))$$

K:
$$A \to (B \to A)$$
$$A \to \forall x^B . A$$

As an example, we can see that the following axiom would be valid only if the universal quantifier is not linear:

---

[47] In [Curry *et al.* (1972)] one already finds some of the parallels listed here, such as, e.g.:

"(PK) $\vdash \alpha \supset .\beta \supset \alpha,$
(PS) $\vdash \alpha \supset .\beta \supset \gamma :\supset: \alpha \supset \beta. \supset .\alpha \supset \gamma,$
(Π_0) $\vdash (\forall x)\alpha x. \supset \alpha U,$
(Π_2) $\vdash \alpha \supset (\forall x)\alpha,$
(ΠP) $\vdash (\forall x)(\alpha x \supset \beta x). \supset .(\forall x)\alpha x \supset (\forall x)\beta x,$"

(Ibid., p. 433.)

Note the parallel between PK and Π_2, as well as between PS and ΠP. Furthermore, following the same line of reasoning, another Π-rule is soon defined mirroring PC — type-scheme for combinator 'C', namely '$\gamma \supset \alpha. \supset \beta :\supset: \alpha \supset \gamma. \supset \beta$' —, which is called Π_1:
"(Π_1) $\vdash (\forall x)(\alpha \supset \beta x). \supset .\alpha \supset (\forall x)\beta x$"   (Ibid., p. 439.)
(In fact, Π_1, Π_2 and ΠP, though not the parallel with the propositional PC, PK and PS, were already presented in Curry's own earlier work [Curry (1963)] (p. 344).)

That would seem to justify why the structural similarity between implication and universal quantification, which was quite explicit in Heyting's intuitionistic predicate calculus [Heyting (1946)], is so naturally reflected in the Curry–Howard interpretation. (And indeed, a form of both ΠP and Π_1 without the leftmost universal quantification already appear in [Heyting (1946)] as 'Rule $(\gamma 1)$' and 'formula (7)', respectively, where it is said they both come from Hilbert & Ackermann's *Grundzüge der theoretischen Logik* [Hilbert and Ackermann (1938)].)

[48] In Quine's (1950) *Methods of Logic* this corresponds to a 'rule of passage' (see [Quine (1950)] (Chapter 23)).

$\forall x^A.(B(x) \to C(x)) \to (\forall x^A.B(x) \to \forall x^A.C(x))$ *(distribution)* (parallel to 'S')

$$\frac{\dfrac{\dfrac{\boxed{[t:A]} \quad [z:\forall x^A.B(x)]}{EXTR(z,t):B(t)} \quad \dfrac{\boxed{[t:A]} \quad [y:\forall x^A.(B(x) \to C(x))]}{EXTR(y,t):B(t) \to C(t)}}{\dfrac{\dfrac{APP(EXTR(y,t),EXTR(z,t)):C(t)}{\boxed{\Lambda t.}\,APP(EXTR(y,t),EXTR(z,t)):\forall t^A.C(t)}}{\lambda z.\Lambda t.APP(EXTR(y,t),EXTR(z,t)):\forall x^A.B(x) \to \forall t^A.C(t)}}}{\lambda y.\lambda z.\Lambda t.APP(EXTR(y,t),EXTR(z,t)):\forall x^A.(B(x) \to C(x)) \to (\forall x^A.B(x) \to \forall t^A.C(t))}$$

Note that the $\boxed{\Lambda t.}$-abstraction is discharging assumptions non-linearly and cancelling more than one free occurrence of the variable 't' in the expression '$APP(EXTR(y,t),EXTR(z,t))$'.

There is room for further extending the Curry–Howard interpretation to deal with modal logics, if one makes a special (and useful) reading of the modal connective '$\Box$'. Looking at implication (universal quantification) as being fundamentally characterised by *modus ponens* (resp. universal extraction), which in the Curry–Howard interpretation is formalised by β-normalisation, one can see that each implication/universal quantifier changes only its *assertability conditions* rule according to the logic, the rules corresponding to the explanation of the (immediate) *consequences* — the proof-theoretic semantical rules, according to a particular semantical standpoint explored in de Queiroz (1989, 1990, 1991) — remaining fixed. Now, looking at the modal '$\Box$' as a sort of second-order universal quantification (quantification over structured collections of formulas), in a way such that:

$$\Box A \equiv \forall \mathbb{W}^{\mathcal{U}}.A(\mathbb{W}),$$

with corresponding:

$\Box$-*introduction*:

$$\underbrace{\frac{\substack{[\mathbb{W}:\mathcal{U}] \\ f(\mathbb{W}):A(\mathbb{W})}}{\Lambda \mathbb{W}.f(\mathbb{W}):\forall \mathbb{W}^{\mathcal{U}}.A(\mathbb{W})}}_{\Box A} \qquad \text{or simply} \qquad \frac{\substack{[\mathbb{W}:\mathcal{U}] \\ f(\mathbb{W}):A(\mathbb{W})}}{\Lambda \mathbb{W}.f(\mathbb{W}):\Box A}$$

where '$\mathcal{U}$' is a collection of 'worlds', '$\mathbb{W}$' a world-variable denoting structured collections — lists, bags, trees, etc. — of labelled formulas, and '$F(\mathbb{W})$' is an expression which may depend on the world-variable '$\mathbb{W}$'.

It is easy to see that the same reasoning made for implication and first-order quantification can be carried through for the case of modal logics. For example, the axiom for the modal logic K finds a parallel in the axioms for implication and first-order universal quantification which correspond to the S (*distribution*) combinator:

Modal Logic K:

$$\Box(A \to B) \to (\Box A \to \Box B),$$

which when rewritten gives us:

$$\forall \mathbb{W}^{\mathcal{U}}.(A \to B)(\mathbb{W}) \to (\forall \mathbb{W}^{\mathcal{U}}.A(\mathbb{W}) \to \forall \mathbb{W}^{\mathcal{U}}.B(\mathbb{W})),$$

and the parallel with the 'S' combinator is quite clear: '$\forall \mathbb{W}^{\mathcal{U}}$' is being distributed over the implication. Similarly to the case of implication and first-order universal quantification one can find a parallel between axioms characterising different modal logics and the type schemes of combinators. For example, the weakest rule characterising the so-called 'standard normative logics' according to [Chellas (1980)], namely:

$$\text{RM} \qquad \frac{A \to B}{\Box A \to \Box B},$$

mirrors the type-scheme of the weakest combinator, namely B (i.e. *left transitivity*), and the parallel with the weakest possible characterisation of implication is immediate.

Moreover, the axiom schema taking us from relevant implication to intuitionistic implication, namely:

$$A \to (B \to A),$$

finds a parallel in the axiom schema taking us from modal logic K to modal logic S4, namely:

$$\Box A \to \Box\Box A,$$

where the reading changes from 'if A is true, then anything implies A' to 'if A is necessary, then A is necessarily necessary'.

We have also been investigating the connections between the proof procedures for the axiom schema taking us from intuitionistic implication to classical implication, namely,

$$((A \to B) \to A) \to A,$$

and the proof procedures for the axiom schema giving the so-called Grzegorczyk's [Grzegorczyk (1967)] extension of S4, namely:

$$(\Box(A \to \Box A) \to A) \to A.$$

Concerning the 'jump' to second-order quantification, observe that unlike the previous cases (implication and first-order universal quantification) where we had only element variables ($x, y, \ldots$, etc.), now we have type variables (e.g. 'W'). And indeed, in modal logics one is dealing with higher-order objects, and therefore some kind of 'higher-order universal extraction (and universal abstraction)' is needed.

Now, for the characterisation of a type-theoretic account of second-order normalisation one can use the seminal results independently obtained by Girard [Girard (1971)] and Reynolds [Reynolds (1974)] on second-order typed λ-calculus and the semantics of polymorphism. Additionally, an attempt at a formulation of the second-order normalisation in a type-theoretic framework has been made with the development of a *'type of types'*-Fix operator described in [de Queiroz and Maibaum (1990)].

Chapter 3

The Existential Quantifier

Preamble

We are concerned with showing how 'labelled' Natural Deduction presentation systems based on an extension of the so-called Curry–Howard functional interpretation can help us understand and generalise most of the deduction calculi designed to deal with the logical notion of existential quantification. We present the labelling mechanism for '$\exists x^D.F(x)$' using what we call 'ε-terms', which have the form of '$\varepsilon x.(f(x), a)$' in a dual form to the '$\Lambda x.f(x)$' terms of '$\forall x^D.F(x)$' in the sense that the 'witness' ('a' in the term above) is chosen (and hidden) at the time of assertion/*introduction*.

With the intention of demonstrating the generality of our framework, we provide a brief comparison with some other calculi of existentials, including the model-theoretic interpretations of Skolem and Herbrand, indicating the way in which some of the classical results on prenex normal forms can be extended to a wide range of logics. The essential ingredient is to look at the conceptual framework of labelled natural deduction as an attempted reconstruction of Frege's functional calculus where the devices of the *Begriffsschrift* (i.e. connectives and quantifiers) work separately from, yet in harmony with, those of the *Grundgesetze* (i.e. abstractors).[1]

---

[1] An earlier version of this chapter was presented at the 1991 European Summer Meeting of the Association for Symbolic Logic, *Logic Colloquium '91*, sections 1–5 and 10 of the (IUHPS) *9th International Congress of Logic, Methodology and Philosophy of Science*, Uppsala, Sweden, August 7–14, 1991. A one-page abstract has appeared in the *Abstracts* of the Congress, and has also appeared in the *JSL*, and a full version was published in a special issue of the *Bulletin of the IGPL* [de Queiroz and Gabbay (1995)].

3.1   Motivation

The functional interpretation of logical connectives, the so-called 'Curry–Howard' [Curry (1934); Howard (1980)] interpretation,[2] provides an adequate framework for the establishment of various calculi of logical inference. Being an 'enriched' system of natural deduction, in the sense that terms representing proof-constructions are carried alongside the formulas, it constitutes an apparatus of great usefulness in the formulation of logical calculi in an operational manner. By uncovering a certain harmony between a functional calculus on the terms (the *labels*) and a logical calculus on the formulas, it proves to be instrumental in giving mathematical foundations for systems of logic presentation designed to handle meta-level features at the object-level via a labelling mechanism, such as, e.g. Gabbay's [Gabbay (1994)] *Labelled Deductive Systems*.

The intention here is to show how such an apparatus can help us understand and generalise some of the deduction calculi designed to deal with the mathematical notion of existential quantification. We present the labelling mechanism for '$\exists x^D.F(x)$' using what we call 'ε-terms', which have the form of '$\varepsilon x.(f(x), a)$' in a dual form to the '$\Lambda x.f(x)$' terms of '$\forall x^D.F(x)$' in the sense that the 'witness' ('a' in the term above) is chosen at the time of assertion/*introduction* as in:

$\exists$-*introduction*

$$\frac{a : D \qquad f(a) : F(a)}{\varepsilon x.(f(x), a) : \exists x^D.F(x)}$$

Notice that the 'witness' 'a' is kept unbound in the term — to the lefthand side of the ':' sign –, whilst in the formula the binding is performed: the formula in the concluding judgement is '$\exists x^D.F(x)$', a closed formula.

The basic intuition of our solution is as follows. Imagine an arbitrary proof system and think of a proof Π_1 leading to $\exists x.F(x)$. Thus, the steps and 'context' of Π_1 derives the above existential formula. We now want to proceed with one more step and introduce a constant a and continue the proof using $F(a)$. We would like to have a discipline telling us what is the logical status (relative to Π_1) of any C which can be proved later, after this 'Skolemisation' step. We decided to answer this question in a general setting. We assume that another proof Π_2 is available (fortunately) to $F(a)$ for some term a. We take a single step of identifying the a in Π_2 with the expected element in $\exists x.F(x)$ of Π_1. We proceed, using

---

[2]As we have previously suggested [Gabbay and de Queiroz (1992)], this might as well be called the 'Curry–Howard–Tait' interpretation, considering the significance of Tait's [Tait (1965)] intensional interpretation of Gödel's [Gödel (1958)] system T via a semantics of convertibility, and the relevance of his proof of Gentzen's results using the framework of the functional interpretation.

the above assumption, to prove the third formula C. We can represent the above diagrammatically as follows:

$$\begin{array}{cc} \Pi_1 & \Pi_2 \\ \exists x.F(x) & F(a) \end{array}$$

$$\cdots \qquad \cdots$$

Identification step

$$\vdots$$

$$C$$

In a labelled deductive system, the context of the proof is given by the label attached to the formula proved, and thus the two formulas prior to identification can be represented as

$$\Pi_1 : \exists x.F(x), \qquad \Pi_2 : F(a)$$

and the rule allows us to get

$$\text{Identify}(\Pi_1, \Pi_2, a, g(t)) : F(g(t))$$

meaning that y is a term which is the result of identification of the two proofs obtained by unskolemising on a in $F(a)$ with a new function symbol g to represent how the proof of $F(a)$ depends on the context Π_2. If we now prove some $d(g(t)) : C$ using $F(g(t))$ we can abstract on t (and on g) to get:

$$\lambda y.\lambda g.d(\text{Identify}(\Pi_1, \Pi_2, a, g(t))) : \forall t(F(g(t)) \rightarrow C)$$

with corresponding 'elimination' procedure as:

$$\frac{\Pi_1 : \exists x.F(x) \qquad \Pi_2 : F(a) \qquad d(g(t)) : F(g(t)) \rightarrow C}{\text{Instantiate}(\Pi_1, \Pi_2, a, g(t), d) : C}$$

The parallel with ordinary Skolemisation procedures is straightforward: the fact that we can abstract on g and t finds its counterpart in the proviso that the symbols chosen for the Skolem-function and the Skolem-constant must be *new*.

Now, in order to formally deal with the above intuitive explanation of Skolemisation within the so-called Curry–Howard functional interpretation, we need to establish a proper functional calculus on the labels to go along in harmony with our logical calculus on the formulas. So, in our solution here, an ε-term of type $\exists x^D.F(x)$ is defined to be a function-term which carries its own (hidden) argument with it.

As will be shown later on in the paper, the identification step briefly described above is made explicit by the (β-type) normalisation step of our labelled natural deduction system for the $\exists$-quantifier:

$\exists$-*reduction*

$$\frac{\dfrac{a:D \qquad f(a):F(a)}{\varepsilon x.(f(x),a):\exists x^D.F(x)}\exists\text{-}intr \qquad \dfrac{[t:D,g(t):F(t)]}{d(g,t):C}}{INST(\varepsilon x.(f(x),a),\acute{g}td(g,t)):C}\exists\text{-}elim$$

$$\triangleright_\beta \quad \begin{array}{c} a:D,f(a):F(a) \\ d(f/g,a/t):C \end{array}$$

Note that, assuming we had a proof Π_2 of $F(a)$ with the Skolem constant 'a' and Skolem function 'f', in the identification step we have that the 'a' replaces the 't', and the 'f' replaces the 'g'.

3.1.1 *The* pairing *interpretation*

Our $\exists$-type differs from both Howard's [Howard (1980)] *strong existence* $\exists$-type[3] and from Martin-Löf's [Martin-Löf (1975b, 1984)] Σ-type in that ε-terms are taken to be 'course-of-values' function-terms carrying their own (hidden) argument, not simply pairs as is the case for Σ-types. The reason for this difference is simple, and the justification can be given by reference to the 'indefinite' nature of existentials. Seen as simply pairs, proof-constructions of an existential statement might give the misleading impression that the witness is readily available by the operations of projection over the ordered pair. This 'availability' does not match with the true spirit of indefiniteness of the existential quantifier.[4] To see better how

---

[3]Howard's definition is as follows:

"(ii) *Strong existence (choice operators)* It is natural to interpret an object of type $\exists y\alpha(y)$ as a pair $\langle t, F^{\alpha(t)}\rangle$. Thus, [its interpretation in Heyting's arithmetic] $(\exists y\alpha(y))^*$ would be defined as the set of all pairs $\langle n, Z\rangle$ such that $Z \in \alpha(n)^*$. Hence introduce projection operators P_1 and P_2 which give the required components t and $F^{\alpha(t)}$ when applied to a pair $\langle t, F^{\alpha(t)}\rangle$ regarded as an object of type $\exists y\alpha(y)$."   [Howard (1980)] (p. 89)

[4]Cf. Martin-Löf's [Martin-Löf (1984)] account of existentials via a Σ-type of ordered pairs. The so-called 'logical interpretation' of a Σ-type into an $\exists$-type gives direct access to the witness via the left projection over the ordered pair as in:

"If we now turn to the logical interpretation [of Σ-types with left and right projections], we see that

$$\frac{c \in (\exists x \in A)B(x)}{p(c) \in A}$$

holds, which means that from a *proof* of $(\exists x \in A)B(x)$ we can obtain an element of A for which the property B holds."   [Martin-Löf (1984)] (p. 45)

Later, in a chapter showing the 'constructive proof of the axiom of choice' the eliminatory operation of (left) projection is explicitly used to access the witness of an existential:

"Let A set, $B(x)$ set $[x \in A]$, $C(x, y)$ set $[x \in A, y \in B(x)]$, and assume $z \in (\Pi x \in A)(\Sigma y \in B(x))C(x, y)$. If x is an arbitrary element of A, i.e. $x \in A$, then, by Π-elimination, we obtain

$$Ap(z, x) \in (\Sigma y \in B(x))C(x, y).$$

We now apply left projection to obtain [by Σ-elimination]

$$p(Ap(z, x)) \in B(x)$$

the 'indefiniteness' of existentials are to be reflected in the calculus, it might be useful to think of the duality between the universal and the existential quantifiers as first-order counterpart to the duality between propositional conjunction and disjunction. With conjunction the respective conjuncts in the proof-construction are readily available for the elimination rules:

$$\frac{\dfrac{a_1 : A_1 \qquad a_2 : A_2}{\langle a_1, a_2 \rangle : A_1 \wedge A_2} \wedge\text{-}intr}{FST(\langle a_1, a_2 \rangle) : A_1} \wedge\text{-}elim \qquad\qquad \frac{\dfrac{a_1 : A_1 \qquad a_2 : A_2}{\langle a_1, a_2 \rangle : A_1 \wedge A_2} \wedge\text{-}intr}{SND(\langle a_1, a_2 \rangle) : A_2} \wedge\text{-}elim,$$

whereas with disjunction, once one of the disjuncts is used to construct a proof of the disjunction, it becomes 'hidden', i.e. the *elimination* rule has to proceed by Skolem-like introduction of new local assumptions:

$$\frac{\dfrac{a_1 : A_1}{inl(a_1) : A_1 \vee A_2} \vee\text{-}intr \qquad \begin{array}{c}[x : A_1]\\ d(x) : C\end{array} \qquad \begin{array}{c}[y : A_2]\\ e(y) : C\end{array}}{CASE(inl(a_1), \acute{x}d(x), \acute{y}e(y)) : C} \vee\text{-}elim$$

$$\frac{\dfrac{a_2 : A_2}{inr(a_2) : A_1 \vee A_2} \vee\text{-}intr \qquad \begin{array}{c}[s_1 : A_1]\\ d(s_1) : C\end{array} \qquad \begin{array}{c}[s_2 : A_2]\\ e(s_2) : C\end{array}}{CASE(inr(a_2), \acute{x}d(x), \acute{y}e(y)) : C} \vee\text{-}elim.$$

Notice that in each case, once the disjunction was introduced, the disjuncts became *inaccessible*[5] to the eliminator '$CASE$', and the introduction of new local assumptions '$s_1 : A_1$' and '$s_2 : A_2$' were necessary in order to eliminate the disjunction.

---

and right projection to obtain

$$q(Ap(z,x)) \in C(x, p(Ap(z,x))).\text{''}$$

[Martin-Löf (1984)] (p. 50)

Note that the use of 'p' and 'q' (left and right projections, respectively) is made explicitly, when one would expect the introduction of Skolem-constant and Skolem-function in the elimination of the existential Σ: in order to eliminate the existential quantifier, given that the existential statement leaves the witness unspecified, one has to introduce a new name, say '$t : D$', to denote a witness from the domain of quantification (meaning 'let it be t'), and a new function-name over the witness name, say '$g(t)$', to state that the witness does satisfy the requirements, say '$F(t)$' (meaning 'let t satisfy the requirements'); arriving at a third statement, say 'C', from those new assumptions, namely '$[t : D]$' and '$[g(t) : F(t)]$', one can conclude that this same statement 'C' can be obtained directly from the original existential statement, no matter which new names one had chosen for the witness and the (Skolem)-function.

[5]In a plain Natural Deduction presentation it is perhaps easier to see why this is the case. In other words, contrast:

whereas the following is OK:

$$\frac{\dfrac{A_1}{A_1 \vee A_2} \vee\text{-}intr \qquad [A_1] \quad [A_2]\quad\quad C \qquad C}{C} \vee\text{-}elim.$$

A similar duality happens with the universal and the existential quantifiers: in the case of universal quantification,

$$\frac{a : D \qquad \dfrac{\begin{array}{c}[x : D]\\ f(x) : F(x)\end{array}}{\Lambda x.f(x) : \forall x^D.F(x)}\text{\textit{$\forall$-intr}}}{EXTR(\Lambda x.f(x), a) : F(a)}\text{\textit{$\forall$-elim}},$$

the 'witness' 'a' is in the scope of the eliminator '$EXTR$', whereas in the case of existential quantification:

$$\frac{\dfrac{a : D \qquad f(a) : F(a)}{\varepsilon x.(f(x), a) : \exists x^D.F(x)}\text{\textit{$\exists$-intr}} \qquad \dfrac{[t : D, g(t) : F(t)]}{d(g, t) : C}}{INST(\varepsilon x.(f(x), a), \acute{g}\acute{t}d(g, t)) : C}\text{\textit{$\exists$-elim}},$$

the witness 'a' is not in the scope of the eliminatory '$INST$', but rather in the scope of the introductory 'ε'. Thus, the introduction of new local assumptions '$t : D$' ('let it be t') and '$g(t) : F(t)$' ('let $g(t)$ be the proof-construction witnessing that t does satisfy $F(t)$') were necessary in order to eliminate the existential quantifier. In other words, arriving at a proof of '$\exists x^D.F(x)$', and knowing that the witness is hidden, then proceed by assuming that it has the (new) name 't' and that it satisfies the property '$F(t)$'; if a third statement can be obtained from these two new assumptions, then proceed by eliminating the existential quantifier (thus discharging the newly introduced assumptions) in favour of such third statement.

Obviously, if the proof-constructions of $\exists x^D.F(x)$ are taken to be ordered pairs, then the Skolem-type procedure of introducing new local assumptions involving a Skolem-constant and a Skolem-function in order to eliminate the existential quantifier would be virtually unnecessary. Take, for example

$$\frac{\dfrac{a : D \qquad f(a) : F(a)}{\langle a, f(a)\rangle : \exists x^D.F(x)}\text{\textit{$\exists$-introduction}}}{SND(\langle a, f(a)\rangle) : F(FST(\langle a, f(a)\rangle))}\text{\textit{$\exists$-elimination}} \quad (??)$$

$$\frac{\underbrace{\hspace{3cm}}_{f(a)} \qquad \underbrace{\hspace{3cm}}_{F(a)}}{}$$

the one below is INVALID:

$$\frac{\dfrac{A_1}{A_1 \vee A_2}\text{\textit{$\vee$-intr}}}{A_1}\text{\textit{$\vee$-elim}} \quad (??)$$

Similarly for the existential quantifier:

OK:

$$\frac{\dfrac{F(a)}{\exists x.F(x)}\text{\textit{$\exists$-intr}} \qquad \dfrac{[F(t)]}{C}}{C}\text{\textit{$\exists$-elim}}$$

INVALID:

$$\frac{\dfrac{F(a)}{\exists x.F(x)}\text{\textit{$\exists$-intr}}}{F(a)}\text{\textit{$\exists$-elim}} \quad (??)$$

(i.e. the witness 'a' is reused in the instantiation step) and the witness would be recoverable without any need to introduce Skolem-type local assumptions, given that:

$$FST(\langle a, b \rangle) = a$$
$$SND(\langle a, b \rangle) = b$$

for any a, b.

Moreover, it seems that thinking of the existential in terms of a 'type of pairs' can lead one to the confusing quest (within the functional interpretation and 'labelled' natural deduction systems) for a 'generalised' type of conjunctions and existential quantification.[6] In order to understand the existential quantifier, instead of thinking in terms of a 'generalised' pairing-type, it appears to be more appropriate to think of the duality between the existential and the universal quantifier concerning the choice of the witness: either in the *introduction* or in the *elimination*, as in Lorenzen's [Lorenzen (1961, 1969)] dialogical games and Hintikka's [Hintikka (1983)] semantical games. So, we find that the Heyting-style [Heyting (1930, 1956)] explanation does not always help.[7] We also want to find appropriate generalising conditions, so that the framework can be parameterised, and is not restricted to the intuitionistic case. Just by looking at the rule of $\exists$-*introduction*, where an (ε-)abstraction is involved, we can see that we should have 'room for manoeuvre', so to speak, as to the classification of distinct existential quantifiers: the abstraction may be binding one, many, none, etc., occurrences of the corresponding name-variable. As we shall see below, the classical existential quantifier, for example, will not require that the term alongside the predicate formula in the premise ('$f(a)$' term, '$F(a)$' formula, in the $\exists$-*introduction* above) have the name-variable ('a' above) occurring free.

---

[6]By following strictly Heyting's semantics of proof-conditions, where both the conjunction and the existential quantifier are explained as 'ordered pairs', the explanation given in [Martin-Löf (1984)] says that '$\wedge$' and '$\exists$' are particular instantiations of a 'more general' type such as:

$$\Sigma\text{-}introduction \quad \frac{a \in A \quad b \in B(a)}{\langle a, b \rangle \in (\Sigma x \in A)B(x)} \qquad \Sigma\text{-}elimination \quad \frac{c \in (\Sigma x \in A)B(x) \quad \overset{[x \in A, y \in B(x)]}{d(x, y) \in C(\langle x, y \rangle)}}{E(c, (x, y)d(x, y)) \in C(c)}$$

And, indeed, in the context of Heyting's proof-based semantics, such as the semantics of *Intuitionistic Type Theory*, one is led to think of conjunction and existential quantification as instantiations of a single 'generalised' type Σ, as '$A \wedge B \equiv (\Sigma x \in A)B$' [Martin-Löf (1984)] (p. 43), and '$\exists x \in A.B(x) \equiv (\Sigma x \in A)B(x)$' (Ibid., p. 42). It seems more sensible to say that in ordinary reasoning the existential quantifier is more likely to be related to the universal quantifier than to conjunction. So, the approach we take here is that if the universal quantifier is the type of functions $x : A \mapsto b(x) : B(x)$, then the existential quantifier is also the type of functions $x : A \mapsto b(x) : B(x)$ with the peculiarity that the argument is not any $x : A$ but a singular $a : A$.

[7]"In accordance with the intuitionistic interpretation of the existential quantifier, the rule of Σ-introduction may be interpreted as saying that a (canonical) proof of $(\exists x \in A)B(x)$ is a pair (a, b)." [Martin-Löf (1984)] (p. 42)

For comparative purposes we briefly look at some of the established calculi (including Hilbert's ε- and η- calculi; Hailperin's [Hailperin (1957)] ν-calculus; Gentzen's [Gentzen (1935)] Natural Deduction and its variations, such as, e.g. Prawitz [Prawitz (1965)], Quine [Quine (1950)], Copi [Copi (1954)], Kalish-Montague [Kalish *et al.* (1964)], and Fine [Fine (1985)]), demonstrating that many of their inherent limitations (such as the need for cumbersome restrictions on the rules of instantiation and generalisation) can be overcome in a more general way by a 'labelled' Natural Deduction framework such as the functional interpretation.[8] We shall also attempt to demonstrate how the proof-calculus reflects the semantical interpretations of Leisenring [Leisenring (1969)], especially with respect to the well-established resolutions *à la* Skolem and *à la* Herbrand.

3.2   Quantifiers and normalisation

By examining the quantifier rules we can observe that in the presentation of both quantifiers '∀' and '∃' the canonical proof involves a 'course-of-values' term, a 'Λ'-term for the universal quantifier, and an 'ε'-term for the existential quantifier. The fundamental difference comes with the 'choice' of witnesses: the 'Λ'-term ('$\Lambda x f(x)$') does not carry a particular witness with it, whereas the 'ε'-term does ('$\varepsilon x.(f(x), a)$'). This crucial difference in the 'choice' of witnesses becomes clear when the normalisation (reduction) rules are laid down: the '*EXTR*action' eliminatory operator, or 'destructor', introduces a new open term ('a') which is not part of the canonical term ('$\Lambda x f(x)$'), indicating the 'arbitrariness' of the choice of the witness; the '*INST*antiation' destructor introduces only closed terms. The open term which plays the rôle of witness, namely 'a', is already part of the canonical term ('$\varepsilon x.(f(x), a)$'), therefore chosen at the time of assertion. This situation, where the name for the individual is bound in the introduction of a universal quantifier, and remains free after the introduction of an existential quantifier is reflected in the meta-language distinction between an 'arbitrary' and a 'specific'

---

[8]We are here referring to the functional interpretation as the framework where the functional calculus on the labels matches the logical calculus on the formulas, but instead of abiding by the tenets of Heyting's [Heyting (1930)] intuitionism (such as, e.g. 'a proposition is defined by laying down what counts as a proof of the proposition', [Prawitz (1980); Martin-Löf (1984)]), it is the functional interpretation based on a semantics of convertibility (normalisation) as defined in [Tait (1965)]. Such a framework has been previously explored to study the various notions of implication and its corresponding deduction calculi [Gabbay and de Queiroz (1992)]. It was shown that systems of implication below and above intuitionistic implication can be obtained from a single general framework where the rule of →-convertibility (normalisation) is kept fixed, while for each system of implication particular conditions on the rule of discharge of assumptions (→-*introduction*) are imposed according to each particular kind of implication (corresponding to weaker or stronger λ-calculi).

individual, as in Martin-Löf's [Martin-Löf (1987)] (p. 411) explanation of the inference rules for the quantifiers:

$$\frac{A(a) \text{ is true}}{\exists x.A(x)} \qquad (\text{'}A(a) \text{ is true of a } \textit{specific} \text{ individual'})$$

$$\frac{A(x) \text{ is true}}{\forall x.A(x)} \qquad (\text{'}A(x) \text{ is true of an } \textit{arbitrary} \text{ individual'})$$

Note that in the case of our labelled natural deduction, the specific versus arbitrary is reflected not simply by means of the typographical convention (initial versus terminal roman letters), but rather in the individual's name remaining unbound or becoming bound in the logical formula. As for the 'functional' side, namely the functional calculus on the labels, the same guidelines should apply, and that is why we have a Λ-term in one case and an ε-term in the other. Only, in our framework we need to make explicit the 'arbitrary' quality of the individual by writing out the assumption 'suppose x is an arbitrary element from the domain...' as in:

$$[\text{Suppose } x \text{ is in } D]$$

$$\vdots$$

$$\frac{A(x) \text{ is true}}{\forall x^D A(x)}$$

and the three dots indicate that some inference steps were made between the supposition that x was an arbitrary element from domain D, and the premise of the inference rule, which says that $A(x)$ is true. And the fact that the introduction of the universal quantifier involves the discharge of assumptions characterises the $\forall$-*introduction* as an improper inference rule (to use a terminology from [Prawitz (1965)]), similarly to the introduction of implication ($\rightarrow$-*introduction*). The introduction of the existential quantifier, however, does not assume that the individual was an arbitrary element from the domain concerned, thus it does not involve any assumption discharge.

3.2.1 *Introducing variables for the Skolem dependency functions*

Now, if the witness (and the proof-construction that confirms it as a 'good' witness) is chosen at the *introduction* inference, the *elimination* inference will have to act on the supposition that new names will have to be introduced in order for the deduction to 'go ahead'. The reasoning would go as follows: if there is an individual which satisfies a certain property ('$\exists x^D.F(x)$'), and I am not supposed to know who it is, I shall proceed by giving it a new name ('let it be t') and by assuming that the confirmation that it is a good one (i.e. that it satisfies the given property) will also be given a new name ('let $g(t)$ be the proof that $F(t)$ indeed holds'). Now, if I can reach a third statement, which may or may not depend on the new name for the witness (note that 'C' is an arbitrary formula), I am entitled to say that whatever choice I might have made of the new names (i.e. 't' and 'g'), such third statement can be inferred straight from the existential statement.

Perhaps the parallel with the propositional disjunction is useful here. Suppose we have arrived at a step in a proof where the disjunction has to be eliminated. We cannot simply continue by inferring one of the disjuncts, so we need to find a way around: if a third statement can be reached from both disjuncts, then we are entitled to say that the latter statement can be inferred straight from the disjunctive statement. This 'way around' seems to be imposed by the fact that from statements involving indefinites (i.e. disjunction, existential quantification, modal possibility, equality, etc.) no direct inference can be made.

For the purpose of constructing proof-trees we shall use:

∃-*elimination*

$$\frac{e : \exists x^D.F(x) \qquad \begin{array}{c}[t : D, g(t) : F(t)]\\ d(g,t) : C\end{array}}{INST(e, \acute{g}\grave{t}d(g,t)) : C}.$$

In terms of the semantical interpretations, the ∃-*elimination* rule can be seen as saying that whatever new 'function symbol' 'g' one chooses, if one can derive a statement 'C' (which may depend on both new variables 'g' and 't') from '$F(t)$' with label '$g(t)$', one can also derive the same statement 'C' from the existential formula '$\exists x^D.F(x)$' directly. Notice that the 'function symbol' 'g' is bound in the term alongside the conclusion 'C'.[9]

---

[9]Cf. Theorem II.26 of [Leisenring (1969)] (p. 55f), and his comments on the informal counterpart of the theorem in terms of mathematical reasoning:

"Theorem II.26 provides a formal justification in terms of the ε-calculus of a type of reasoning which is commonly used in mathematics. Suppose that we are trying to prove some statement C and in the course of the proof we prove a statement of the form

(1)   'for all x, there exists a y such that $B(x, y)$',

where $B(x, y)$ asserts some relationship between x and y. It is often convenient to have, for each x, a way of denoting some y such that $B(x, y)$. Since the notation must express the fact that y depends on x, we introduce a new function symbol g and say

(2)   'for all x, $B(x, g(x))$',

thus using $g(x)$ to denote an appropriate y. Theorem II.26 shows that if we can deduce C using statement (2), then we can deduce C directly from statement (1) without using the function g." [Leisenring (1969)] (p. 55)

Notice the parallel between Leisenring's informal explanation of his Theorem II.26 and the formal presentation of our ∃-*elimination*, where the counterpart of his 'g' is present in a term of our

Abstraction on function variables, yet keeping the logic first-order. The perspicuous reader will have noticed that we are here using abstraction over variables which may denote second-order objects (i.e. functions): we ´-abstract on 'g' over the expression '$d(g, t)$'. The question may arise as to whether the logic will remain first-order. And here once more we recall that the two-dimensional aspect of our calculus (i.e. a functional calculus on the labels, in harmony with a logical calculus on the formulas) means that we may have variables which will only appear in the functional calculus, no formulas having access to them. This is the case of our Skolem-dependency variable 'g' above.

Restrictions. We have been explaining the framework of our labelled natural deduction in terms of a certain harmony between a functional calculus on the labels and a logical calculus on the formulas. This can be expressed as saying that for every assumption discharge of the logical calculus there corresponds an abstraction on the functional calculus. Now, observe that the rule of $\exists$-*elimination* discharges the two 'Skolem-type' new assumptions, whose counterpart will correspond to abstractions on the respective label-variables in the functional calculus. So, when going from

$$d(g, t) : C$$

(i.e. the formula C with a label expression which depends on both t and g) in the premise, to

$$INST(e, \acute{g}\acute{t}d(g, t)) : C$$

in the conclusion, we are discharging the assumptions '$t : D$' and '$g(t) : F(t)$' and binding the respective variables on the label ('$\acute{t}$' and '$\acute{g}$', respectively). Thus, we really need to make sure that the new variables t and g do indeed occur free in the label alongside the third statement we want to draw straight from the existential after the application of the rule.

Furthermore, notice that the formula 'C' is unaffected by the application of the rule: even if it depends on the new variables, it will remain so. The new variables are only touched upon by the functional calculus. That is a feature of our 'two-dimensional' framework: a functional calculus on the labels in harmony with, yet independent of, a logical calculus on the formulas.

---

proof-calculus as the 'label' of an assumption which is then discharged. This 'discharging' is reflected in the label of the conclusion because this Skolem-type assumption has the variable corresponding to its label bound by the ´-abstractor, indicating the 'arbitrariness' of the choice of the particular 'g'. In other words, we can interpret the last sentence 'we can deduce C directly from statement (1) *without using the function g*' as saying that there is a generality statement being made about g when passing from the premise to the conclusion of the rule of $\exists$-*elimination*.

In order to see how the dependencies are eventually resolved, let us look again at the rule of β-reduction for the existential quantifier:

$\exists$-*reduction*

$$\dfrac{\dfrac{a:D \quad f(a):F(a)}{\varepsilon x.(f(x),a):\exists x^D.F(x)}\exists\text{-}intr \qquad \dfrac{[t:D,g(t):F(t)]}{d(g,t):C}}{INST(\varepsilon x.(f(x),a),\acute{g}\acute{t}d(g,t)):C}\exists\text{-}elim \quad \rhd_\beta \quad \dfrac{[a:D,f(a):F(a)]}{d(f/g,a/t):C}$$

Notice that the statement 'C' is not affected by the substitution performed on the label expression which results from the normalised proof. For example, the following proof fragment:

$$\dfrac{\dfrac{a:D \qquad f(a):F(a)}{\varepsilon x.(f(x),a):\exists x^D.F(x)}\exists\text{-}intr \qquad \dfrac{[t:D,g(t):F(t)]}{g(t):F(t)}(\dagger)}{INST(\varepsilon x.(f(x),a),\acute{g}\acute{t}g(t)):F(t)}\exists\text{-}elim$$

will be β-normalised to

$$[a:D,f(a):F(a)]$$
$$f(a):F(t)$$

and here we have the formula '$F(t)$' unaffected by the substitution performed on the label alongside it.[10] That this in no disagreement with the intended interpretation of existential statements is corroborated by what might be called 'the hiding principle' of indefinites: from an existential statement it should not be allowed a reference to the particular witness which supports the claim.[11] Thus, once

---

[10]Notice that the inference performed at point marked with '($\dagger$)' is simply to repeat the premise in the conclusion. Strictly speaking, this would involve going through the equality law of *reflexivity* on the label expressions such as, e.g.:

$$\dfrac{\dfrac{g(t):F(t)}{g(t)=_r g(t):F(t)}}{r(g(t)):F(t)}(*).$$

In [Gabbay and de Queiroz (1992)] we already explained this aspect of our labelled natural deduction when constructing a proof of $\lambda x.r(x):A\to A$.

The step marked '$(*)$' is justified by the following rules of inference discussed in more detail in the context of the interpretation of equality within labelled natural deduction:

$$\dfrac{a=_s b:A}{s(a):A} \qquad \dfrac{a=_s b:A}{s(b):A},$$

where the kind of equality 's' (which could be $\beta,\eta,\xi,\zeta,$ r) is carried along in the label.

[11]Contrary to Girard *et al.*'s view that the Natural Deduction rules for the so-called 'sums' (i.e. disjunction and existential quantification) are wrong, we feel that the use of a third statement is in perfect agreement with the intended meaning of 'indefinites'. About those elimination rules Girard *et al.* say:

"The elimination rules are very bad. What is catastrophic about them is the parasitic presence of a formula C which has no structural link with the formula which is eliminated. C plays the rôle of a context, and the writing of these rules is a concession to the sequent calculus."

[Girard *et al.* (1989)] (Chapter 'Sums in Natural Deduction', Section 'Defects of the System') Thinking of disjunction and existential quantification as indefinites we see nothing 'parasitic' about

the existential quantifier is introduced (and the witness appropriately 'hidden' in the label), no subsequent reference to the witness is allowed in the logical calculus. (Cf. the case of disjunction shown earlier on, and that of the connective of propositional equality, which is dealt with in [de Queiroz and Gabbay (1992)].)

3.2.2 *The hiding principle*

When defining the general pattern of an inference rule it is expected that it should be, in some sense, 'context-free': the only data available for the inference to be made is that available from the formulas in the premises and assumptions only.

When arriving at a disjunction (or indeed at an existential) one should not rely on the information: the *elimination* rule has to be applicable whether or not any other information that might be relevant for the inference has already been obtained. Although in cases involving assumptions (i.e. the *improper* inference rules), there may be an explicit reference to the history of the deduction, in general one wishes to have the inferences as free from the context as possible. Now, what happens with the indefinites, i.e. formulas like disjunctions and existentials, where at the point of *elimination* nothing may be known about which disjunct (resp. which witness) was used to arrive at such statement: it should only be assumed that simply one disjunct (resp. witness) would have made the statement true. Indeed, with *introduction* rules which 'forget' the singular witness which allowed the connective to be introduced,

$$\frac{A}{A \vee B} \qquad \frac{B}{A \vee B}$$

and

$$\frac{F(a_1)}{\exists x.F(x)} \qquad \frac{F(a_2)}{\exists x.F(x)} \qquad \frac{F(a_3)}{\exists x.F(x)} \quad \cdots \quad \frac{F(a_n)}{\exists x.F(x)}$$

(for any n) one needs to proceed by starting with new assumptions concerning the 'hidden' information.

---

the third statement, rather on the contrary: it is helping to keep the witness in hiding, so to speak. It is not quite clear whether the third statement really plays the rôle of the context. Assuming that the reference to the context in the above passage concerns the sequent calculus inference:

$$\frac{\Gamma, P(a) \vdash \Delta}{\Gamma, \exists x.P(x) \vdash \Delta}$$

(where a does not occur in the conclusion, i.e. the *eigenvariable* condition holds) then the rôle of the context (understood as a collection of formulas) is played by sets of formulas 'Γ' and 'Δ', rather than by a formula 'C', which is meant to be a single formula.

In other words, at points like:

$$\begin{array}{cc} \vdots & \vdots \\ A \vee B & \exists x.F(x) \\ \vdots & \vdots \end{array}$$

a general procedure for eliminating the logical connective is needed in order to keep the inference general enough so that it does not depend on the information obtained in previous steps, unless one of the premises is depending explicitly on assumptions that will be discharged by the inference rule (which, clearly, is not the case for either of the two kinds of premises above).

Closed formulas, the reuse of label-bound variables, and vacuous ε-abstractions. As the reader may have realised, we are here dealing with closed formulas, which will be theorems if and only if we can obtain closed labelled expressions alongside them in a deduction using the given rules of inference (plus the appropriate discipline of handling assumptions).

Let us construct a derivation for the following formula:
$(\forall x^D.F(x) \to C) \to \exists y^D.(F(y) \to C)$

$$\cfrac{\cfrac{[t:D]}{\cfrac{[f(t):F(t)]}{\cfrac{\Lambda t.f(t):\forall t^D F(t)}{\cfrac{\cfrac{APP(z,\Lambda t.f(t)):C}{\lambda f.APP(z,\Lambda t.f(t)):F(t)\to C}\;(*)}{\varepsilon y.(\lambda f.APP(z,\Lambda t.f(t)),t):\exists y^D.(F(y)\to C)}\;(**)}}\quad [z:\forall x^D.F(x)\to C]}}{\cfrac{\lambda z.\varepsilon y.(\lambda f.APP(z,\Lambda t.f(t)),t):(\forall x^D.F(x)\to C)\to\exists y^D.(F(y)\to C)}{\lambda t.\lambda z.\varepsilon y.(\lambda f.APP(z,\Lambda t.f(t)),t):D\to((\forall x^D.F(x)\to C)\to\exists y^D.(F(y)\to C))}}\;\;[t:D]'}$$

where at '$(*)$' we obtained a closed label-expression alongside an open formula. Notice that we are making full use of the 'two-dimensional' aspect of our labelled calculus: we performed a λf.-abstraction at point '$(*)$' to obtain an open formula '$F(t) \to C$'. Now, in order to obtain a closed label alongside a closed formula it was necessary to reuse the free variable of the formula (which was previously bound in the label alongside the formula) to obtain an existential closed formula via a 'vacuous' ε-abstraction (notice that at point '$(**)$' the expression '$\lambda f.APP(z,\Lambda t.f(t))$' no longer depends on t). Given that t was reintroduced as the witness of an ε-expression, the side effect was that the label expression was no longer closed, and the abstraction from t over the expression, introducing the implication with the domain D as antecedent, took us to a point where a closed label expression was obtained alongside a closed formula. Only, what we obtained was a proof of a slightly weakened version of the original formula: if the domain is non-empty then the formula is true. Fortunately, this is in no disagreement with the intended meaning of the formula.

In some cases we may not need to perform λ-abstractions on function variables to introduce implications,[12] and yet we may arrive at a point where a closed label expression is alongside an open formula. Again, as in the previous case, we need to perform further deductions to achieve 'closed label expression alongside closed formula'. For that we may have to use vacuous ε-abstractions as before.

Let us take the formula below:

$$\exists y^D.(\exists x^D.F(x) \to F(y))$$

$$
\cfrac{
[t:D] \quad \cfrac{
\cfrac{
[z:\exists x^D.F(x)] \quad \cfrac{
\cfrac{
\cfrac{
\cfrac{[t:D]}{[g(t):F(t)]}
}{g(t) =_r g(t) : F(t)}
}{r(g(t)):F(t)}
}{INST(z,\acute{g}\acute{t}r(g(t))):F(t)}
}{\lambda z.INST(z,\acute{g}\acute{t}r(g(t))):\exists x^D.F(x)\to F(t)}
}{\varepsilon y.(\lambda z.INST(z,\acute{g}\acute{t}r(g(t))),t):\exists y^D.(\exists x^D.F(x)\to F(y))} \, (*)
}{\lambda t.\varepsilon y.(\lambda z.INST(z,\acute{g}\acute{t}r(g(t))),t):D\to\exists y^D.(\exists x^D.F(x)\to F(y))}
$$

Notice that at point marked '$(*)$' we had a premise with a closed label expression alongside an open formula.

3.3 Other approaches to existential quantification

The study of existential statements has been systematically pursued by logicians, mathematicians and linguists alike. Since the early days of formal logic when formalisations of the notion of definite description were proposed, e.g. in Frege's (1893) *Grundgesetze*,[13] the notion of singular terms has been addressed by several mathematical logicians, beginning with no one less than David Hilbert.

---

[12] Although the example above refers to a theorem of classical predicate logic, notice that the abstraction on function variables is also needed in proving some theorems of intuitionistic logic, such as, e.g. '$\forall x^D.(F(x) \to \exists y^D.F(y))$' (see the example A.9 of Section 3.6).

[13] When attempting to explain the notion of 'definite article' (*Grundgesetze I*, §11), Frege defines the notation '$\backslash\acute{e}\Phi(\epsilon)$' to mean "the object falling under the concept $\Phi(\xi)$ if $\Phi(\xi)$ is a concept under which falls one and only one object".

3.3.1 *Systems of natural deduction based on direct existential instantiation*

There are essentially two ways of eliminating the existential quantifier in the systems of natural deduction discussed in the literature, namely *direct* and *indirect* existential instantiation. The one we have been studying in this chapter is regarded as 'indirect' existential instantiation, and is more commonly known as existential *elimination*. It is called *indirect* because the step from an existential formula to any other statement in a deduction is to be made dependent on the instantiated formula being inserted as a new local assumption. So, when arriving at a formula such as $\exists x.F(x)$, one does not proceed by directly instantiating with a new constant, say t, to obtain the instantiated formula $F(t)$. One has to introduce the latter as a new local assumption, as in:

$$\frac{\exists x.F(x) \qquad \begin{array}{c} [F(t)] \\ C \end{array}}{C}$$

The other way of eliminating the existential quantifier is to make a *direct* instantiation, as in:

$$\frac{\exists x.F(x)}{F(t)}$$

where, in order to prevent fallacious deductions, extra devices must be introduced to make sure that the newly introduced variable is indeed new, and that the same t is not universally quantified. Though varying in terms of the extra conditions to be added to the rule of existential instantiation (and eventually to the rule of universal generalisation), the approach of direct instantiation is used in a number of (plain) natural deduction systems described in the literature.

The systems of Natural Deduction based on direct existential instantiation were used by Quine [Quine (1950)] and, subsequently, by Copi [Copi (1954)], Kalish & Montague [Kalish *et al.* (1964)], Fine [Fine (1985)], among many others. They consisted essentially of Gentzen-style inference rules for the propositional connectives, and the following schemata of rules for quantifiers:

$$\text{UI} \quad \frac{\forall x F(x)}{F(t)} \qquad\qquad \text{UG} \quad \frac{F(a)}{\forall x F(x)}$$

$$\text{EI} \quad \frac{\exists x.F(x)}{F(a)} \qquad\qquad \text{EG} \quad \frac{F(t)}{\exists x.F(x)}$$

where 'I' and 'G' denoted 'instantiation' and 'generalisation', respectively, for both the Universal and the Existential quantifiers.

In this section we wish to give an overview of some of these systems with the aim of pointing out how and to what extent our labelled natural deduction system copes better with the peculiarities of the existential quantifier.

3.3.1.1 *Quine's system*

Perhaps the first system to use the meta-level device of annotating the instantial variables used in a deduction was Quine's system described in, e.g. Quine's [Quine (1950)] *Methods of Logic*. The idea was that in order to prevent fallacious deductions, such as:

PREMISE: $\exists x.F(x)$

DEDUCTION: $F(y)$ (EI)

              $\forall x F(x)$ (UG)   (wrong)

appropriate restrictions had to be imposed in the application of both EI and UG. To use Quine's own words, two neat restrictions would prove sufficient: the instantial variables of EI and UG must be different for each step (called *Flagging*), and the instantial variable of each such step must be alphabetically later than all free variables of the generic line of that step (named *Ordering*). Quine's way to making sure that the deductions did conform to the restrictions was by means of the device of 'flagging' variables. The idea was basically to record the instantial variable off the right of the line of each step of the deduction, so that the restrictions could be enforced.

As remarked by Fine [Fine (1985)], further to *Flagging* and *Ordering* an additional *Local Restriction* would also be required: the instantial variable does not occur in the formula $F(x)$ in any application of the rules UG and EI.

3.3.1.2 *Fine's generalised system*

In his monograph on *Reasoning with Arbitrary Objects* [Fine (1985)], K. Fine puts forward the idea that a generalised system of natural deduction with a rule of existential generalisation is to be defined by a set of restrictions on the instantial variables used by the rules UG and EI. Central to the whole theory is the idea that instantial variables (also called *A-letters*) denote *arbitrary objects* which get defined as they are introduced in the course of a deduction.

Considering what would count as an admissible restriction, Fine then asks what information from the derivation might properly be made use of in formulating the restrictions on the A-letters. Recognising the need for 'keeping track' of what variables and rules of inference were used in each step of a deduction, he gives the definition of auxiliary devices meant to allow the 'keeping-track of proof steps' to be effective. One of these devices is that of *line-type*: a *line-type* is a quintuple $\langle q, t, R, S, T \rangle$, where

q is a quantifier ($\forall$ or $\exists$)
t is an A-letter
R, S, T are sets of A-letters subject only to the condition that $T \supseteq R \cup S \cup \{t\}$.

The notion of *derivation-type* is then defined as a finite sequence $\tau_i, \ldots, \tau_n$, $n \geq 0$, of *line-types*, subject to the condition that, if T_i is the last term of τ_i and T_j the last term of τ_j, for $j > i$, then $T_j \supseteq T_i$. As Fine points out, a *derivation-type* encodes the information that can be extracted from the derivation as a whole. Here there is the question of what should be considered as relevant information.

Dependency diagrams.   Based on the (meta-level) information collected in the *line-types*, and with the help of 'dependency diagrams', Fine is then able to spell out the conditions under which a derivation is to be accepted as correct.

Instantiation or Elimination: check with ordinary language.   Defending the use of existential instantiation as 'closer to ordinary reasoning' than existential elimination, Fine argues that "the rules of UG and EI correspond to procedures of ordinary reasoning". In the case of the second rule, namely EI, the ordinary reasoning procedure given as an example goes as following: "having established that there exists a bisector to an angle, we feel entitled to give it a name and declare that it is a bisector to the angle". Now, the step of giving it a (previously inexistent) name and declaring it a bisector of the angle is indeed in accord with ordinary reasoning. It is not, however, in accord with rules of the EI form. It seems perfectly arguable that the Gentzen-Prawitz style *∃-elimination* reflects the procedure in a more 'natural' way. The instantial term in the latter is provided with the status of an 'assumption', such as 'let 'b' be the bisector', where 'b' is a 'new' name. In ordinary reasoning the assumption 'b' gets discharged when, given that some third statement, say 'C', is obtained from the assumption that 'b' is a bisector, one concludes that the same statement 'C' can be inferred directly from the original existential statement, regardless of whichever 'b' was chosen in the instantial deduction.

3.3.1.3   *Incorporating annotations into the object language*

We have seen that the information as to how and when instantial variables were used in a derivation involving the existential quantifier is of paramount importance in deciding whether a certain derivation constitutes a valid one in the semantics. Now, let us recall that in a labelled natural deduction system we carry the information concerning the proof steps as already an 'official' part of the calculus: for each logical connective and quantifier there corresponds a specific functional device on the labels. Moreover, by introducing abstractors to bind free variables of the labels whenever there is an assumption discharge on the logical calculus, we obtain a suitable harmony between a functional calculus on the labels (where the 'meta-level' information is recorded) and a logical calculus on the formulas. We have abstractors, whose scopes are precisely defined, working together with logical connectives and quantifiers. Needless to say, we are here in possession of a much more 'natural' calculus to handle existential statements, given that we are capable of handling relevant information from the meta-level in the object-level. For example, the conditions on the dependency diagrams become scope conditions on the abstractors of the calculus on the labels. Instead of handling instantial objects dependency in terms of diagrams, and then 'interpret dependence in terms of scope'[14] we can actually handle dependency conditions solely in terms of abstractor-scoping.

The right restrictions. Instead of setting out a generalised system based on ∃-instantiation and then start figuring out what the right restrictions should be imposed on the instantial variables, we want to keep as close as possible to the original analysis of deduction made by Gentzen where we have *introduction* and *elimination* rules (standing in a certain harmony) for each connective and quantifier. With the addition of labels alongside the formulas, we have the general mechanisms of introduction of new variables (as labels of formulas), as well as the devices of abstraction and substitution. With this additional dimension added to a logical calculus with connectives and quantifiers, we are able to meet the minimum soundness conditions discussed in Fine's monograph (i.e. *Flagging, Ordering, Irreflexivity* of syntactic dependence, *Local Restriction*, etc.) in a natural fashion simply by making use of the properties of (Fregean) 'course-of-values' functional terms.

3.3.2 *Axiomatic systems based on the notion of 'such that'*

It is fair to say that the formal counterpart to the notion of 'the object such that it has a given property' finds its origins in Frege's device of 'course-of-values' presented in the *Grundgesetze*. Nevertheless, given that Frege's course-of-values terms could also denote truth-values (not merely 'objects such that ...'), the calculus presented in the *Grundgesetze* does not give us a 'pure' formalisation of the

---

[14]Fine, op. cit., p. 163.

notion of 'such that'. The logical concept of 'singular' terms, i.e. those terms denoted by the 'such that'-expressions, found in Hilbert and Bernays [Hilbert and
Bernays (1934, 1939)] proof-theoretic studies a major development. By placing
truth-values at the 'meta-level', so to speak, and being primarily concerned with a
calculus of expressions and logical deductions, Hilbert and Bernays undertook the
task of formalising the notion of 'such that' in its full generality. The main device
was essentially the same as Frege's *Wertverlauf*, i.e. a device allowing the binding
free occurrences of variables in a formula of the predicate calculus to obtain a
term which would denote any object falling under the concept determined by the
formula. The resulting terms would be the so-called singular terms. For example,
an ε-singular term of the sort '$\varepsilon_x P(x)$' denotes one individual (if it exists) *such
that* it has the property P. (The formalisation of mathematics given in Bourbaki's
Theorie des Ensembles involves the so-called τ-terms, which are defined in the
same spirit as Hilbert's ε-terms.[15])

3.3.2.1   *Hilbert's η- and ε-calculi*

As we have seen above, the origins of Hilbert's calculi of singular terms are to be
found in Frege's *Grundgesetze*. There we find the basic concepts of predicates as
propositional functions, the variable binding device of forming course-of-values
expressions, the techniques of substitution and instantiation, etc. What we do not
find in Frege's original treatise is a systematic account of the meta-mathematical
properties of the predicate calculus with the addition of course-of-values expressions. The essential ingredient, one may fairly say, is already given in the first
volume of the *Grundgesetze*.

   An axiomatic calculus of singular terms, designed with a view to prove meta-
mathematical properties of logical calculi, such as for example the consistency of
analysis, was first formulated and studied systematically by Hilbert in his early
paper on a τ-calculus (Hilbert, 1923). Soon afterwards came Ackermann's doc-
toral dissertation written under Hilbert, which attempted to prove the consistency
of analysis via an ε-calculus. It was, according to Leisenring [Leisenring (1969)],
the first published work in which the ε-symbol was used.

   As we have already mentioned, the essential difference between Hilbert's calculi and ours is that we handle ε-like terms in the functional calculus alongside

---

[15]Bourbaki explains the intuitive meaning of τ-terms as:
"(...) if B is an assertion and x a letter, then $\tau_X(B)$ is an object. Let us consider the assertion B as
expressing a property of the object X; then, if there exists an object which has the property in question,
$\tau_X(B)$ represents a distinguished object which has this property; if not, $\tau_X(B)$ represents an object
about which nothing can be said."

[p. 20]

the formulas. Our ε-symbol is not part of the logical calculus, and only appears alongside existential formulas.

3.3.2.2 *Hailperin's ν-calculus*

Devised by T. Hailperin [Hailperin (1957)] in the tradition of Frege's *Wertverlauf* calculus, the ν-calculus consisted of an attempt to generalise Hilbert and Bernays' [Hilbert and Bernays (1934, 1939)] calculi of singular terms by replacing the usual variable letters bound by the ε- (or η-) operators in the respective calculus by *restricted variables*. Roughly speaking a *restricted variable* was defined as any expression of the form '$\nu x Q$', x being an individual variable and Q being a formula of the predicate calculus. Instead of attempting to use the ν-terms to replace the usual quantifiers, as is the case of Hilbert and Bernays' calculi,[16] Hailperin's calculus consisted of an attempt to combine a formal counterpart to the notion of 'x such that Q' *à la* Frege with the usual quantifiers. So, for example, instead of having '$\forall x^D P(x)$' (meaning 'for all x from domain D, $P(x)$ is true'), Hailperin would write '$\forall(\nu x D)P(x)$' where '$\nu x D$' would stand for 'any x such that D', and D is any formula of the calculus.

In our labelled natural deduction every variable introduced is, in some sense, restricted: to every variable (introduced as a new atomic label) it is associated a formula. In this chapter we have only considered unanalysed domains 'D', but the framework is general enough to allow for complex formulas to appear as the domain of quantification.

3.4 Model-theoretic semantics

The purpose here is simply to indicate how the proof calculus reflects the model-theoretical semantic interpretations. There are important lessons one can learn from this exercise, not least how to uncover shortcomings in the plain Natural Deduction systems in terms of expressiveness. For example, we have seen from previous sections that only the Skolem-constant appears in the rule of ∃-*elimination* presented in Gentzen-Prawitz style of Natural Deduction:

$$\exists\text{-}elimination \quad \frac{\exists x.F(x) \qquad \begin{array}{c}[F(t)]\\ C\end{array}}{C}$$

the Skolem-function not having any rôle to play.

---

[16]The ε-calculus allows the replacement of an existential formula like $\exists x.P(x)$ by $P(\varepsilon_x P(x))$, as well as the replacement of universally quantified formulas such as $\forall x P(x)$ by $P(\varepsilon_x \neg P(x))$.

The model-theoretic interpretations have shown us how to obtain proof procedures by looking at the model-theoretic aspects of the quantifiers. The framework of labelled Natural Deduction systems, on the other hand, demonstrates that the proof procedures and the key properties of quantifiers are better attributed to the individual quantifier one is dealing with, neither to the axiomatic nor to the usual 'universal as a series of conjunctions' and 'existential as a series of disjunctions' semantical interpretations. The quantifiers themselves are characterised by their functionality given by their corresponding *introduction* and *reduction* rules.

3.4.1 *Constants versus variables revisited*

So far we have not discussed as to the way in which 'constants' are distinguished from 'variables' within our labelled natural deduction. The implicit assumption being that an assumption of the form '$x : D$' is to mean 'let x be an element from domain D', only the labelling mechanisms attached to quantifiers tell us which individuals are being taken as constants (or variables) at a given point in a deduction: whereas in the label alongside an universal quantifier being introduced the corresponding individual name becomes Λ-bound, the introduction of an existential quantifier leaves the individual name free (not bound) in its corresponding label expression. There is no definite typographical convention, no subscripting of dependent variables, and no special meta-level proviso stating which names denote constant and which names denote variables.

3.4.1.1 *Skolem resolution*

The rule of $\exists$-*elimination* has a well-known theorem-counterpart in the model-theoretic semantics, namely Skolem's resolution (Theorem II.26 of [Leisenring (1969)]). It says that when trying to prove a statement 'C', and in the course of the proof we get a proof of a statement like 'for all x, there exists a y such that $F(x, y)$', and we can show that the same 'C' can be derived from the skolemised 'forall x, $F(x, g(x))$', then we can deduce 'C' directly from the previous formula, regardless of the choice of 'g':

$\exists$-*elimination*

$$
\cfrac{e : \exists x^D.F(x) \qquad \cfrac{[t : D, g(t) : F(t)]}{d(g,t) : C}}{INST(e, \acute{g}\acute{t}d(g,t)) : C} \qquad \text{(instead of simply: } \cfrac{\exists x.F(x) \qquad \cfrac{[F(t)]}{C}}{C} \text{)}
$$

where '$d(g, t)$' is a term which depends on both on 't' and 'g'. (See examples of proofs below.) Observe that we have a much heavier syntax than the plain Natural

Deduction system, but this is the price we pay if we want to keep track not only of our proof steps but also of our introduction of new arbitrary names for individuals during our process of making deductions.

One might say that the $\exists$-*elimination* rule is actually saying that whatever new 'function symbol' 'g' one chooses, if one can derive a statement 'C' from '$F(t)$' with term '$g(t)$', one can also derive the same statement 'C' directly from the existential formula '$\exists x^D.F(x)$', no matter which 'g' was chosen. Notice that the 'function symbol' 'g' is no longer a free variable in the label expression of the conclusion: it is bound by the ´-abstractor in the term alongside the conclusion 'C'. (Notice also that in plain Natural Deduction the 'function symbol' 'g' does not even appear.)

3.4.1.2 *Herbrand resolution*

A general procedure for the elimination of universal quantifiers in the scope of existential quantifiers finds in Herbrand's resolution its semantical justification (Theorem II.25 of [Leisenring (1969)]). The idea can be summarised as follows:

Suppose we have the following prenex formula:

$$\exists x.\forall y.P(x, y)$$

and we want to get rid of the universal quantifier occurring inside the scope of the existential quantifier to obtain a $\exists$-prenex formula. Herbrand's theorem states that the formula above is equivalent to the following formula:

$$\forall u.(\exists x.\forall y.P(x, y) \rightarrow \exists w.P(w, u))$$

where the new individual symbol 'u' is called Herbrand constant, and 'w' is constructed from any Herbrand function 'g' as dependent on the constant eventually replacing 'x' in a deduction. No matter which Herbrand-constant 'u' and Herbrand function 'g' we choose, the universal quantifier can be eliminated in the way prescribed.

Now, if we construct a proof of the latter formula starting from the former using the functional interpretation (with explicit mention of the domain of quantification 'D'), we get:

$$\frac{\Lambda u.\lambda e.INST(e, \acute{g}t\varepsilon w.(EXTR(g(w), u), t)) : \forall u^D.(\exists x^D.\forall y^D.P(x, y) \rightarrow \exists w^D.P(w, u))}{}$$

$$\cfrac{\cfrac{[e : \exists x^D.\forall y^D.P(x, y)] \quad \cfrac{[t : D] \quad \cfrac{[u : D] \quad [g(t) : \forall y^D P(t, y)]}{EXTR(g(t), u) : P(t, u)}}{\varepsilon w.(EXTR(g(w), u), t) : \exists w^D.P(w, u)}}{INST(e, \acute{g}t\varepsilon w.(EXTR(g(w), u), t)) : \exists w^D.P(w, u)}}{\lambda e.INST(e, \acute{g}t\varepsilon w.(EXTR(g(w), u), t)) : \exists x^D.\forall y^D.P(x, y) \rightarrow \exists w^D.P(w, u)}$$

Observe that here we have the explicit account of the Herbrand resolution: whichever 'u' we choose from the domain 'D' (if it is empty we can introduce a new Herbrand constant), whichever Herbrand function 'g' we choose, we can eliminate the universal quantifier in the scope of the existential quantifier in a way that any instantiation of 'y' will depend on the particular instantiation of 'x'. The 'arbitrariness' of the choice of 'u' is reflected both in the 'functional' side (i.e. the abstractors on the label) and the 'logical' side (the universal quantification on the formula), whereas the arbitrariness of 'g' is reflected directly in the functional calculus on the labels. The functional calculus shows that they do not have to be particular constants or function-constants. They are both bound variables in the final label, 'u' being Λ-bound and 'g' being '-bound.

3.4.2 *Eliminability and conservative extensions*

One of the primary motivations of Gentzen in his analysis of deduction was to develop a formal system that could come as close as possible to actual mathematical reasoning. To every logical symbol belongs precisely one class of inference figures which *introduce* the symbol, and one which *eliminates* it. Instead of starting from basic logical propositions (the axioms), Natural Deduction starts from assumptions. The framework of Natural Deduction is such that for the introduction of a new symbol into the language, one has to provide *independent* inference rules which determine the rules of how to deal with that symbol in particular. In the Hilbert-style 'axioms plus rules of inference', the introduction of a new symbol had to be proved a 'conservative extension', given that the new axioms introduced could affect existing connectives because of the lack of independence enjoyed by Natural Deduction formal systems. For example, in the case of the ε-calculus, where a new logical symbol '$\varepsilon x A$' was introduced to represent singular terms, Hilbert and Bernays [Hilbert and Bernays (1934, 1939)] had to prove the so-called 'eliminability' theorems, showing the new calculus to be a conservative extension of first-order predicate calculus. They had to do so, mainly because the introduction of the new logical symbol was not made via logically *independent* inference rules, but by the introduction of a new axiom such as '$\exists x A \rightarrow A(\varepsilon x A)$' in Leisenring's [Leisenring (1969)] axiomatic. Clearly the theorems involving '$\exists$' might be affected by the introduction of the new symbol. The point of the elimination theorems is to show that the introduction of the new symbol is well-disciplined, in the sense that all theorems not involving the new symbol can still be proved, and that no new theorems not involving the new symbol can be proved in the enriched calculus.

This way of enriching first-order predicate calculus is certainly effective, as Hilbert and Bernays have indeed proved. It is not, however, the only way to handle the introduction of new logical symbols in a given calculus, in particular the introduction of singular terms in first-order predicate logic. We are here engaged in an investigation into how benefit from a reassessment of Frege's 'two-dimensional' calculus by producing a functional interpretation of the existential quantifier within the framework of a labelled Natural Deduction system, where we demonstrate that Gentzen's criterion of *independence* can be provided even for the apparently problematic ε-like calculi. The difficulties of proving the 'new' calculus a conservative extension of the previous one is transformed into a much simpler task, given that the methodology itself imposes the introduction of new signs to be made in a well-behaved fashion.

At the heart of the analysis of deduction into Natural Deduction systems was the principle of inversion: the *introduction* rule for a logical connective would have to be the exact inverse of its *elimination* rule, so that the extension of the system with a new connective would be 'naturally' conservative. Instead of considering any system of inference rules with premises and conclusions, the central to the idea of a Natural Deduction formulation of a logic was the pattern of *introduction/elimination* rules, together with the possibility of normalising proofs which contained 'redundant' steps without leaving any *junk* behind. The conservativeness of extensions is seen from a rather different perspective: in order to make sure extensions are indeed conservative, one has to prove a normalisation theorem for the extended system. In other words, the definition of a new connective is made in such a way that the *elimination* and the *introduction* rules stand to one another in a symmetrical relation as guaranteed by the *inversion principle*.

3.5   Finale

From what we have seen above, it would not seem unreasonable to claim that the use of labels alongside the formulas is crucial in making the connections between the various proof calculi and the model-theoretic semantics much clearer and much more explicit than with unlabelled systems. We have also seen that the 'two-dimensional' feature of formal logic (i.e. a functional calculus on the terms, and a logical calculus on the formulas), already apparent in Frege's *Grundgesetze*, finds in the functional interpretation a concrete representative.

Furthermore, some interesting, even if often overlooked, aspects of the corresponding logical connective show up in a rather 'natural' fashion. In [Gabbay and de Queiroz (1992)], we have seen that what distinguishes one implication from another (linear, relevant, intuitionistic, classical, etc.) is the way in which assumptions are discharged, all implications having *modus ponens*. From [de Queiroz and Gabbay (1997)], this particular feature (i.e. the discipline of handling assumptions as the distinguishing factor) is shown to be applicable to other kinds of conditionals, including the modal *necessity*.

Here, it seems that an interesting feature of the existential quantifier is unmistakably showing up, and that is the kind of generality involved in the introduction of Skolem-functions.

The fact that the variable for the Skolem function is bound by an abstractor (' '' in this case), may be an indicator that the well-known Skolem equivalence should not be between

$$\forall x \exists y. R(x, y) \text{ and } \exists g. \forall x R(x, g(x)),$$

but rather between

$$\forall x \exists y. R(x, y) \text{ and } \forall g. \forall x R(x, g(x)),$$

for any *newly* chosen Skolem-function 'g'.

The latter indicates more explicitly that the choice of 'g' is indeed irrelevant: in other words, instead of 'there exists a function g ...', one ought to say 'for any (newly chosen) function g ...'.[17]

---

[17]It is customary to use the existential quantification over the Skolem functions rather than the universal quantification.

For example, when discussing two sorts of counterexamples to the view that constructive mathematics is a part of classical mathematics, Tait [Tait (1983)] refers to this sort of equivalence by means of the usual 'there is a Skolem-function ...:', viz.:

"(...) The first [counterexample] concerns the axiom of choice

$$\forall x^A \exists y^B C(x, y) \rightarrow \exists z^A \rightarrow B. \forall x : A.C(x, zx),$$

which is constructively valid but which is regarded as an object of controversy in classical mathematics."

[Tait (1983)] (p. 183)

It seems that the functional interpretation of the existential quantifier makes us feel more inclined to have the '$\exists z^A \rightarrow B. \ldots$' replaced by '$\boxed{\forall} z : A \rightarrow B \ldots$' with the special proviso that z must be a *new* function.

Another example of the use of '$\exists$' instead of '$\forall$' comes from a recent paper on the free variable first-order predicate calculus, which begins with:

"The intuitive idea that existential quantifiers are appropriately expressed by function symbols, using an equivalence such as

$$(\forall x)(\exists y), \ldots, x, y, \ldots \Leftrightarrow (\exists f)(\forall x), \ldots, x, f(x), \ldots$$

has played a crucial role in logical investigations and in certain aspects of computer science. These "Skolem functions" were introduced [previously] to prove the famous Skolem-Löwenheim theorem." [Davis and Fechter (1991)]

However, we need to be a little more careful here. First of all, with our labelled natural deduction we manage to make the function variable 'lose its identity', so to speak, without using any quantifiers of the logical calculus. Secondly, the need for the special proviso *newly chosen* may indicate that the generality of the Skolem functions lie somewhere between existential and universal, leaving perhaps enough room for divergent interpretations of the notion of *choice* in mathematical logic.

3.5.1 *Extensions to higher-order existentials*

A natural extension to the analysis of first-order existential quantification via the functional interpretation which we have demonstrated here is the study of a notion of higher-order existential quantification best known as modal *possibility*. As a sequel to our analysis of the notion of modal *necessity* which is also being carried out within the framework of the functional interpretation [de Queiroz and Gabbay (1997)], we are currently working on an application of the methods and techniques described here to the characterisation of some of the notions of modal *possibility* described in the literature.

3.5.2 *Further connections to model-theoretic interpretations*

What we have given here is in this respect is merely a glimpse of a whole area to explore. It is our intention to further pursue the investigation into a more detailed account of the connections between the model-theoretic interpretations of existential statements and our functional interpretation. For example, we would like to see strengthened the connections between our framework and the model-theoretic framework used by Shoenfield (1967), by characterising the counterpart of the latter's notions of *special constants*, *Henkin's theories*, etc.

3.6 Examples of deduction

For the sake of illustrating the framework we have just described, we should like to present a few examples of proofs of propositions involving the existential quantifier in our labelled natural deduction.

3.6.1 *Generic examples*

The proof of many propositions relating the existential quantifier with the universal quantifier, some of which are presented as axioms, some as 'rules of passage', and others as theorems,[18] can be readily constructed by the use of the rules given for the existential quantifier in our functional interpretation:

Example 3.1. $\forall x^D(F(x) \to C) \to (\exists x^D.F(x) \to C)$

$$
\cfrac{
  \cfrac{
    \cfrac{
      [e : \exists x^D.F(x)] \quad
      \cfrac{
        [g(t) : F(t)] \quad
        \cfrac{[t : D] \quad [z : \forall x^D.(F(x) \to C)]}{EXTR(z,t) : F(t) \to C}
      }{APP(EXTR(z,t), g(t)) : C}
    }{INST(e, \acute{g}tAPP(EXTR(z,t), g(t))) : C}
  }{\lambda e.INST(e, \acute{g}tAPP(EXTR(z,t), g(t))) : \exists x^D.F(x) \to C}
}{\lambda z.\lambda e.INST(e, \acute{g}tAPP(EXTR(z,t), g(t))) : \forall x^D.(F(x) \to C) \to (\exists x^D.F(x) \to C)}
$$

A proof of $\forall x^D.\neg F(x) \to \neg \exists x^D.F(x)$ becomes a special case of the above proof if we put '$\neg P \equiv P \to \mathcal{F}$', rewriting it to $\forall x^D.(F(x) \to \mathcal{F}) \to (\exists x^D.F(x) \to \mathcal{F})$.

Example 3.2. $(\exists x^D.F(x) \to C) \to \forall x^D.(F(x) \to C)$

$$
\cfrac{
  \cfrac{
    \cfrac{
      \cfrac{[t : D] \quad [f(t) : F(t)]}{\varepsilon x.(f(x),t) : \exists x^D.F(x)} \quad [z : \exists x^D.F(x) \to C]
    }{APP(z, \varepsilon x.(f(x),t)) : C}
  }{\lambda f.APP(z, \varepsilon x.(f(x),t)) : F(t) \to C}
}{
  \cfrac{\Lambda t.\lambda f.APP(z, \varepsilon x.(f(x),t)) : \forall x^D(F(x) \to C)}{\lambda z.\Lambda t.\lambda f.APP(z, \varepsilon x.(f(x),t)) : (\exists x^D.F(x) \to C) \to \forall x^D.(F(x) \to C)}
}
$$

We can prove '$\neg \exists x^D.F(x) \to \forall x^D.\neg F(x)$' by putting '$\neg P \equiv P \to \mathcal{F}$', rewriting it to '$(\exists x^D.F(x) \to \mathcal{F}) \to \forall x^D.(F(x) \to \mathcal{F})$'.

Example 3.3. $\forall x^D F(x) \to (\exists x^D(F(x) \to C) \to C)$.

$$
\cfrac{
  \cfrac{
    \cfrac{
      [z : \exists x^D(F(x) \to C)] \quad
      \cfrac{
        \cfrac{[t : D] \quad [u : \forall x^D F(x)]}{EXTR(u,t) : F(t)} \quad [g(t) : F(t) \to C]
      }{APP(g(t), EXTR(u,t)) : C}
    }{INST(z, \acute{g}tAPP(g(t), EXTR(u,t))) : C}
  }{\lambda z INST(z, \acute{g}tAPP(g(t), EXTR(u,t))) : \exists x^D(F(x) \to C) \to C}
}{\lambda u \lambda z INST(z, \acute{g}tAPP(g(t), EXTR(u,t))) : \forall x^D F(x) \to (\exists x^D(F(x) \to C) \to C)}
$$

---

[18]Cf., for example, [Heyting (1956)], [Quine (1950)], [Leisenring (1969)].

The proof of $\forall x^D F(x) \to \neg\exists x^D \neg F(x)$ is just a special case of the above proof when we put '$\neg P \equiv P \to \mathcal{F}$', rewriting the original proposition into $\forall x^D F(x) \to (\exists x^D (F(x) \to \mathcal{F}) \to \mathcal{F})$.

Example 3.4. $\exists x^D F(x) \to (\forall x^D (F(x) \to C) \to C)$

$$\cfrac{\cfrac{[z : \exists x^D F(x)] \qquad \cfrac{[g(t):F(t)] \qquad \cfrac{[t:D] \quad [u:\forall x^D(F(x)\to C)]}{EXTR(u,t):F(t)\to C}}{APP(EXTR(u,t),g(t)):C}}{INST(z,\acute{g}\grave{t}APP(EXTR(u,t),g(t))):C}}{\cfrac{\lambda uINST(z,\acute{g}\grave{t}APP(EXTR(u,t),g(t))):\forall x^D(F(x)\to C)\to C}{\lambda z\lambda uINST(z,\acute{g}\grave{t}APP(EXTR(u,t),g(t))):\exists x^D F(x)\to(\forall x^D(F(x)\to C)\to C)}}$$

Similarly to the previous case, the proof of $\exists x^D F(x) \to \neg\forall x^D \neg F(x)$ becomes a special case of the above proof when we put '$\neg P \equiv P \to \mathcal{F}$', rewriting the original proposition into $\exists x^D F(x) \to (\forall x^D (F(x) \to \mathcal{F}) \to \mathcal{F})$.

Example 3.5. $\exists x^D (F(x) \to C) \to (\forall x^D F(x) \to C)$

$$\cfrac{\cfrac{[z:\exists x^D(F(x)\to C)] \qquad \cfrac{\cfrac{[t:D] \quad [u:\forall x^D F(x)]}{EXTR(u,t):F(t)} \qquad [g(t):F(t)\to C]}{APP(g(t),EXTR(u,t)):C}}{INST(z,\acute{g}\grave{t}APP(g(t),EXTR(u,t))):C}}{\cfrac{\lambda uINST(z,\acute{g}\grave{t}APP(g(t),EXTR(u,t))):\forall x^D F(x)\to C}{\lambda z\lambda uINST(z,\acute{g}\grave{t}APP(g(t),EXTR(u,t))):\exists x^D(F(x)\to C)\to(\forall x^D F(x)\to C)}}$$

Again, similarly to previous cases, the proof of '$\exists x^D \neg F(x) \to \neg\forall x^D F(x)$' becomes a special case of the proof just given, if we put '$\neg P \equiv P \to \mathcal{F}$', rewriting the original formula as $\exists x^D (F(x) \to \mathcal{F}) \to (\forall x^D F(x) \to \mathcal{F})$.

Example 3.6. $(\forall x^D.F(x) \to C) \to \exists y^D.(F(y) \to C)$ (classical)

$$\cfrac{\cfrac{\cfrac{[t:D] \qquad \cfrac{\cfrac{[f(t):F(t)]}{\Lambda t f(t):\forall t^D F(t)} \qquad [z:\forall x^D F(x)\to C]}{APP(z,\Lambda t.f(t)):C}}{\lambda fAPP(z,\Lambda t.f(t)):F(t)\to C}}{\cfrac{\varepsilon y.(\lambda f.APP(z,\Lambda t.f(t)),t):\exists y^D(F(y)\to C)}{\lambda z\varepsilon y.(\lambda f.APP(z,\Lambda t.f(t)),t):(\forall x^D F(x)\to C)\to\exists y^D(F(y)\to C)}}}{\lambda t.\lambda z.\varepsilon y.(\lambda f.APP(z,\Lambda t.f(t)),t):D\to((\forall x^D F(x)\to C)\to\exists y^D(F(y)\to C))}$$

Note that here we have a weakened version of the classical case[19] where the full classical version is conditional to the domain 'D' being non-empty.

Example 3.7. $\exists x^D(F \to G(x)) \to (F \to \exists x^D G(x))$

$$
\cfrac{
  \cfrac{
    [z : \exists x^D(F \to G(x))] \quad
    \cfrac{
      \cfrac{
        [t : D] \quad
        \cfrac{
          [v : F] \quad [h(t) : F \to G(t)]
        }{
          APP(h(t), v) : G(t)
        }
      }{
        \varepsilon x(APP(h(x), v), t) : \exists x^D G(x)
      }
    }{
      INST(z, \acute{h}t\varepsilon x.(APP(h(x), v), t)) : \exists x^D G(x)
    }
  }{
    \lambda v.INST(z, \acute{h}t\varepsilon x.(APP(h(x), v), t)) : F \to \exists x^D G(x)
  }
}{
  \lambda z.\lambda v.INST(z, \acute{h}t\varepsilon x.(APP(h(x), v), t)) : \exists x^D(F \to G(x)) \to (F \to \exists x^D G(x))
}
$$

Example 3.8. $(F \to \exists x^D G(x)) \to \exists x^D(F \to G(x))$ (classical)

$$
\cfrac{
  \cfrac{
    [t : D] \quad
    \cfrac{
      \cfrac{
        \cfrac{
          [z : F] \quad [y : F \to \exists x^D G(x)]
        }{
          APP(y, z) : \exists x^D G(x)
        } \quad
        \cfrac{
          \cfrac{
            \cfrac{[t : D]}{[g(t) : G(t)]} \quad g(t) =_r g(t) : G(t)
          }{
            r(g(t)) : G(t)
          }
        }{}
      }{
        INST(APP(y, z), \acute{g}tr(g(t))) : G(t)
      }
    }{
      \lambda z.INST(APP(y, z), \acute{g}tr(g(t))) : F \to G(t)
    }
  }{
    \varepsilon x(\lambda z.INST(APP(y, z), \acute{g}tr(g(t))), t) : \exists x^D(F \to G(x))
  } \;(*)
}{
  \cfrac{
    \lambda y.\varepsilon x.(\lambda z.INST(APP(y, z), \acute{g}tr(g(t))), t) : (F \to \exists x^D G(x)) \to \exists x^D.(F \to G(x))
  }{
    \lambda t.\lambda y.\varepsilon x.(\lambda z.INST(APP(y, z), \acute{g}tr(g(t))), t) : D \to ((F \to \exists x^D G(x)) \to \exists x^D(F \to G(x)))
  }
}
$$

and here, similarly to previous cases of intuitionistic non-theorems, we have a vacuous εx-abstraction, obtaining a weakened version of the classical theorem, which states that the theorem holds if the domain is non-empty.

Example 3.9. $\forall x^D(F(x) \to \exists y^D F(y))$[20]

$$
\cfrac{
  \cfrac{
    \cfrac{
      [x : D] \quad [f(x) : F(x)]
    }{
      \varepsilon y.(f(y), x) : \exists y^D F(y)
    }
  }{
    \lambda f.\varepsilon y.(f(y), x) : F(x) \to \exists y^D F(y)
  }
}{
  \Lambda x.\lambda f.\varepsilon y.(f(y), x) : \forall x^D(F(x) \to \exists y^D F(y))
}
$$

Example 3.10. $\exists x^D(F(x) \to \forall t^D F(t))$ (classical)

$$
\cfrac{
  \cfrac{
    [t : D] \quad
    \cfrac{
      \cfrac{
        \cfrac{
          \cfrac{[t : D]}{[f(t) : F(t)]}
        }{
          \Lambda t.f(t) : \forall t^D F(t)
        }
      }{
        \lambda f.\Lambda t.f(t) : F(t) \to \forall t^D F(t)
      }
    }{
      \varepsilon x.(\lambda f.\Lambda t.f(t), t) : \exists x^D(F(x) \to \forall t^D F(t))
    }
  }{}
}{
  \lambda t.\varepsilon x.(\lambda f.\Lambda t.f(t), t) : D \to \exists x^D(F(x) \to \forall t^D F(t))
}
$$

---

[19] By following the same procedure used in the previous cases, we can show that this is a general presentation of the classical '$\neg\forall x^D F(x) \to \exists x^D \neg F(x)$'.

[20] In [Martin-Löf (1984)] (p. 47)

and we obtain a 'weakened' version of the classically valid schema above, namely, '$D \rightarrow \exists x^D (F(x) \rightarrow \forall t^D F(t))$', which says that the schema we started off with is valid if the domain 'D' over which one is existentially quantifying is not empty.[21]

Example 3.11. $\forall x^D (F(x) \rightarrow G(x)) \rightarrow (\exists x^D F(x) \rightarrow \exists y^D G(y))$

$$
\frac{\dfrac{\dfrac{[h(t) : F(t)] \quad \dfrac{[t : D] \quad [z : \forall x^D (F(x) \rightarrow G(x))]}{EXTR(z,t) : F(t) \rightarrow G(t)}}{APP(EXTR(z,t), h(t)) : G(t)}}{\dfrac{[u : \exists x^D F(x)] \quad \varepsilon y.(APP(EXTR(z,y), h(y)), t) : \exists y^D G(y)}{\dfrac{INST(u, \hat{h} \varepsilon y.(APP(EXTR(z,t), h(y)), t)) : \exists y^D G(y)}{\lambda u.INST(u, \hat{h} \varepsilon y.(APP(EXTR(z,t), h(y)), t)) : \exists x^D F(x) \rightarrow \exists y^D G(y)}}}}{\lambda z.\lambda u.INST(u, \hat{h} \varepsilon y.(APP(EXTR(z,t), h(y)), t)) : \forall x^D (F(x) \rightarrow G(x)) \rightarrow (\exists x^D F(x) \rightarrow \exists y^D G(y))}
$$

Example 3.12. $(\exists x^D F(x) \rightarrow \forall x^D G(x)) \rightarrow \forall x^D (F(x) \rightarrow G(x))$

$$
\frac{\dfrac{\dfrac{\dfrac{[y : D] \quad [f(y) : F(y)]}{\varepsilon x.(f(x), y) : \exists x^D F(x)} \quad [u : \exists x^D F(x) \rightarrow \forall x^D G(x)]}{APP(u, \varepsilon x.(f(x), y)) : \forall x^D G(x)}}{[y : D] \quad \dfrac{EXTR(APP(u, \varepsilon x.(f(x), y)), y) : G(y)}{\lambda f.EXTR(APP(u, \varepsilon x.(f(x), y)), y) : F(y) \rightarrow G(y)}}}{\dfrac{\Lambda y.\lambda f.EXTR(APP(u, \varepsilon x.(f(x), y)), y) : \forall y^D (F(y) \rightarrow G(y))}{\lambda u.\Lambda y.\lambda f.EXTR(APP(u, \varepsilon x.(f(x), y)), y) : (\exists x^D F(x) \rightarrow \forall x^D G(x)) \rightarrow \forall y^D (F(y) \rightarrow G(y))}}
$$

Example 3.13. $\forall y^D F(y)(\forall x^D F(x) \rightarrow F(y))$[22]

$$
\frac{\dfrac{\dfrac{[y : D] \quad [u : \forall x^D F(x)]}{EXTR(u, y) : F(y)}}{\lambda u.EXTR(u, y) : \forall x^D F(x) \rightarrow F(y)}}{\Lambda y.\lambda u.EXTR(u, t) : \forall t^D (\forall x^D F(x) \rightarrow F(y))}
$$

Example 3.14. $\exists y^D (\exists x^D F(x) \rightarrow F(y))$[23] (classical)

$$
\frac{\dfrac{\dfrac{[z : \exists x^D F(x)] \quad \dfrac{\dfrac{\dfrac{[g(t) : F(t)]}{g(t) =_r g(t) : F(t)}}{r(g(t)) : F(t)}}{INST(z, \acute{g} tr(g(t))) : F(t)}}{\dfrac{[t : D] \quad \lambda z.INST(z, \acute{g} tr(g(t))) : \exists x^D F(x) \rightarrow F(t)}{\varepsilon y.(\lambda z.INST(z, \acute{g} tr(g(t))), t) : \exists y^D (\exists x^D F(x) \rightarrow F(y))}}}{\lambda t.\varepsilon y.(\lambda z.INST(z, \acute{g} tr(g(t))), t) : D \rightarrow \exists y^D (\exists x^D F(x) \rightarrow F(y))} (*)
$$

---

[21] In the system of Kalish–Montague [Kalish *et al.* (1964)] (p. 146) this example is given. From the difficulties shown by that system, which uses boxes to delimit scopes but does not 'keep track of proof steps' as we do here with labels, it is clear that the lack of the extra dimension given by the label, forces one to use the method of indirect proof by means of negation (Ibid., p. 165).

[22] [Quine (1950)] (p. 180, schema (11))

[23] Id. Ibid., schema (14).

and, similarly to some previous cases, by allowing a vacuous 'ε'-abstraction at step '$(*)$', we obtain a weakened version of the classically valid schema, and the precondition is that the domain 'D' is not empty.

3.6.2   *Specific examples*

3.6.2.1   *Inclusive logics*

Some of the difficulties of other systems of natural deduction can be easily over-come. For example, the handling of *inclusive* logics (cf. [Fine (1985)] (Chapter 21)),[24]:

Example 3.15. $\forall x F(x) \rightarrow \exists x F(x)$

$$\frac{\dfrac{\dfrac{[\forall x F(x)]}{F(t)}}{\exists x F(x)}}{\forall x F(x) \rightarrow \exists x F(x)}$$

Using the functional interpretation, where the presence of terms and of the domains of quantification make the framework a much richer instrument for deduction calculi, we have:

$$\frac{\dfrac{[t:D] \qquad [z:\forall x^D.F(x)]}{EXTR(z,t):F(t)}}{\dfrac{\varepsilon x.(EXTR(z,x),t):\exists x^D F(x)}{\lambda z.\varepsilon x.(EXTR(z,x),\boxed{t}):\forall x^D F(x) \rightarrow \exists x^D F(x)}}$$

Here the presence of a free variable (namely 't') indicates that the assumption '$[t:D]$' remains to be discharged. By making the domain of quantification explicit one does not have the antecedent (vacuously) true and the consequent

---

[24]"An *inclusive* logic is one that is meant to be correct for both empty and non-empty domains. There are certain standard difficulties in formulating a system of inclusive logic. If, for example, we have the usual rules of UI, EG and conditional proof, then the following derivation of the theorem $\forall x F x \supset \exists x F x$ goes through (...) But the formula $\forall x F x \supset \exists x F x$ is not valid in the empty domain; the antecedent is true, while the consequent is false."   [Fine (1985)] (p. 205)

  Here the difficulty of formulating a system of inclusive logic does not exist simply because the individuals are taken to be part of the calculus: recall that the labelled natural deduction presentation system is made of a functional calculus on the terms, and a logical calculus of deductions on the formulas side. It requires that the names of individuals be introduced in the functional part in order for the quantifiers to be introduced and eliminated.

trivially false in the case of empty domain: the proof of the proposition is still depending on the assumption 'let t be an element from D', i.e. that the type 'D' is presumably non-empty. To be categorical the above proof would still have to proceed one step, as in:

$$\cfrac{\cfrac{\cfrac{\cfrac{[t:D] \qquad [z:\forall x^D F(x)]}{EXTR(z,t):F(t)}}{\varepsilon x.(EXTR(z,x),t):\exists x^D F(x)}}{\lambda z.\varepsilon x.(EXTR(z,x),\boxed{t}):\forall x^D F(x) \to \exists x^D F(x)}}{\underbrace{\lambda t.\lambda z.\varepsilon x.(EXTR(z,x),t)}:D \to (\forall x^D F(x) \to \exists x^D F(x))}$$

no free variable

Now we look at the proof-construction ('$\lambda t\lambda z\varepsilon x.(EXTR(z,x),t)$') we can see no free variables, thus the corresponding proof is categorical, i.e. does not rely on any assumption.

3.6.2.2 *Classical logic and Skolem functions*

Here we attempt to reassess the connections between the classical Skolem-type proof calculus involving existentials and our functional interpretation, by constructing proofs for the examples used in [Davis and Fechter (1991)]:

Example 3.16. $\exists x^D(P(x) \vee Q(x)) \to (\exists x^D P(x) \vee \exists x^D Q(x))$

$$\cfrac{\cfrac{\cfrac{[g(t):P(t)\vee Q(t)] \quad \cfrac{\cfrac{[t:D] \quad [m(t):P(t)]}{\varepsilon y.(m(y),t):\exists y^D P(y)}}{inl(\varepsilon y.(m(y),t)):\exists y^P(y)\vee\exists y Q(y)} \quad \cfrac{\cfrac{[t:D] \quad [n(t):Q(t)]}{\varepsilon y.(n(y),t):\exists y^D Q(y)}}{inr(\varepsilon y.(n(y),t)):\exists y^P(y)\vee\exists y^Q(y)}}{CASE(g(t),\hat{m}inl(\varepsilon y.(m(y),t)),\hat{n}inr(\varepsilon y.(n(y),t))):\exists y^D P(y)\vee\exists y^D Q(y)}}{[z:\exists x^D(P(x)\vee Q(x))] \quad INST(z,CASE(g(t),\hat{m}inl(\varepsilon y.(m(y),t)),\hat{n}inr(\varepsilon y(n(y),t)))):\exists y^D P(y)\vee\exists y^D Q(y)}}{\lambda z.INST'(z,CASE(g(t),\hat{m}inl(\varepsilon y.(m(y),t)),\hat{n}inr(\varepsilon y.(n(y),t)))):\exists x^D(P(x)\vee Q(x))\to(\exists y^D P(y)\vee\exists y^D Q(y))}$$

Example 3.17. $\exists x^D\exists y^D P(x,y) \to \exists y^D\exists x^D.P(x,y)$

$$\cfrac{\cfrac{\cfrac{[g(u):\exists y^D P(u,y)] \quad \cfrac{[t:D] \quad \cfrac{[u:D] \quad [f(u,t):P(u,t)]}{\varepsilon w.(f(w,t),u):\exists w^D P(w,t)}}{\varepsilon s.(\varepsilon w.(f(w,s),u),t):\exists s^D\exists w^D P(w,s)}}{INST(g(u),\hat{f}t\varepsilon s.(\varepsilon w.(f(w,s),u),t)):\exists s^D\exists w^D P(w,s)}}{[z:\exists x^D\exists y^D P(x,y)] \quad INST(z,\hat{g}\hat{u}INST(g(u),\hat{f}t\varepsilon s.(\varepsilon w.(f(w,s),u),t))):\exists s^D\exists w^D P(w,s)}}{\lambda z.INST(z,\hat{g}\hat{u}INST(g(u),\hat{f}t\varepsilon s.(\varepsilon w.(f(w,s),u),t))):\exists x^D\exists y^D P(w,s)\to\exists s^D\exists w^D P(w,s)}$$

Example 3.18. $\exists x^D \forall y^D P(y,x) \rightarrow \forall y^D \exists x^D P(y,x)$

$$\cfrac{[z:\exists x^D \forall y^D P(y,x)] \quad \cfrac{\cfrac{[t:D] \quad \cfrac{[u:D] \quad [g(t):\forall y^D P(y,t)]}{EXTR(g(t),u):P(u,t)}}{\varepsilon w.(EXTR(g(w),u),t):\exists w^D P(u,w)}}{\Lambda u.\varepsilon w.(EXTR(g(w),u),t):\forall u^D \exists w^D P(u,w)}}{\cfrac{INST(z,\acute{g}t\Lambda u.\varepsilon w.(EXTR(g(w),u),t)):\forall u^D \exists w^D P(u,w)}{\lambda z.INST(z,\acute{g}t\Lambda u.\varepsilon w.(EXTR(g(w),u),t)):\exists x^D \forall y^D P(y,x) \rightarrow \forall u^D \exists w^D P(u,w)}}.$$

Chapter 4

Normalisation

Preamble

We are concerned with the question as to why the η-conversions are not usually taken into account as *reductions* in the proofs of the normalization theorems, and we offer one possible answer to such a question: *confluence* requires a new type of proof transformation, which we have here defined and called ι-reduction. By applying the so-called Knuth–Bendix completion procedure, we demonstrate that the problem of non-confluence caused by η-reductions is finally solved.

4.1 Introduction

Deductive systems based on the so-called Curry–Howard isomorphism [Howard (1980)] have an interesting feature: normalization and strong normalization (Church–Rosser property) theorems can be proved by reductions on the terms of the functional calculus. Exploring this important characteristic, we have proved these theorems for the *labelled natural deduction* - (*LND*) via a term rewriting system (*TRS*) constructed from the *LND*-terms of the functional calculus [de Oliveira and de Queiroz (1994)]. The *LND* system is an instance of Gabbay's *labelled deductive systems* - (*LDS*) [Gabbay (1994)], which is based on a generalization of the functional interpretation of logical connectives [Gabbay and de Queiroz (1992)] (i.e. Curry–Howard isomorphism).

Proving the *termination* and *confluence* properties for the *TRS* associated to the *LND* system (*LND-TRS*) [de Oliveira and de Queiroz (1994)], we have in fact proved the normalization and strong normalization theorems for the *LND* system, respectively. The *termination* property guarantees the existence of a normal form of the *LND*-terms, while the *confluence* property its uniqueness. Thus, because of the Curry–Howard isomorphism, we have that every *LND* derivation converts

111

to a normal form (normalization theorem) and it is unique (strong normalization theorem).

The significance of applying this technique in the proof of the normalization theorems lies in the presentation of a simple and computational method, which allowed the discovery of a new basic set of transformations between proofs, which we baptized as "ι (iota)-reductions" [de Oliveira (1995)]. With this result, we obtained a confluent system which contains the η-reductions. Traditionally, the η-reductions have not been given an adequate status, as rightly pointed out by Girard in [Girard *et al.* (1989)] (p. 16), when he defines the *primary* equations, which correspond to the β-equations and the *secondary* equations, which are the η-equations. Girard [Girard *et al.* (1989)] says that the system given by these equations is consistent and decidable, however he notes the following:

> "Although this result holds for the whole set of equations, one only ever considers the first three. It is a consequence of the *Church–Rosser property* and the *normalization theorem* (...)"

The first three equations, referred to by Girard, are the *primary* ones, i.e. β-equations.

Applying the so-called *completion procedure*, proposed by Knuth and Bendix in [Knuth and Bendix (1970)], to *LND-TRS*, the following term, which causes a non-confluence in the system, is produced (i.e. a divergent critical pair is generated):

$$w(CASE(c, \acute{a}_1 inl(a_1), \acute{a}_2 inr(a_2))).$$

This term rewrites in two different ways[1]:

    1. $\triangleright_\eta w(c)$ **2.** $\triangleright_\zeta CASE(c, \acute{a}_1 w(inl(a_1)), \acute{a}_2 w(inr(a_2)))$

The method of Knuth and Bendix says that when a terminating system is not confluent it is possible to add rules in such a way that the resulting system becomes

---

[1] The η-reduction for the $\vee$ connective is framed as follows:

$$\cfrac{c : A_1 \vee A_2 \quad \cfrac{[a_1 : A_1]}{inl(a_1) : A_1 \vee A_2} \vee\text{-}intr \quad \cfrac{[a_2 : A_2]}{inr(a_2) : A_1 \vee A_2} \vee\text{-}intr}{CASE(c, \acute{a}_1 inl(a_1), \acute{a}_2 inr(a_2)) : A_1 \vee A_2} \vee\text{-}elim \quad \triangleright_\eta \quad c : A_1 \vee A_2$$

and the ζ-reduction is defined as follows:

$$\cfrac{w\left(\cfrac{p : A_1 \vee A_2 \quad \cfrac{[s_1 : A_1]}{d(s_1) : C} \quad \cfrac{[s_2 : A_2]}{e(s_2) : C}}{CASE(p, \acute{s}_1 d(s_1), \acute{s}_2 e(s_2)) : C} \right) : W}{} r$$

$$\triangleright_\zeta$$

$$\cfrac{p : A_1 \vee A_2 \quad \cfrac{\cfrac{[s_1 : A_1]}{d(s_1) : C}}{w(d(s_1)) : W} r \quad \cfrac{\cfrac{[s_2 : A_2]}{e(s_2) : C}}{w(e(s_2)) : W} r}{CASE(p, \acute{s}_1 w(d(s_1)), \acute{s}_2 w(e(s_2))) : W}$$

Whilst 'r' is usually restricted to an *elimination* rule, we have relaxed this condition: it is only required that 'r' does not discharge any assumptions from the other (independent) branch, i.e. that the auxiliary branches do not interfere with the main branch.

confluent. Thus applying this procedure to *LND-TRS*, a new rule is added to the system:

$$CASE(c, \acute{a_1}w(inl(a_1)), \acute{a_2}w(inr(a_2))) \rhd_\iota w(c)$$

Since terms represent proof-constructions in the *LND* system, this rule defines a new transformation between proofs:

ι-reduction-$\vee$

$$\cfrac{c : A_1 \vee A_2 \quad \cfrac{\cfrac{[a_1 : A_1]}{inl(a_1) : A_1 \vee A_2}\vee\text{-}intr \quad w(inl(a_1)) : W}{} \quad \cfrac{\cfrac{[a_2 : A_2]}{inr(a_2) : A_1 \vee A_2}\vee\text{-}intr \quad w(inr(a_2)) : W}{}}{CASE(c, \acute{a_1}w(inl(a_1)), \acute{a_2}w(inr(a_2))) : W}\vee\text{-}elim \quad \rhd_\iota \quad \cfrac{c : A_1 \vee A_2}{w(c) : W}r.$$

Similarly, the ι reduction for the existential quantifier is defined [de Oliveira (1995)], since, similarly to $\vee$, the quantifier $\exists$ is a "Skolem type" connective (i.e. in the *elimination* inference for this type of connective is necessary to open local assumptions):[2]

ι-reduction-$\exists$

$$\cfrac{c : \exists x^D.P(x) \quad \cfrac{\cfrac{[t : D] \quad [g(t) : P(t)]}{\varepsilon y.(g(y), t) : \exists y^D.P(y)}\exists\text{-}intr \quad w(\varepsilon y.(g(y), t)) : W}{}}{INST(c, \acute{g}tw(\varepsilon y.(g(y), t))) : W}\exists\text{-}elim \quad \rhd_\iota \quad \cfrac{c : \exists x^D.P(x)}{w(c) : W}r,$$

where 'ε' is an abstractor.

With this result, we believe that we have answered the question as to why the η-reductions are not considered in the proofs of the normalization theorems (*confluence* requires ι-reductions). However, by applying a computational and well-defined method, the completion procedure, it seems that this problem of the non-confluence caused by η-reductions are solved.

4.2 Proof transformations in labelled deduction

By approaching the normalisation theorem via *LND* makes it possible to look at η rules as rules of reduction, rather than expansions, and normalization may be proved via term rewriting. It turns out that the use of term rewriting (and Knuth completion) brought about the need for a new kind of rewriting rule, and, thus, a new proof transformation rule. the reductions here called "ι"-rules. Together with

---

[2]For more details on the treatment of the existential quantifier in labelled natural deduction see [de Queiroz and Gabbay (1995)].

the ζ rules, they offer a kind of explanation for the non-interference between the main branches and the secondary branches in 'Skolem-type' logical connectives, such as $\vee$, $\exists$ and $\doteq$.[3]

The relevance of the proposal hereby presented is due to basically two reasons:

(1) Well-known results for Natural Deduction are proved via a computationally meaningful method: by proving the properties of termination and confluence for the *TRS* associated to *LND*, we prove the normalization theorem and the strong normalization theorem, respectively. Thus, we define a new method for studying the problem of normalization between proofs in the context of labelled natural deduction: transformation between proofs via term rewriting [de Oliveira and de Queiroz (1994)].

(2) The discovery of a new basic set of transformation between proofs: the ι-reductions.

We shall first present the syntax of the *TRS* associated to *LND*, by means of an equational system with ordered types. Next, the *TRS* defined in [de Oliveira and de Queiroz (1994)] is presented, and the proofs of termination and confluence are given.

---

[3] In [de Queiroz and Gabbay (1992); de Queiroz and de Oliveira (2008)] the study of natural deduction for equality yields an equational fragment of *Labelled Natural Deduction*, with proof rules framed as:

$\doteq$-*introduction*

$$\frac{a =_s b : D}{s(a,b) :\doteq_D (a,b)}$$

$\doteq$-*reduction*

$$\frac{\dfrac{a =_s b : D}{s(a,b) :\doteq_D (a,b)} \doteq\text{-}intr \quad \dfrac{[a =_t b : D]}{d(t) : C}}{REW\,R(s(a,b), \acute{t}d(t)) : C} \doteq\text{-}elim \quad \triangleright_\beta \quad \dfrac{[a =_s b : D]}{d(s/t) : C}$$

$\doteq$-*induction*

$$\frac{e :\doteq_D (a,b) \quad \dfrac{[a =_t b : D]}{t(a,b) :\doteq_D (a,b)} \doteq\text{-}intr}{REW\,R(e, \acute{t}t(a,b)) :\doteq_D (a,b)} \doteq\text{-}elim \quad \triangleright_\eta \quad e :\doteq_D (a,b)$$

where '´' is an abstractor which binds the occurrences of the (new) variable 't' introduced with the local assumption '$[a =_t b : D]$' as a kind of 'Skolem'-type constant denoting the (presumed) 'reason' why 'a' was assumed to be equal to 'b'.

The properties of confluence and termination for the equational fragment are proved in [de Oliveira and de Queiroz (1999)].

4.3 Equivalences between proofs in *LND*

Here we shall present the notion of equivalence between proofs in *LND*, based in the formalism adopted in equational logic with ordered types. Such a formalism suit our purposes here basically for two reasons:

- the terms of *LND* have types (i.e. the formulas)
- via equational logic with ordered types, the correct domain of the operators can be specified. For example, the operator FST (which is a *destructor*) cannot have as an argument an operator such as inl or inr: the terms $FST(inl(x))$ and $FST(inr(x))$ are not *LND*-terms, as it shall be seen in this section.

Definition 4.1 (Signature). The signature of our LND Σ_{LND} is the set formed by the definitions of subtypes and operators of LND (ol). The subtypes are grouped in the set S_{LND}:

- $S_{\text{LND}} = \{T_\wedge,\ T_\vee,\ T_\rightarrow,\ T_\exists,\ T_\forall,\ T_{\doteq},\ T_{atomic},\ T^\cdot,\ T_{wff},\ D,\ D'\}$
- *Subtype declaration:*

  — $T_\wedge < T_{wff}$
  — $T_\vee < T_{wff}$
  — $T_\rightarrow < T_{wff}$
  — $T_\exists < T_{wff}$
  — $T_\forall < T_{wff}$
  — $T_{\doteq} < T_{wff}$
  — $T_{atomic} < T_{wff}$
  — $T^\cdot < T_{wff}$
  — $D' < D$

- *Operators:*

  — $FST : T_\wedge \rightarrow T_{wff},\ \ SND : T_\wedge \rightarrow T_{wff},\ \ \langle\,\rangle : T_{wff} \times T_{wff} \rightarrow T_\wedge$
  — $APP : T_\rightarrow \times T_{wff} \rightarrow T_{wff},\ \ \lambda : T_{wff} \times T_{wff} \rightarrow T_\rightarrow$
  The λ-abstractor is a special operator. It constructs terms of the form $\lambda x.f(x)$, whose type is $T_\rightarrow$. The first argument of λ must be a variable of type T_{wff} and the second argument, also of type T_{wff}, may or may not contain a variable of first argument.
  — $' : T_{wff} \times T_{wff} \rightarrow T^\cdot$,
  This abstraction operator is similar to operator λ. They construct terms of the form $\acute{x}f(x)$.
  — $CASE : T_\vee \times T^\cdot \times T^\cdot \rightarrow T_{wff},\ \ inl : T_{wff} \rightarrow T_\vee,\ \ inr : T_{wff} \rightarrow T_\vee$

- $INST : T_\exists \times T^\cdot \to T_{wff}, \quad \varepsilon : T_{wff} \times D \to T_\exists$
  The ε-abstractor is also a special operator, since it constructs terms of the
  form $\varepsilon x.(f(x), a)$ of type $T_\exists$.
- $EXTR : T_\forall \times D \to T_{wff}, \quad \Lambda : T_{wff} \times T_{wff} \to T_\forall$
  The Λ-abstractor is similar to λ, except for the fact that it constructs terms
  of the form $\Lambda x.f(x)$ of type $T_\forall$.
- $REWR : T_{\doteq} \times T^\cdot \to T_{wff}, \quad r : D \times D \to T_{\doteq}, \quad r : T_{wff} \times T_{wff} \to T_{\doteq}$

According to the *principle of inversion*, the type of a term in the functional side
is determined by the main connective of the formula in the logical calculus. Thus,
the types $T_\wedge$, $T_\vee$, $T_\to$, $T_{\doteq}$, $T_\forall$ and $T_\exists$ are associated to the logical connectives
$(\wedge, \vee, \to, \doteq, \forall, \exists)$. The type T_{wff} refers to any well-formed formula. When a
term has an atomic formula as its type, it has the type T_{atomic}. Since in *LND* there
exist two types of information in the so-called meta-level — terms which represent
proofs and terms which represent objects from a certain domain (see, for example,
the rule of $\forall$-introduction) —, the terms may also have a certain domain D as its
type and possibly a subdomain D' of D.

   A subset of operators of *LND* can be grouped according to the type of proof
that they represent.

Definition 4.2 ('constructor' and 'DESTRUCTOR'). A subset of operators
of *LND* determines two types of signature: $\Sigma_{\text{constructor}}$ and $\Sigma_{\text{DESTRUCTOR}}$.
The set $\Sigma_{\text{constructor}}$ is constituted by the declaration of the operators which
form the *canonical* proofs corresponding subtype declarations, whereas the set
$\Sigma_{\text{DESTRUCTOR}}$ is formed by the declaration of the elimination operators together with the corresponding subtype declarations, which form the *non-canonical*
proofs.

$$Operators\ of\ \Sigma_{constructor} = \{\ \lambda\ ,\langle\ \rangle,\ \acute{},\ inl,\ inr,\ \Lambda,\ \varepsilon,\ r\}$$

$$Operators\ of\ \Sigma_{DESTRUCTOR} = \{\ APP, FST, SND, CASE, INST,$$
$$REWR\ \}$$

   The set of terms of *LND* is denoted by $T_{\Sigma-LND}$. The terms of *LND* are defined
as follows:

Definition 4.3 (*LND*-terms). An LND-term of type s is defined as:

- Variables and constants, the so-called atomic labels (ra), of type $s \in S_{LND}$
  are *LND*-terms. So, if $ra \in \Sigma_{LND}$, then $ra \in T_{\Sigma-LND}$ with *arity(ra)* = 0.

- $ol(u_1, \ldots, u_n)$ is an *LND//*-term, given that there exists a declaration of ol: $s_1, \ldots, s_n \rightarrow s \in \Sigma_{LND}$, such that the type t of $ol(u_1, \ldots, u_n)$ is $\leq_\Sigma s$, u_i is an *LND*-term of type s_i for $i = 1, \ldots n$ with arity$(ol) = n$.

Since terms represent proof constructions in *LND*, the transformations between proofs form an equational system defined as follows.

Definition 4.4 (equational system for *LND*). The equational system of *LND* (E_{LND}) is made of definitional equalities which reflect the proof transformations, and for this reason the equational axioms are classified into four groups:

(1) β
(2) η
(3) ζ
(4) ι *(cf. Section 4.4.5)*

The group of β-equations is the following:

- $FST(\langle a_1, a_2 \rangle) =_\beta a_1$
- $SND(\langle a_1, a_2 \rangle) =_\beta a_2$
- $CASE(inl(a_1), \acute{s_1}d(s_1), \acute{s_2}e(s_2)) =_\beta d(a_1/s_1)$
- $CASE(inr(a_2), \acute{s_1}d(s_1), \acute{s_2}e(s_2)) =_\beta e(a_2/s_2)$
- $APP(\lambda x.b(x), a) =_\beta b(a/x)$
- $EXTR(\Lambda x.f(x), a) =_\beta f(a/x)$
- $INST(\varepsilon x.(f(x), a), \acute{g}\acute{t}d(g, t)) =_\beta d(f/g, a/t)$
- $REWR(s(a, b), \acute{t}d(t)) =_\beta d(s/t)$

The η-equations are:

- $\langle FST(c), SND(c) \rangle =_\eta c$
- $CASE(c, \acute{a_1}inl(a_1), \acute{a_2}inr(a_2)) =_\eta c$
- $\lambda x.APP(c, x) =_\eta c$ (where c does not depend on x)
- $\Lambda x.EXTR(c, x) =_\eta c$ (where x does not occur free in c).
- $INST(c, \acute{g}\acute{t}\varepsilon y.(g(y), t)) =_\eta c$
- $REWR(c, \acute{t}t(a, b)) =_\eta c$

The ζ-equations are:

- $w(CASE(p, \acute{s_1}d(s_1), \acute{s_2}e(s_2))) =_\zeta CASE(p, \acute{s_1}w(d(s_1)), \acute{s_2}w(e(s_2)))$
- $w(INST(c, \acute{g}\acute{t}d(g, t))) =_\zeta INST(c, \acute{g}\acute{t}w(d(g, t)))$
- $w(REWR(e, \acute{t}d(t))) =_\zeta REWR(e, \acute{t}w(d(t)))$

The ι-equations are:

- $CASE(c, \acute{a_1}w(inl(a_1)), \acute{a_2}w(inr(a_2))) =_\iota w(c)$
- $INST(c, \acute{g}tw(\varepsilon y.(g(y), t))) =_\iota w(c)$
- $REWR(c, \acute{t}w(t(a, b))) =_\iota w(c)$

The pair $\langle \Sigma_{LND}, E_{LND} \rangle$ defines a specification of the set of all equivalences between proofs in *LND*.

An equational specification $\langle \Sigma, E \rangle$, where E is a set of equations between Σ-terms, defines a class of Σ-algebras A, such that $A \models E$. Birkhoff (1935) proved completeness for the equational logics, thus making sure that for all terms u, v belonging to a set $T_\Sigma(X)$, $\langle \Sigma, E \rangle \vdash u = v \Leftrightarrow \langle \Sigma, E \rangle \models u = v$. Given an equational specification $\langle \Sigma, E \rangle$, the validity problem (or the uniform word problem) for $\langle \Sigma, E \rangle$ is easily solved, in case it is possible to construct a canonical *TRS* equivalent to $\langle \Sigma, E \rangle$. After constructing a canonical *TRS*, to decide whether $E \vdash u = v$, it suffices to reduce u and v to their corresponding normal forms u' and v' and compare them. However, in general, the main results of the theory of standard rewriting systems do not naturally extend to *TRS*'s with ordered types. In order to get all the results, it is necessary that the *TRS* has the *sort decreasing* property.

The next section presents the *TRS*, whose syntax was defined by means of the specification $\langle \Sigma_{LND}, E_{LND} \rangle$, and proves the main properties of the system: *sort decreasing*, *termination* and *confluence*.

4.4 The term rewriting system for *LND*

We shall present the rewriting system associated to *LND* (*LND-TRS*), and prove *sort decreasing*, *termination* and *confluence*.

4.4.1 *Defining the* LND-TRS

LND-TRS computes the normal form of proofs in *LND*. Thus, the rules of such a system are based on the transformations between proofs (β, η and ζ), and a new set of proof transformations arise, namely the ι rules, uncovered during the construction of the proof of confluence of the system as originally defined with only β, η and ζ. This way, the equations of the *LND* equational system, defined in the previous section, are oriented according to these transformations. *LND-TRS* is defined as follows:

Definition 4.5 (*LND-TRS*). LND-TRS is a term rewriting system with ordered types which computes the normal form of proofs in LND. The rules are the following:

(1) $FST(\langle a_1, a_2 \rangle) \triangleright_\beta a_1$

(2) $SND(\langle a_1, a_2 \rangle) \triangleright_\beta a_2$

(3) $CASE(inl(a_1), \acute{s_1}d(s_1), \acute{s_2}e(s_2)) \triangleright_\beta d(a_1/s_1)$

(4) $CASE(inr(a_2), \acute{s_1}d(s_1), \acute{s_2}e(s_2)) \triangleright_\beta e(a_2/s_2)$

(5) $APP(\lambda x.b(x), a) \triangleright_\beta b(a/x)$

(6) $EXTR(\Lambda x.f(x), a) \triangleright_\beta f(a/x)$

(7) $INST(\varepsilon x.(f(x), a), \acute{g}td(g, t)) \triangleright_\beta d(f/g, a/t)$

(8) $REWR(s(a, b), \acute{t}d(t)) \triangleright_\beta d(s/t)$

(9) $\langle FST(c), SND(c) \rangle \triangleright_\eta c$

(10) $CASE(c, \acute{a_1}inl(a_1), \acute{a_2}inr(a_2)) \triangleright_\eta c$

(11) $\lambda x.APP(c, x) \triangleright_\eta c$ (where c does not depend on x)

(12) $\Lambda x.EXTR(c, x) \triangleright_\eta c$ (where x does not occur free in c).

(13) $INST(c, \acute{g}t\varepsilon y.(g(y), t)) \triangleright_\eta c$

(14) $REWR(c, \acute{t}t(a, b)) \triangleright_\eta c$

(15) $w(CASE(p, \acute{s_1}d(s_1), \acute{s_2}e(s_2))) \triangleright_\zeta CASE(p, \acute{s_1}w(d((s_1)), \acute{s_2}w(e(s_2)))$

(16) $w(INST(c, \acute{g}td(g, t))) \triangleright_\zeta INST(c, \acute{g}tw(d(g, t)))$

(17) $w(REWR(e, \acute{t}d(t))) \triangleright_\zeta REWR(e, \acute{t}w(d(t)))$

(18) $CASE(c, \acute{a_1}w(inl(a_1)), \acute{a_2}w(inr(a_2))) \triangleright_\iota w(c)$

(19) $INST(c, \acute{g}tw(\varepsilon y.(g(y), t))) \triangleright_\iota w(c)$

(20) $REWR(c, \acute{t}w(t(a, b))) \triangleright_\iota w(c)$

The rules of *LND-TRS* come with a subscript which informs the type of proof transformation (β, η, ζ or ι) which the rule is supposed to represent. This way, one has the notion of β-redex, η-redex, ζ-redex and ι-redex, which characterizes the type of reduction:

Definition 4.6 (β (η, ζ, ι)-redex). A β-redex is a redex belonging to an *LND*-term which is reducible according to a β rule. Similarly for η-redex and ζ-redex.

An *LND*-term in the normal form is defined as follows:

Definition 4.7 (normal *LND*-term). An *LND*-term is in the normal form if it does not contain any β-redex, η-redex, ζ-redex or ι-redex.

4.4.2 *The* sort decreasing *property*

In order to make sure that the results for standard *TRS*'s get naturally extended to *TRS*'s with ordered types, it is necessary for the *TRS* to be *sort decreasing*. In the case of *LND* this property is trivially verified, since all rewriting rules representing proof transformations keep the type of the rewritten terms. For example:

$$FST(\langle a_1, a_2 \rangle) : A \quad \triangleright_\beta \quad a_1 : A$$

Thus, the *LND-TRS* is *sort decreasing*.

4.4.3 *Defining an order*

When comparing the left-hand side and the right-hand side of the reductions β, η and ι, one easily checks that the right-hand side is syntactically simpler term. By defining an order which guarantees the property of termination for a system with only those three sets of equations is a rather trivial task: it would suffice to use a complexity measure on terms, based, for example, on the number of operators. However, this does not apply to the ζ transformations. The set of ζ transformations is characterized as being a set of permutative reductions. A permutative reduction is one such that both sides of the equality contain the same symbols, such as the associative and commutative laws. According the observation by Peterson and Stickel in [Peterson and Stickel (1981)], the permutative reductions represent an additional difficulty in determining the confluence and termination properties. For this reason, one has to be more careful with those kinds of reduction.

For proving the termination property for *LND-TRS* we might adopt a methodology used for cases in which subsets of the rules of the rewriting system have common features, giving rise to various subsystems of rewriting. In this situation, the termination property is proved separately for each subsystem. For the *LND-TRS*, this technique would be useful, since the system could be split into two 'modules': one subsystem formed by the rules of β, η and ι reduction and the other one by the rules of ζ reduction. However, the termination property is not modular for the general case: it is necessary that the subsystems of rewriting possess certain properties [Klop (1990)]. Furthermore, the subsystems of rewriting must be disjoint, i.e. the function symbols and the constant symbols need to be different. In case the alphabet is not distinct, one takes renamed copies of the subsystems [Klop (1990)]. This way, it is preferable to use an order which suits the whole system.

The proof method adopted here does not use such device of splitting the *TRS* into modules, since the chosen order may be applied to the system as a whole. Due to the difficulty presented by the so-called 'permutative' we shall first take a careful look at the set of ζ reductions, and then establish an adequate order for the whole system.

4.4.3.1 *Analyzing the ζ-reductions*

The set of ζ reductions, as previously presented, does not illustrate all possible ζ transformations between proofs. The operator w may have arity 1, 2 or 3, so that the set of ζ reductions ζ is extended in the following way:

- $w(CASE(p, \acute{s}_1 d(s_1), \acute{s}_2 e(s_2))) \triangleright_\zeta CASE(p, \acute{s}_1 w(d(s_1), \acute{s}_2 w(e(s_2))))$
- $w(CASE(p, \acute{s}_1 d(s_1), \acute{s}_2 e(s_2)), u) \triangleright_\zeta$
$$CASE(p, \acute{s}_1 w(d(s_1), u), \acute{s}_2 w(e(s_2), u))$$
- $w(CASE(p, \acute{s}_1 d(s_1), \acute{s}_2 e(s_2)), u_1, u_2) \triangleright_\zeta$
$$CASE(p, \acute{s}_1 w(d(s_1), u_1, u_2), \acute{s}_2 w(e(s_2), u_1, u_2))$$
- $w(INST(c, \sigma g.\acute{t} d(g, t))) \triangleright_\zeta INST(c, \acute{g} t w(d(g, t)))$
- $w(INST(c, \acute{g} t d(g, t)), u) \triangleright_\zeta INST(c, \acute{g} t w(d(g, t), u))$
- $w(INST(c, \acute{g} t d(g, t)), u_1, u_2) \triangleright_\zeta INST(c, \acute{g} t w(d(g, t), u_1, u_2))$
- $w(REWR(e, \acute{t} d(t))) \triangleright_\zeta REWR(e, \acute{t} w(d(t)))$
- $w(REWR(e, \acute{t} d(t)), u) \triangleright_\zeta REWR(e, \acute{t} w(d(t), u))$
- $w(REWR(e, \acute{t} d(t)), u_1, u_2) \triangleright_\zeta REWR(e, \acute{t} w(d(t), u_1, u_2))$

Among this set of reductions, the first three rules, which concern permutation of $CASE$, represent an additional difficulty in determining the order, since, besides the fact that they are supposed to be a commutative reduction, the size of its right-hand side is greater than its left-hand side, and the operator w is duplicated.

This leads to the use of recursive path ordering.

Definition 4.8 (recursive path ordering). Let $>$ be a partial order over the set of operators Σ. The recursive path ordering $>^*$ over the set $T_\Sigma(X)$ of terms over Σ is defined recursively as follows:

$$s = f(s_1, \ldots, s_m) >^* g(t_1, \ldots, t_n) = t$$

if and only if one of the following conditions holds:

(1) $f = g$ and $\{s_1, \ldots, s_m\} \gg^* \{t_1, \ldots, t_n\}$
(2) $f > g$ and $\{s\} \gg^* \{t_1, \ldots, t_n\}$
(3) $f \not\geq g$ and $\{s_1, \ldots, s_m\} \gg^*$ *ou* $= \{t\}$

where $\gg^*$ is the extension of $>^*$ for multisets.

By using a precedence order in which $w > CASE$, when proving that

$$w(CASE(p, \acute{s}_1 d(s_1), \acute{s}_2 e(s_2))) >^* CASE(p, \acute{s}_1 w(d(s_1), \acute{s}_2 w(e(s_2))))$$

the following situations would have to be analyzed:

- If $w > CASE$ one has to show that

$$\{w(CASE(p, \acute{s_1}d(s_1), \acute{s_2}e(s_2)))\} \gg^* \{p, \acute{s_1}w(d(s_1)), \acute{s_2}w(e(s_2))\}:$$

— $w(CASE(p, \acute{s_1}d(s_1), \acute{s_2}e(s_2))) >^* p$ by the subterm property.
— $w(CASE(p, \acute{s_1}d(s_1), \acute{s_2}e(s_2))) >^* \acute{s_1}w(d(s_1))$. In this case, one has an abstraction-term (i.e. $\acute{s_1}w(d(s_1)))^4$; every time a λ-term is compared with some other term, the function from which the variable is being abstracted will be used in the comparison. This way, the term $w(d(s_1))$ will be used in the comparison. Since $w = w$, one has to prove that
$\{CASE(p, \acute{s_1}d(s_1), \acute{s_2}e(s_2)))\} \gg^* \{d(s_1)\}.$
This relation is true by the subterm property.

Thus, when $w > CASE$ the recursive path ordering is rather adequate.

- If $w = CASE$, to prove that

$$CASE(CASE(p, \acute{s_1}d(s_1), \acute{s_2}e(s_2)), u_1, u_2)) >^*$$
$$CASE(p, \acute{s_1}CASE(d(s_1), u_1, u_2), \acute{s_2}CASE(e(s_2), u_1u_2))$$

it is necessary to show that
$\{CASE(p, \acute{s_1}d(s_1), \acute{s_2}e(s_2)), u_1, u_2\} \gg^*$
$\quad\quad\quad\quad\quad \{p, \acute{s_1}CASE(d(s_1), u_1, u_2), \acute{s_2}CASE(e(s_2), u_1, u_2)\}.$
It is not possible to guarantee that this relation holds since, although by the subterm property one has that $CASE(p, \acute{s_1}d(s_1), \acute{s_2}e(s_2)) >^* p$, by this same property the following relations also hold:

— $u_1 <^* \acute{s_1}CASE(d(s_1), u_1, u_2)$
— $u_1 <^* \acute{s_2}CASE(e(s_2), u_1, u_2)$
— $u_2 <^* \acute{s_1}CASE(d(s_1), u_1, u_2)$
— $u_2 <^* \acute{s_2}CASE(e(s_2), u_1, u_2)$

This relation could still be guaranteed if $CASE(p, \acute{s_1}d(s_1), \acute{s_2}e(s_2))$ were $>^*$ than $\acute{s_1}CASE(d(s_1), u_1, u_2)$ and $\acute{s_2}CASE(e(s_2), u_1, u_2)$:

— When comparing $CASE(p, \acute{s_1}d(s_1), \acute{s_2}e(s_2))$ and $\acute{s_1}CASE(d(s_1), u_1, u_2)$ one has that $CASE = CASE$, thus it is necessary to prove that
$\{p, \acute{s_1}d(s_1), \acute{s_2}e(s_2)\} \gg^* \{d(s_1), u_1, u_2\}.$
It is not possible to guarantee that this relation holds without previous knowledge about p, u_1 and u_2.
— When comparing $CASE(p, \acute{s_1}d(s_1), \acute{s_2}e(s_2))$ and $\acute{s_2}CASE(e(s_2), u_1, u_2)$ a situation similar to the previous case occurs.

---
[4]Every term with abstractors λ, ´ is here called simply λ-term.

From this analysis, the use of recursive path ordering was disconsidered. The same situation occurs with the other permutative equations. In fact, this is the same problem that occurs when trying to apply this ordering to rules which deal with the law of associativity (e.g. $+(+(a, b), c) \rightarrow +(a, +(b, c))$). Plaisted [Plaisted (1994)] had already noticed that to use recursive path ordering in rules of this kind it is necessary to resort to other mechanisms. Thus, it is not difficult to notice that in all permutative reductions the first argument of the term on the right-hand side of the reduction is smaller than the first argument of the term on the left-hand side of the reduction. So, one must use an order in which the first argument of the terms of the equation could be compared. It is exactly this "trick" which Dershowitz proposes in [Dershowitz (1982)]. He proposes a recursive path ordering in which an operand may be used as operator:

"One way of extending the recursive path ordering is to allow some function of a term $f(t_1, \ldots, t_n)$ to serve the role of the operator f. For example, we can consider the kth operand t_k to be the operator, and compare two terms by first recursively comparing their kth operands. This yields a simplification ordering for the same reasons that the original definition does."

By means of the example below this recursive path ordering is illustrated [Dershowitz (1982)]:

Example 4.1. The recursive path ordering given in this example uses an operand as operator. Let us take the *TRS* $R = \{if(if(a, b, c), d, e) \rightarrow if(a, if(b, d, e), if(c, d, e))\}$. The conditional expression "$if(a, b, c)$" represents an "$if - then - else$": "$if \ a \ then \ b \ else \ c$". This system "normalizes" conditional expressions by removing nested if's from the condition "a".

To prove that this system is terminating, the condition is considered as an operator. The condition statement "$if(a, b, c)$" on the left-hand side of the rule is greater (via the subterm property) than the condition "a" on the right-hand side. Thus, it is necessary to show that the left-hand side is greater than both the operands at the right-hand side, "$if(b, d, e)$" and "$if(c, d, e)$": $if(if(a, b, c), d, e) > if(b, d, e)$ and $if(if(a, b, c), d, e) > if(c, d, e)$, since $if(a, b, c)$ is greater than the operators "b" and "c" (subterm property). It suffices then to prove that the left hand side is greater than the operands "d" and "e", which is easily proved, once again via the subterm property.

This example fits perfectly within the first three ζ reductions of the set here defined. Indeed, $CASE$ is very similar to if. The operator $CASE$ denotes case-analysis. So, by taking the first argument as an operator and the remaining arguments as operands, the recursive path ordering, adopted in the above example, applies not only to the ζ reductions, but also to the remaining set of reductions in the system. Next, we prove termination using this order.

4.4.4 *Proving the termination property*

As it is well-known, in order to prove the property of termination for a *TRS*, after one adopts a recursive path ordering $>^*$, it suffices to show that for every rule $e \to d$ of the system, $e >^* d$. This way, the proof of termination for *LND-TRS* will be done by showing that the lefthand side of every rule in the system is greater than its right-hand side.

The proof shown here is based on, apart from the properties of the recursive path ordering, the lemma of embedded homeomorphic relation, given by Dershowitz [Dershowitz (1979)]:

Lemma 4.1. *Let s and t be terms in $T_\Sigma(X)$. If $s \trianglelefteq t$, then $s \leq t$ in any simplification order $>$ over $T_\Sigma(X)$.*

This lemma establish that if a term s is syntactically simpler than a term t, then $t > s$ in the simplification order. Therefore, $t >^* s$ in the recursive path ordering, which is a simplification ordering.

4.4.4.1 *Proof for the subset of β reductions*

(1)(a) $FST(\langle a_1, a_2 \rangle) \triangleright_\beta a_1$
  (b) $SND(\langle a_1, a_2 \rangle) \triangleright_\beta a_2$

- One has that $FST(\langle a_1, a_2 \rangle) >^* a_1$ and $SND(\langle a_1, a_2 \rangle) >^* a_2$ by the subterm property.

(2)(a) $CASE(inl(a_1), \acute{s}_1 d(s_1), \acute{s}_2 e(s_2)) \triangleright_\beta d(a_1/s_1)$
  (b) $CASE(inr(a_2), \acute{s}_1 d(s_1), \acute{s}_2 e(s_2)) \triangleright_\beta e(a_2/s_2)$

- $inl(a_1) >^* a_1$ and $inr(a_2) >^* a_2$, by the subterm property.
- $\{CASE(inl(a_1), \acute{s}_1 d(s_1), \acute{s}_2 e(s_2))\} \gg^* \emptyset$ and $\{CASE(inr(a_2), \acute{s}_1 d(s_1), \acute{s}_2 e(s_2))\} \gg^* \emptyset$ vacuously.

(3) $APP(\lambda x.b(x), a) \triangleright_\beta b(a/x)$

- Comparing $\lambda x.b(x)$ and a:
  Here we have a λ-term. In this case, we compare the function that is being abstracted from (i.e. $b(x)$):

  — $b(x)$ and a: the value of x is a, so, by the subterm property we have that $b(a) >^* a$; and, $\{APP(\lambda x.b(x), a)\} \gg^* \emptyset$.

(4) $EXTR(\Lambda x.f(x), a) \rhd_\beta f(a/x)$

- Comparing $\Lambda x.f(x)$ and a:
  Again, here we have a λ-term. In this case, we compare the function which is being abstracted from (i.e. $f(x)$):

  — $f(x)$ and a: the value of x is a, so $f(a) >^* a$ by the subterm property; and, $\{EXTR(\Lambda x.f(x), a)\} \gg^* \emptyset$.

(5) $INST(\varepsilon x.(f(x), a), \acute{g}\grave{t}d(g, t)) \rhd_\beta d(f/g, a/t)$

- $\varepsilon x.(f(x), a) >^* f$ by the lemma of embedded homeomorphic relation.

- $\{INST(\varepsilon x.(f(x), a), \acute{g}\grave{t}d(g, t))\} \gg^* \{a\}$ by the subterm property.

(6) $REWR(s(a, b), \grave{t}d(t)) \rhd_\beta d(s/t)$

- $s(a, b) >^* s$ by the lemma of embedded homeomorphic relation.

- $\{REWR(s(a, b), \grave{t}d(t))\} \gg^* \emptyset$ vacuously.

■

4.4.4.2 *Proof for the subset of η reductions*

(1) $\langle FST(c), SND(c) \rangle \rhd_\eta c$

- $\langle FST(c), SND(c) \rangle >^* c$ by the subterm property.

(2) $CASE(c, \acute{a_1}inl(a_1), \acute{a_2}inr(a_2)) \rhd_\eta c$

- $CASE(c, \acute{a_1}inl(a_1), \acute{a_2}inr(a_2)) >^* c$ by the subterm property.

(3) $\lambda x.APP(c, x) \rhd_\eta c$

- $\lambda x.APP(c, x) >^* c$ by the subterm property.

(4) $\Lambda t.EXTR(c, t) \rhd_\eta c$

- $\Lambda t.EXTR(c, t) >^* c$ by the subterm property.

(5) $INST(c, \acute{g}\grave{t}\varepsilon y.(g(y), t)) \rhd_\eta c$

- $INST(c, \acute{g}\grave{t}\varepsilon y.(g(y), t)) >^* c$ by the subterm property.

(6) $REWR(c, \grave{t}t(a, b)) \rhd_\eta c$

- $REWR(c, \grave{t}t(a, b)) >^* c$ by the subterm property.

■

4.4.4.3 *Proof for the subset of ζ reductions:*

(1) $w(CASE(p, \acute{s}_1d(s_1), \acute{s}_2e(s_2))) \triangleright_\zeta ; CASE(p, \acute{s}_1w(d((s_1))), \acute{s}_2w(e(s_2)))$

- $w(CASE(p, \acute{s}_1d(s_1), \acute{s}_2e(s_2))) >^* p$ by the subterm property.
- $\{w(CASE(p, \acute{s}_1d(s_1), \acute{s}_2e(s_2)))\} \gg^* \{\acute{s}_1w(d((s_1))), \acute{s}_2w(e(s_2))\}$:

  — $w(CASE(p, \acute{s}_1d(s_1), \acute{s}_2e(s_2))) >^* \acute{s}_1w(d(s_1))$
    since $CASE(p, \acute{s}_1d(s_1), \acute{s}_2e(s_2)) >^* d(s_1)$ by the subterm property.
  — $w(CASE(p, \acute{s}_1d(s_1), \acute{s}_2e(s_2))) >^* \acute{s}_2w(e(s_2))$
    since $CASE(p, \acute{s}_1d(s_1), \acute{s}_2e(s_2)) >^* e(s_2)$ by the subterm property.

(2) $w(INST(c, \acute{g}\acute{t}d(g,t))) \triangleright_\zeta INST(c, \acute{g}\acute{t}w(d(g,t)))$

- $INST(c, \acute{g}\acute{t}d(g,t)) >^* c$ by the subterm property.

- $\{w(INST(c, \acute{g}\acute{t}d(g,t)))\} \gg^*$
  $\{\acute{g}\acute{t}w(d(g,t))\}$ pois $INST(c, \acute{g}\acute{t}d(g,t)) > d(g,t)$ by the subterm property.

(3) $w(REWR(e, \acute{t}d(t))) \triangleright_\zeta REWR(e, \acute{t}w(d(t)))$

- $REWR(e, \acute{t}d(t))) >^* e$ by the subterm property.
- $\{w(REWR(e, \acute{t}d(t)))\} \gg^* \{\acute{t}w(d(t))\}$ pois $REWR(e, \acute{t}d(t)) >^* d(t)$
  by the subterm property.

The proof for the rules in which w has arity greater than 1 is done analogously, since the additional arguments are not placed as first operator, and they are added in the same way, be it on the left-hand side or the right-hand side of the rules.

The proof for the set of ι reductions is shown in the Appendix.

From the proof of termination of *LND-TRS*, the following result follows:

Theorem 4.1 (normalization). *Every derivation in LND converts to a normal form.* ∎

4.4.5 *Proving the confluence property*

Once we have proved the termination property for the *LND-TRS*, it suffices to apply the superposition algorithm to check if there exist divergent critical pairs.

The superposition algorithm tests each rule of the system with a view towards verifying if there are divergent critical pairs. In case there is any divergent critical pair, one needs to give it an orientation as illustrated by the Knuth–Bendix completion procedure.

The proof of confluence, shown in Appendix A, had as a result the incorporation of three new rules to *LND-TRS*:

18. $CASE(c, á_1 w(inl(a_1)), á_2 w(inr(a_2))) \rhd_\iota w(c)$
19. $INST(c, ǵtw(\varepsilon y.(g(y), t))) \rhd_\iota w(c)$
20. $REWR(c, \acute{t}w(t(a, b))) \rhd_\iota w(c)$

This new set of rules defines new basic transformations between proofs, which do not seem to have appeared in the literature.

ι-$\vee$-*reduction*

$$
\cfrac{c : A_1 \vee A_2 \qquad \cfrac{\cfrac{[a_1 : A_1]}{inl(a_1) : A_1 \vee A_2} \vee\text{-}intr}{w(inl(a_1)) : W} r \qquad \cfrac{\cfrac{[a_2 : A_2]}{inr(a_2) : A_1 \vee A_2} \vee\text{-}intr}{w(inr(a_2)) : W} r}{CASE(c, á_1 w(inl(a_1)), á_2 w(inr(a_2))) : W} \vee\text{-}elim \quad \rhd_\iota \quad \cfrac{c : A_1 \vee A_2}{w(c) : W}
$$

ι-$\exists$-*reduction*

$$
\cfrac{c : \exists x^D.P(x) \qquad \cfrac{\cfrac{[t : D] \quad [g(t) : P(t)]}{\varepsilon y.(g(y), t) : \exists y^D.P(y)} \exists\text{-}intr}{w(\varepsilon y.(g(y), t)) : W} r}{INST(c, ǵtw(\varepsilon y.(g(y), t))) : W} \exists\text{-}elim \quad \rhd_\iota \quad \cfrac{c : \exists x^D.P(x)}{w(c) : W} r
$$

ι-$\doteq$-*reduction*

$$
\cfrac{c :\doteq_D (a, b) \qquad \cfrac{\cfrac{[a =_t b : D]}{t(a, b) :\doteq_D (a, b)} \doteq\text{-}intr}{w(t(a, b)) : W} r}{REWR(c, \acute{t}w(t(a, b))) : W} \doteq\text{-}elim \quad \rhd_\iota \quad \cfrac{c :\doteq_D (a, b)}{w(c) : W} r
$$

These transformations, together with the ζ transformations, offer a formal justification for the non-interference of the other branches with the main branch in the *elimination* rules of $\vee$, $\exists$ and $\doteq$: the rule r (an interference in the form of an *introduction*) may be applied to each secondary branches or in the main branch. The example below shows the application of such transformations:

Example 4.2. This example illustrates the ι transformation:

$$
\cfrac{c : A_1 \vee A_2 \qquad \cfrac{\cfrac{[a_1 : A_1]}{inl(a_1) : A_1 \vee A_2} \quad b : B}{\langle inl(a_1), b \rangle : (A_1 \vee A_2) \wedge B} \qquad \cfrac{\cfrac{[a_2 : A_2]}{inr(a_2) : A_1 \vee A_2} \quad b : B}{\langle inr(a_2), b \rangle : (A_1 \vee A_2) \wedge B}}{CASE(c, á_1 \langle inl(a_1), b \rangle, á_2 \langle inr(a_2), b \rangle) : (A_1 \vee A_2) \wedge B}
$$

$$
\rhd_\iota \cfrac{c : A_1 \vee A_2 \qquad b : B}{\langle c, b \rangle : (A_1 \vee A_2) \wedge B}
$$

Another consequence of the confluence property of *LND-TRS* is the strong normalization theorem for *LND*:

Theorem 4.2 (strong normalization). *Every derivation in LND converts to a unique normal form.*

Proof.   The proof of this theorem is based on the same argument of the proof of the normalization theorem.   ∎

4.5   Examples of transformations between proofs

This section illustrates, by means of two examples, the method of transformation between proofs based on term-rewriting defined for the functional calculus of *LND*.

Example 4.3.   Suppose we have the following non-normal proof in *LND*:

$$\cfrac{a:G\wedge(A\wedge B)\qquad \cfrac{\cfrac{[x:G\wedge(A\wedge B)]}{SND(x):A\wedge B}}{\lambda x.SND(x):G\wedge(A\wedge B)\to A\wedge B}}{\cfrac{APP(\lambda x.SND(x),a):A\wedge B\qquad\qquad y:C}{\langle APP(\lambda x.SND(x),a),y\rangle:(A\wedge B)\wedge C}}$$

The following rewritings transform this proof, represented by the term $\langle APP(\lambda x.SND(x),a),y\rangle$, to its normal form:

$$\triangleright_\beta\langle SND(a),y\rangle$$

The normal derivation constructed from this term is the following:

$$\cfrac{\cfrac{a:G\wedge(A\wedge B)}{SND(a):A\wedge B}\qquad y:C}{\langle SND(a),y\rangle:(A\wedge B)\wedge C}$$

Example 4.4.

$$\cfrac{\cfrac{a_1:A_1\to B_1}{inl(a_1):(A_1\to B_1)\vee(A_1\to B_1)}\quad \cfrac{\cfrac{[x:A_1]\quad [s_1:A_1\to B_1]}{APP(s_1,x):B_1}}{\lambda x.APP(s_1,x):A_1\to B_1}\quad \cfrac{\cfrac{[x_1:A_1]\quad [s_2:A_1\to B_1]}{APP(s_2,x):B_1}}{\lambda x.APP(s_2,x):A_1\to B_1}}{CASE(inl(a_1),\acute{s_1}\lambda x.APP(s_1,x),\acute{s_2}\lambda x.APP(s_2,x)):A_1\to B_1}$$

This proof can be transformed to its normal form by means of one of the following rewritings over the term $CASE(inl(a_1),\acute{s_1}\lambda x.APP(s_1,x),\acute{s_2}\lambda x.APP(s_2,x))$:

(1)

$$\triangleright_\beta \lambda x.APP(a_1, x)$$
$$\triangleright_\eta a_1$$

(2)

$$\triangleright_\eta CASE(inl(a_1), s'_1 s_1, s'_2 \lambda x.APP(s_2, x))$$
$$\triangleright_\eta CASE(inl(a_1), s'_1 s_1, s'_2 s_2)$$
$$\triangleright_\beta a_1$$

(3)

$$\triangleright_\eta CASE(inl(a_1), s'_1 s_1, s'_2 \lambda x.APP(s_2, x))$$
$$\triangleright_\beta a_1$$

Thus, the normal form is, as expected, the following:

$$a_1 : A_1 \to B_1$$

4.6 Final remarks

We have defined a *TRS* associated to *LND* (*LND-TRS*), which computes the normal form of proofs in *LND*. The proof of termination for *LND-TRS* had as a consequence the normalization theorem for the *LND* system, whereas the proof of confluence led to the strong normalization theorem. The termination of *LND-TRS* was proved via a special kind of order called recursive path ordering, proposed by Dershowitz [Dershowitz (1979)]. Confluence was established via the mechanism of superposition among the rules of the system.

The results presented here seem to be relevant not only to the *LND* system, but seems to be of general interest since they offer a possible answer to the question as to why normalization proofs for natural deduction systems do not take into account the so-called η-rules as reduction rules, but rather as expansion rules. The uncovering of the need for a new set of reduction rules (i.e. the 'ι' rules) in order to obtain confluence brings in new information into the discussion.

Appendix: Proof of confluence of *LND-TRS*

Here comes the proof of confluence of *LND-TRS* via the application of the Knuth–Bendix procedure. The result of such a completion procedure is the addition of the following three new rules to the initial rewriting system:

18. $CASE(c, \acute{a_1}w(inl(a_1)), \acute{a_2}w(inr(a_2))) \triangleright_\iota w(c)$
19. $INST(c, \acute{g}tw(\varepsilon y.(g(y), t))) \triangleright_\iota w(c)$
20. $REWR(c, \acute{t}w(t(a, b))) \triangleright_\iota w(c)$.

With the addition of those three new rules, *LND-TRS* becomes confluent.
The rules of the initial *LND-TRS* are as follows:

(1) $FST(\langle a_1, a_2 \rangle) \triangleright_\beta a_1$

(2) $SND(\langle a_1, a_2 \rangle) \triangleright_\beta a_2$

(3) $CASE(inl(a_1), \acute{s_1}d(s_1), \acute{s_2}e(s_2)) \triangleright_\beta d(a_1/s_1)$

(4) $CASE(inr(a_2), \acute{s_1}d(s_1), \acute{s_2}e(s_2)) \triangleright_\beta e(a_2/s_2)$

(5) $APP(\lambda x.b(x), a) \triangleright_\beta b(a/x)$

(6) $EXTR(\Lambda x.f(x), a) \triangleright_\beta f(a/x)$

(7) $INST(\varepsilon x.(f(x), a), \acute{g}td(g, t)) \triangleright_\beta d(f/g, a/t)$

(8) $REWR(s(a, b), \acute{t}d(t)) \triangleright_\beta d(s/t)$

(9) $\langle FST(c), SND(c) \rangle \triangleright_\eta c$

(10) $CASE(c, \acute{a_1}inl(a_1), \acute{a_2}inr(a_2)) \triangleright_\eta c$

(11) $\lambda x.APP(c, x) \triangleright_\eta c$ (where c does not depend on x).

(12) $\Lambda x.EXTR(c, x) \triangleright_\eta c$ (where x does not occur free in c).

(13) $INST(c, \acute{g}t\varepsilon y.(g(y), t)) \triangleright_\eta c$

(14) $REWR(c, \acute{t}t(a, b)) \triangleright_\eta c$

(15) $w(CASE(p, \acute{s_1}d(s_1), \acute{s_2}e(s_2))) \triangleright_\zeta CASE(p, \acute{s_1}w(d(s_1)), \acute{s_2}w(e(s_2)))$

(16) $w(INST(c, \acute{g}td(g, t))) \triangleright_\zeta INST(c, \acute{g}tw(d(g, t)))$

(17) $w(REWR(e, \acute{t}d(t))) \triangleright_\zeta REWR(e, \acute{t}w(d(t)))$.

The analysis of superpositions is shown below.

- Analyzing rules 1 and 3:
  Since d and $e \in T_{\Sigma-LND}$ then the following superposition may occur:

$$CASE(inl(a_1), \acute{s_1}FST(\langle s_1, x \rangle), \acute{s_2}FST(\langle s_2, y \rangle)).$$

This term rewrites in the following ways:
Rewriting sequence 1:
  $\triangleright_\beta CASE(inl(a_1), \acute{s_1}s_1, \acute{s_2}FST(\langle s_2, y \rangle))$
  $\triangleright_\beta CASE(inl(a_1), \acute{s_1}s_1, \acute{s_2}s_2)$
  $\triangleright_\beta a_1$
Rewriting sequence 2:
  $\triangleright_\beta CASE(inl(a_1), \acute{s_1}s_1, \acute{s_2}FST(\langle s_2, y \rangle))$
  $\triangleright_\beta a_1$

Rewriting sequence 3:

$\triangleright_\beta FST(\langle a_1, x\rangle)$

$\triangleright_\beta a_1$

Such a superposition does not yield any divergent critical pair, since the term resulting from the superposition between the rules rewrites to the same term a_1. By analyzing rules 1 and 4, 2 and 3 and 2 and 4 a similar situation occurs. The same happens when one alternates d and e in the rule 3 (or 4), i.e. d may be FST and e may be SND.

- Analyzing rule 3 with itself:

$$CASE(inl(a_1), \acute{s_1}CASE(inl(s_1), \acute{b_1}f(b_1), \acute{b_2}g(b_2)), \acute{s_2}CASE(inl(s_2), \acute{b_1}f(b_1), \acute{b_2}g(b_2)))$$

This term rewrites in the following ways:

Rewriting sequence 1:

$\triangleright_\beta CASE(inl(a_1), \acute{b_1}f(b_1), \acute{b_2}g(b_2))$

$\triangleright_\beta f(a_1)$

Rewriting sequence 2:

$\triangleright_\beta CASE(inl(a_1), \acute{s_1}f(s_1), \acute{s_2}CASE(inl(s_2), \acute{b_1}f(b_1), \acute{b_2}g(b_2)))$

$\triangleright_\beta f(a_1)$

- Analyzing rules 5 and 3:

The following superposition may occur:

$$CASE(inl(a_1), \acute{s_1}APP(\lambda x.b(x), s_1), \acute{s_2}APP(\lambda x.b(x), s_2)).$$

The rewriting alternatives for such a term are as follows:

Rewriting sequence 1:

$\triangleright_\beta CASE(inl(a_1), \acute{s_1}b(s_1), \acute{s_2}APP(\lambda x.b(x), s_2))$

$\triangleright_\beta CASE(inl(a_1), \acute{s_1}b(s_1), \acute{s_2}b(s_2))$

$\triangleright_\beta b(a_1)$

Rewriting sequence 2:

$\triangleright_\beta CASE(inl(a_1), \acute{s_1}b(s_1), \acute{s_2}APP(\lambda x.b(x), s_2))$

$\triangleright_\beta b(a_1)$

Rewriting sequence 3:

$\triangleright_\beta APP(\lambda x.b(x), a_1)$

$\triangleright_\beta b(a_1)$

Analogously, rules 5 and 4 do not generate divergent critical pairs.

- Analyzing rules 6 and 3:

The following term resulting from the superposition is generated:

$$CASE(inl(a_1), \acute{s_1}EXTR(\Lambda x.f(x), s_1), \acute{s_2}EXTR(\Lambda x.f(x), s_2))$$

Alternatives of rewriting for this term:

Rewriting sequence 1:

$\triangleright_\beta EXTR(\Lambda x.f(x), a_1)$

$\triangleright_\beta f(a_1)$

Rewriting sequence 2:
$\triangleright_\beta CASE(inl(a_1), \acute{s_1}f(s_1), \acute{s_2}EXTR(\Lambda x.f(x), s_2))$
$\triangleright_\beta CASE(inl(a_1), \acute{s_1}f(s_1), \acute{s_2}f(s_2))$
$\triangleright_\beta f(a_1)$

Rewriting sequence 3:
$\triangleright_\beta CASE(inl(a_1), \acute{s_1}f(s_1), \acute{s_2}EXTR(\Lambda x.f(x), s_2))$
$\triangleright_\beta f(a_1)$

In a similar way, the superposition of 6 and 4 does not generate any critical pair.

- Analyzing rules 7 and 3:

$$CASE(inl(a_1), \acute{s_1}INST(\varepsilon x.(f(x), s_1), \acute{g}td(g, t)), \acute{s_2}INST(\varepsilon x.(f(x), s_2), \acute{g}td(g, t)))$$

Rewriting sequence 1:
$\triangleright_\beta CASE(inl(a_1), \acute{s_1}d(f, s_1), \acute{s_2}INST(\varepsilon x.(f(x), s_2), \acute{g}td(g, t)))$
$\triangleright_\beta CASE(inl(a_1), \acute{s_1}d(f, s_1), \acute{s_2}d(f, s_2))$
$\triangleright_\beta d(f, a_1)$

Rewriting sequence 2:
$\triangleright_\beta CASE(inl(a_1), \acute{s_1}d(f, s_1), \acute{s_2}INST(\varepsilon x.(f(x), s_2), \acute{g}td(g, t)))$
$\triangleright_\beta d(f, a_1)$

Rewriting sequence 3:
$\triangleright_\beta INST(\varepsilon x.(f(x), a_1), \acute{g}td(g, t))$
$\triangleright_\beta d(f, a_1)$

Similarly, the superposition of rules 7 and 4 does not generate any divergent critical pair.

- Analyzing rules 8 and 3:

$$CASE(inl(a_1), \acute{s_1}REWR(s_1(a, b), \acute{t}h(t)), \acute{s_2}REWR(s_2(a, b), \acute{t}h(t)))$$

Rewriting sequence 1:
$\triangleright_\beta REWR(a_1(a, b), \acute{t}h(t))$
$\triangleright_\beta h(a_1)$

Rewriting sequence 2:
$\triangleright_\beta CASE(inl(a_1), \acute{s_1}h(s_1), \acute{s_2}REWR(s_2(a, b), \acute{t}h(t)))$
$\triangleright_\beta CASE(inl(a_1), \acute{s_1}h(s_1), \acute{s_2}h(s_2))$
$\triangleright_\beta h(a_1)$

Rewriting sequence 3:
$\triangleright_\beta CASE(inl(a_1), \acute{s_1}h(s_1), \acute{s_2}REWR(s_2(a, b), \acute{t}h(t)))$
$\triangleright_\beta h(a_1)$

Similarly, rules 8 and 4 do not yield any divergent critical pair.

- Analyzing rules 9 and 3:

$$CASE(inl(a_1), \acute{s_1}\langle FST(s_1), SND(s_1)\rangle, \acute{s_2}\langle FST(s_2), SND(s_2)\rangle)$$

Rewriting sequence 1:

$\triangleright_\beta \langle FST(a_1), SND(a_1) \rangle$

$\triangleright_\eta a_1$

Rewriting sequence 2:

$\triangleright_\eta CASE(inl(a_1), s'_1 s_1, s'_2 \langle FST(s_2), SND(s_2) \rangle)$

$\triangleright_\eta CASE(inl(a_1), s'_1 s_1, s'_2 s_2)$

$\triangleright_\beta a_1$

Rewriting sequence 3:

$\triangleright_\eta CASE(inl(a_1), s'_1 s_1, s'_2 \langle FST(s_2), SND(s_2) \rangle)$

$\triangleright_\beta a_1$

Thus, the superposition of rules 9 and 4 does not generate any divergent critical pair.

- Analyzing rules 10 and 3:

$CASE(inl(a_1), s'_1 CASE(s_1, b'_1 inl(b_1), b'_2 inr(b_2)), s'_2 CASE(s_2, b'_1 inl(b_1), b'_2 inr(b_2)))$

Rewriting sequence 1:

$\triangleright_\beta CASE(a_1, b'_1 inl(b_1), b'_2 inr(b_2))$

$\triangleright_\eta a_1$

Rewriting sequence 2:

$\triangleright_\eta CASE(inl(a_1), s'_1 s_1, s'_2 CASE(s_2, b'_1 inl(b_1), b'_2 inr(b_2)))$

$\triangleright_\eta CASE(inl(a_1), s'_1 s_1, s'_2 s_2)$

$\triangleright_\beta a_1$

Rewriting sequence 3:

$\triangleright_\eta CASE(inl(a_1), s'_1 s_1, s'_2 CASE(s_2, b'_1 inl(b_1), b'_2 inr(b_2)))$

$\triangleright_\beta a_1$

Thus, the superposition of rules 10 and 4 does not yield any divergent critical pair.

- Analyzing rules 11 and 3:

$$CASE(inl(a_1), s'_1 \lambda x.APP(s_1, x), s'_2 \lambda x.APP(s_2, x))$$

Rewriting sequence 1:

$\triangleright_\beta \lambda x.APP(a_1, x)$

$\triangleright_\eta a_1$

Rewriting sequence 2:

$\triangleright_\eta CASE(inl(a_1), s'_1 s_1, s'_2 \lambda x.APP(s_2, x))$

$\triangleright_\eta CASE(inl(a_1), s'_1 s_1, s'_2 s_2)$

$\triangleright_\beta a_1$

Rewriting sequence 3:

$\triangleright_\eta CASE(inl(a_1), s'_1 s_1, s'_2 \lambda x.APP(s_2, x))$

$\triangleright_\beta a_1$

Similarly, after the superposition of rules 11 and 4 no divergent critical pair is generated.

- Analyzing rules 12 and 4:

$$CASE(inr(a_2), \acute{s_1}\Lambda x.EXTR(s_1, x), \acute{s_2}\Lambda x.EXTR(s_2, x))$$

Rewriting sequence 1:
$\triangleright_\beta \Lambda x.EXTR(a_2, x)$

$\triangleright_\eta a_2$

Rewriting sequence 2:
$\triangleright_\eta CASE(inr(a_2), \acute{s_1}s_1, \acute{s_2}\Lambda x.EXTR(s_2, x))$

$\triangleright_\eta CASE(inr(a_2), \acute{s_1}s_1, \acute{s_2}s_2)$

$\triangleright_\beta a_2$

Rewriting sequence 3:
$\triangleright_\eta CASE(inr(a_2), \acute{s_1}\Lambda x.EXTR(s_1, x), \acute{s_2}s_2)$

$\triangleright_\beta a_2$

Thus, the superposition of rules 12 and 3 does not generate any divergent critical pairs.

- Analyzing rules 13 and 4:

$$CASE(inr(a_2), \acute{s_1}INST(s_1, \acute{g}t\varepsilon y.(g(y), t)), \acute{s_2}INST(s_2, \acute{g}t\varepsilon y.(g(y), t)))$$

Rewriting sequence 1:
$\triangleright_\beta INST(a_2, \acute{g}t\varepsilon y.(g(y), t))$

$\triangleright_\eta a_2$

Rewriting sequence 2:
$\triangleright_\eta CASE(inr(a_2), \acute{s_1}s_1, \acute{s_2}INST(s_2, \acute{g}t\varepsilon y.(g(y), t)))$

$\triangleright_\eta CASE(inr(a_2), \acute{s_1}s_1, \acute{s_2}s_2)$

$\triangleright_\beta a_2$

Rewriting sequence 3:
$\triangleright_\eta CASE(inr(a_2), \acute{s_1}INST(s_1, \acute{g}t\varepsilon y.(g(y), t)), \acute{s_2}s_2)$

$\triangleright_\beta a_2$

Similarly, rules 13 and 3 do not generate any divergent critical pair.

- Analyzing rules 15 and 3:

$$CASE(inl(a_1), \acute{s_1}w(CASE(s_1, \acute{b_1}d(b_1), \acute{b_2}e(b_2))), \acute{s_2}w(CASE(s_2, \acute{b_1}d(b_1), \acute{b_2}e(b_2))))$$

Rewriting sequence 1:
$\triangleright_\beta w(CASE(a_1, \acute{b_1}d(b_1), \acute{b_2}e(b_2)))$

$\triangleright_\zeta CASE(a_1, \acute{b_1}w(d(b_1)), \acute{b_2}w(e(b_2)))$

Rewriting sequence 2:
$\triangleright_\zeta CASE(inl(a_1), \acute{s_1}CASE(s_1, \acute{b_1}w(d(b_1)), \acute{b_2}w(e(b_2))),$
$\qquad\qquad \acute{s_2}w(CASE(s_2, \acute{b_1}d(b_1), \acute{b_2}e(b_2))))$

$\triangleright_\zeta CASE(inl(a_1), \acute{s_1}CASE(s_1, \acute{b_1}w(d(b_1)), \acute{b_2}w(e(b_2))),$
$\qquad\qquad \acute{s_2}CASE(s_2, \acute{b_1}w(d(b_1)), \acute{b_2}w(d(b_2))))$

$\triangleright_\beta CASE(a_1, \acute{b_1}w(d(b_1)), \acute{b_2}w(e(b_2)))$

Rewriting sequence 3:

$\triangleright_\zeta CASE(inl(a_1), \acute{s_1}CASE(s_1, \acute{b_1}w(d(b_1)), \acute{b_2}w(e(b_2)),$
$\qquad\qquad \acute{s_2}w(CASE(s_2, \acute{b_1}d(b_1), \acute{b_2}e(b_2))))$
$\triangleright_\beta CASE(a_1, \acute{b_1}w(d(b_1)), \acute{b_2}w(e(b_2)))$

Analogously, the superposition of rules 15 and 4 does not generate any critical pair.

- Analyzing rules 16 and 3:

$$CASE(inl(a_1), \acute{s_1}w(INST(s_1, \acute{g}td(g,t))), \acute{s_2}w(INST(s_2, \acute{g}td(g,t))))$$

Rewriting sequence 1:

$\triangleright_\beta w(INST(a_1, \acute{g}td(g,t)))$
$\triangleright_\zeta INST(a_1, \acute{g}tw(d(g,t)))$

Rewriting sequence 2:

$\triangleright_\zeta CASE(inl(a_1), \acute{s_1}INST(s_1, \acute{g}tw(d(g,t))), \acute{s_2}w(INST(s_2, \acute{g}td(g,t))))$
$\triangleright_\zeta CASE(inl(a_1), \acute{s_1}INST(s_1, \acute{g}tw(d(g,t))), \acute{s_2}INST(s_2, \acute{g}tw(g,t))))$
$\triangleright_\beta INST(a_1, \acute{g}tw(d(g,t)))$

Rewriting sequence 3:

$\triangleright_\zeta CASE(inl(a_1), \acute{s_1}INST(s_1, \acute{g}tw(d(g,t))), \acute{s_2}w(INST(s_2, \acute{g}td(g,t))))$
$\triangleright_\beta INST(a_1, \acute{g}tw(d(g,t)))$

Analogously, the superposition of rules 16 and 4 does not generate any critical pair.

- Analyzing rules 17 and 3:

$$CASE(inl(a_1), \acute{s_1}w(REWR(s_1, \acute{t}d(t))), \acute{s_2}w(REWR(s_2, \acute{t}d(t))))$$

Rewriting sequence 1:

$\triangleright_\beta w(REWR(a_1, \acute{t}d(t)))$
$\triangleright_\zeta REWR(a_1, \acute{t}w(d(t)))$

Rewriting sequence 2:

$\triangleright_\zeta CASE(inl(a_1), \acute{s_1}REWR(s_1, \acute{t}w(d(t))), \acute{s_2}w(REWR(s_2, \acute{t}d(t))))$
$\triangleright_\zeta CASE(inl(a_1), \acute{s_1}REWR(s_1, \acute{t}w(d(t))), \acute{s_2}REWR(s_2, \acute{t}w(d(t))))$
$\triangleright_\beta REWR(a_1, \acute{t}w(d(t)))$

Rewriting sequence 3:

$\triangleright_\zeta CASE(inl(a_1), \acute{s_1}REWR(s_1, \acute{t}w(d(t))), \acute{s_2}w(REWR(s_2, \acute{t}d(t))))$
$\triangleright_\beta REWR(a_1, \acute{t}w(d(t)))$

Analogously, the superposition of rules 17 and 4 does not generate any critical pair.

- Analyzing rules 1 and 5:

$$APP(\lambda x.FST(\langle x, y\rangle), a)$$

Rewriting sequence 1:
$\triangleright_\beta APP(\lambda x.x, a)$

$\triangleright_\beta a$

Rewriting sequence 2:
$\triangleright_\beta FST(\langle a, y \rangle)$

$\triangleright_\beta a$

In this term, the abstractor operator λ binds the first argument of the term FST. It could be a vacuous abstraction, and in this case, it would pose no problems, the term generated by the superposition will converge. The term thus generated is the following:

$$APP(\lambda x.FST(\langle z, x \rangle), a)$$

Rewriting sequence 1:
$\triangleright_\beta APP(\lambda x.z, a)$

$\triangleright_\beta z$

Rewriting sequence 2:
$\triangleright_\beta FST(\langle z, a \rangle)$

$\triangleright_\beta z$

- Analyzing rules 3 and 5:

$$APP(\lambda x.CASE(inl(x), \acute{s}_1 d(s_1), \acute{s}_2 d(s_2)), a).$$

Rewriting sequence 1:
$\triangleright_\beta CASE(inl(a), \acute{s}_1 d(s_1), \acute{s}_2 d(s_2))$

$\triangleright_\beta d(a)$

Rewriting sequence 2:
$\triangleright_\beta APP(\lambda x.d(x), a)$

$\triangleright_\beta d(a)$

Analogously, the superposition of rules 4 and 5 does not generate any critical pair.

- Analyzing rules 5 and 5:

$$APP(\lambda x.APP(\lambda y.b(y), x), a)$$

Rewriting sequence 1:
$\triangleright_\beta APP(\lambda x.b(x), a)$

$\triangleright_\beta b(a)$

Rewriting sequence 2:
$\triangleright_\beta APP(\lambda y.b(y), a)$

$\triangleright_\beta b(a)$

- Analyzing rules 6 and 5:

$$APP(\lambda x.EXTR(\Lambda y.f(y), x), a)$$

Rewriting sequence 1:

$\triangleright_\beta EXTR(\Lambda y.f(y), a)$

$\triangleright_\beta f(a)$

Rewriting sequence 2:

$\triangleright_\beta APP(\lambda x.f(x), a)$

$\triangleright_\beta f(a)$

Similarly to the previous case, the abstraction Λ may be vacuous, and even so no divergent critical pairs will be generated.

- Analyzing rules 7 and 5:

$$APP(\lambda x.INST(\varepsilon y.(f(y), x), \acute{g}td(g, t)), a)$$

Rewriting sequence 1:

$\triangleright_\beta INST(\varepsilon y.(f(y), a), \acute{g}td(g, t))$

$\triangleright_\beta d(f, a)$

Rewriting sequence 2:

$\triangleright_\beta APP(\lambda x.d(f, x), a)$

$\triangleright_\beta d(f, a)$

- Analyzing rules 8 and 5:

$$APP(\lambda s.REWR(s(a, b), \acute{t}d(t)), u)$$

Rewriting sequence 1:

$\triangleright_\beta REWR(u(a, b), \acute{t}d(t))$

$\triangleright_\beta d(u)$

Rewriting sequence 2:

$\triangleright_\beta APP(\lambda s.d(s), u)$

$\triangleright_\beta d(u)$

- Analyzing rules 9 and 5:

$$APP(\lambda x.\langle FST(x), SND(x)\rangle, a)$$

Rewriting sequence 1:

$\triangleright_\beta \langle FST(a), SND(a)\rangle$

$\triangleright_\eta a$

Rewriting sequence 2:

$\triangleright_\eta APP(\lambda x.x, a)$

$\triangleright_\beta a$

- Analyzing rules 10 and 5:

$$APP(\lambda x.CASE(x, \acute{a_1}inl(a_1), \acute{a_2}inr(a_2)), a)$$

Rewriting sequence 1:

$\triangleright_\beta CASE(a, \acute{a_1}inl(a_1), \acute{a_2}inr(a_2))$

$\triangleright_\eta a$

Rewriting sequence 2:
$\triangleright_\eta APP(\lambda x.x, a)$
$\triangleright_\beta a$

- Analyzing rules 11 and 5:

$$APP(\lambda x.\lambda y.APP(c, y), a)$$

Rewriting sequence 1:
$\triangleright_\beta \lambda y.APP(c, y)$
$\triangleright_\eta c$
Rewriting sequence 2:
$\triangleright_\eta APP(\lambda x.c, a)$
$\triangleright_\beta c$

- Analyzing rules 12 and 5:

$$APP(\lambda x.\Lambda y.EXTR(c, y), a)$$

Rewriting sequence 1:
$\triangleright_\beta \Lambda y.EXTR(c, y)$
$\triangleright_\eta c$
Rewriting sequence 2:
$\triangleright_\eta APP(\lambda x.c, a)$
$\triangleright_\beta c$

- Analyzing rules 13 and 5:

$$APP(\lambda x.INST(c, \acute{g}t\varepsilon y.(g(y), t)), a)$$

Rewriting sequence 1:
$\triangleright_\beta INST(c, \acute{g}t\varepsilon y.(g(y), t))$
$\triangleright_\eta c$
Rewriting sequence 2:
$\triangleright_\eta APP(\lambda x.c, a)$
$\triangleright_\beta c$

(See remarks on vacuous abstractions given previously.)

- Analyzing rules 14 and 5:

$$APP(\lambda x.REWR(x, \acute{t}t(a, b)), a)$$

Rewriting sequence 1:
$\triangleright_\beta REWR(a, \acute{t}t(a, b))$
$\triangleright_\eta a$
Rewriting sequence 2:
$\triangleright_\eta APP(\lambda x.x, a)$
$\triangleright_\beta a$

- Analyzing rules 15 and 5:

$$APP(\lambda x.w(CASE(x, \acute{s_1}d(s_1), \acute{s_2}e(s_2))), a)$$

Rewriting sequence 1:
$\rhd_\beta w(CASE(a, \acute{s_1}d(s_1), \acute{s_2}e(s_2)))$
$\rhd_\zeta CASE(a, \acute{s_1}w(d(s_1)), \acute{s_2}w(e(s_2)))$
Rewriting sequence 2:
$\rhd_\zeta APP(\lambda x.CASE(x, \acute{s_1}w(d(s_1)), \acute{s_2}w(e(s_2))), a)$
$\rhd_\beta CASE(a, \acute{s_1}w(d(s_1)), \acute{s_2}w(e(s_2)))$

- Analyzing rules 16 and 5:

$$APP(\lambda x.w(INST(x, \acute{g}td(g, t))), a)$$

Rewriting sequence 1:
$\rhd_\beta w(INST(a, \acute{g}td(g, t)))$
$\rhd_\zeta INST(a, \acute{g}tw(d(g, t)))$
Rewriting sequence 2:
$\rhd_\zeta APP(\lambda x.INST(x, \acute{g}tw(d(g, t))), a)$
$\rhd_\beta INST(a, \acute{g}tw(d(g, t)))$

- Analyzing rules 17 and 5:

$$APP(\lambda x.w(REWR(x, \acute{t}d(t))), a)$$

Rewriting sequence 1:
$\rhd_\beta w(REWR(a, \acute{t}d(t)))$
$\rhd_\zeta REWR(a, \acute{t}w(d(t)))$
Rewriting sequence 2:
$\rhd_\zeta APP(\lambda x.REWR(x, \acute{t}w(d(t))), a)$
$\rhd_\beta REWR(a, \acute{t}w(d(t)))$

- Superposition with rule 6:

— All cases of superposition with rule 6 are done in a similar way to rule 5:

$$(5) \qquad APP(\lambda x.b(x), a) \rhd_\beta b(a/x)$$
$$(6) \qquad EXTR(\Lambda x.f(x), a) \rhd_\beta f(a/x)$$

- Analyzing rules 1 and 9:

$$\langle FST(\langle x, y \rangle), SND(\langle x, y \rangle) \rangle$$

Rewriting sequence 1:
$\rhd_\eta \langle x, y \rangle$
Rewriting sequence 2:
$\rhd_\beta \langle x, SND(\langle x, y \rangle) \rangle$
$\rhd_\beta \langle x, y \rangle$

Rewriting sequence 3:
$\triangleright_\beta \langle FST(\langle x, y \rangle), y \rangle$
$\triangleright_\beta \langle x, y \rangle$
This subsumes the superposition of rules 2 and 9, due to the structure of rule 9.

- Analyzing rules 3 and 10:

$$CASE(inl(b), \acute{a_1}inl(a_1), \acute{a_2}inr(a_2))$$

Rewriting sequence 1:
$\triangleright_\eta inl(b)$
Rewriting sequence 2:
$\triangleright_\beta inl(b)$

- Analyzing rules 4 and 10:

$$CASE(inr(b), \acute{a_1}inl(a_1), \acute{a_2}inr(a_2))$$

Rewriting sequence 1:
$\triangleright_\eta inr(b)$
Rewriting sequence 2:
$\triangleright_\beta inr(b)$

- Analyzing rules 15 and 10:

$$w(CASE(c, \acute{a_1}inl(a_1), \acute{a_2}inr(a_2)))$$

Rewriting sequence 1:
$\triangleright_\eta w(c)$
Rewriting sequence 2:
$\triangleright_\zeta CASE(c, \acute{a_1}w(inl(a_1)), \acute{a_2}w(inr(a_2)))$
In this case there was no confluence, therefore by the Knuth-Bendix procedure, a new rule must be added according to the order previously chosen:

- $CASE(c, \acute{a_1}w(inl(a_1)), \acute{a_2}w(inr(a_2))) >^* w(c)$
  since $c = c$ and $\{w(inl(a_1)), w(inr(a_2))\} \gg^* \emptyset$ vacuously.

Thus, the following rule, "baptized" as "ι", is added to the system:
18. $CASE(c, \acute{a_1}w(inl(a_1)), \acute{a_2}w(inr(a_2))) \triangleright_\iota w(c)$

- Analyzing rules 5 and 11:

$$\lambda x.APP(\lambda y.b(y), x)$$

Rewriting sequence 1:
$\triangleright_\eta \lambda y.b(y)$
Rewriting sequence 2:
$\triangleright_\beta \lambda x.b(x)$

In this case there was confluence since the terms "$\lambda y.b(y)$" and "$\lambda x.b(x)$" are the same. The bounded variable has no identity. In the theory of λ-calculus these terms are equal due to α-reduction, which renames variables. Here such a reduction is unnecessary, since in the theory of term rewriting, the name of the variable does not distinguish terms.

- Analyzing rules 6 and 12:

$$\Lambda t.EXTR(\Lambda x.f(x), t)$$

Rewriting sequence 1:
$\triangleright_\eta \Lambda x.f(x)$
Rewriting sequence 2:
$\triangleright_\beta \Lambda t.f(t)$
As before, here we have confluence.

- Analyzing rules 7 and 13:

$$INST(\varepsilon x.(f(x), a), \acute{g}t\varepsilon.y(g(y), t))$$

Rewriting sequence 1:
$\triangleright_\eta \varepsilon x.(f(x), a)$
Rewriting sequence 2:
$\triangleright_\beta \varepsilon y.(f(y), a)$
As before, the terms resulting from rewriting sequences 1 and 2 are the same.

- Analyzing rules 16 and 13:

$$w(INST(c, \acute{g}t\varepsilon y.(g(y), t)))$$

Rewriting sequence 1:
$\triangleright_\eta w(c)$
Rewriting sequence 2:
$\triangleright_\zeta INST(c, \acute{g}tw(\varepsilon y.(g(y), t)))$
This case is similar to the one involving rules 15 and 10.

  - $INST(c, \acute{g}tw(\varepsilon y.(g(y), t))) >^* w(c)$, since $c = c$ and $\{w(\varepsilon y.(g(y), t))\} \gg^* \emptyset$

Thus, the following new rule is added to the rewriting system:
19. $INST(c, \acute{g}tw(\varepsilon y.(g(y), t))) \triangleright_\iota w(c)$

- Analyzing rules 8 and 14:

$$REWR(s(a, b), \acute{t}t(a, b))$$

Rewriting sequence 1:
$\triangleright_\eta s(a, b)$
Rewriting sequence 2:
$\triangleright_\beta s(a, b)$

- Analyzing rules 17 and 14:

$$w(c, REWR(c, \acute{t}t(a, b)))$$

Rewriting sequence 1:
$$\triangleright_\eta w(c)$$
Rewriting sequence 2:
$$\triangleright_\zeta REWR(c, \acute{t}w(t(a, b)))$$
No confluence, thus, similarly to the case involving 15 and 10, as well as 16 and 13, a new rule must be added to the rewriting system, respecting the previously chosen order:

- $REWR(c, \acute{t}w(t(a, b))) >^* w(c)$, since $c = c$ and $\{w(t(a, b))\} \gg^* \emptyset$ vacuously.

Thus, the following new rule is added to the system:
20. $REWR(c, \acute{t}w(t(a, b))) \triangleright_\iota w(c)$
- Analyzing rules 1 and 15:

$$w(CASE(p, \acute{s_1}FST(\langle s_1, t_1 \rangle), \acute{s_2}FST(\langle s_2, t_2 \rangle)))$$

Rewriting sequence 1:
$$\triangleright_\zeta CASE(p, \acute{s_1}w(FST(\langle s_1, t_1 \rangle)), \acute{s_2}w(FST(\langle s_2, t_2 \rangle)))$$
$$\triangleright_\beta CASE(p, \acute{s_1}w(s_1), \acute{s_2}w(s_2))$$
Rewriting sequence 2:
$$\triangleright_\beta w(CASE(p, \acute{s_1}s_1, \acute{s_2}s_2))$$
$$\triangleright_\zeta CASE(p, \acute{s_1}w(s_1), \acute{s_2}w(s_2))$$
When we look at rules 2 and 15 we see a similar situation. The same happens after the superposition of rules 1 and 2 simultaneously, i.e. the second argument of the CASE having the external symbol FST and the third argument the symbol SND, and vice-versa.
- Analyzing rules 3 and 15:
$$w(CASE(p, \acute{s_1}CASE(inl(s_1), \acute{t_1}d(t_1), \acute{t_2}e(t_2)),$$
$$\acute{s_2}CASE(inl(s_2), \acute{u_1}d(u_1), \acute{u_2}.e(u_2)))))$$

Rewriting sequence 1:
$$\triangleright_\zeta CASE(p, \acute{s_1}w(CASE(inl(s_1), \acute{t_1}d(t_1), \acute{t_2}e(t_2))),$$
$$\acute{s_2}w(CASE(inl(s_2), \acute{u_1}d(u_1), \acute{u_2}e(u_2))))$$
$$\triangleright_\beta CASE(p, \acute{s_1}w(d(s_1)), \acute{s_2}w(CASE(inl(s_2), \acute{u_1}d(u_1), \acute{u_2}e(u_2))))$$
$$\triangleright_\beta CASE(p, \acute{s_1}w(d(s_1)), \acute{s_2}w(d(s_2)))$$
Rewriting sequence 2:
$$\triangleright_\beta w(CASE(p, \acute{s_1}d(s_1), \acute{s_2}CASE(inl(s_2), \acute{u_1}d(u_1), \acute{u_2}e(u_2))))$$
$$\triangleright_\beta w(CASE(p, \acute{s_1}d(s_1), \acute{s_2}d(s_2))) \triangleright_\zeta CASE(p, \acute{s_1}w(d(s_1)), \acute{s_2}w(d(s_2)))$$

These rules may have a different superposition:

$$w(CASE(inl(a_1), \acute{s_1}d(s_1), \acute{s_2}d(s_2)))$$

Rewriting sequence 1:
$\rhd_\beta w(d(a_1))$

Rewriting sequence 2:
$\rhd_\zeta CASE(inl(a_1), \acute{s_1}w(d(s_1)), \acute{s_2}w(d(s_2)))$
$\rhd_\beta w(d(a_1))$

Similar to the superposition of rules 4 and 15.

- Analyzing rules 5 and 15:

$$w(CASE(p, \acute{s_1}APP(\lambda x.b(x), s_1), \acute{s_2}APP(\lambda x.b(x), s_2)))$$

Rewriting sequence 1:
$\rhd_\zeta CASE(p, \acute{s_1}w(APP(\lambda x.b(x), s_1)), \acute{s_2}w(APP(\lambda x.b(x), s_2)))$
$\rhd_\beta CASE(p, \acute{s_1}w(b(s_1)), \acute{s_2}w(CASE(inl(s_2), \acute{u_1}d(u_1), \acute{u_2}e(u_2))))$
$\rhd_\beta CASE(p, \acute{s_1}w(d(s_1)), \acute{s_2}w(d(s_2)))$

Rewriting sequence 2:
$\rhd_\beta w(CASE(p, \acute{s_1}b(s_1), \acute{s_2}APP(\lambda x.b(x), s_2)))$
$\rhd_\beta w(CASE(p, \acute{s_1}b(s_1), \acute{s_2}b(s_2)))$
$\rhd_\zeta CASE(p, \acute{s_1}w(b(s_1)), \acute{s_2}w(b(s_2)))$

Since rules 5 and 6 are very similar, the superposition of rules 6 and 15 is done in an analogous way.

- Analyzing rules 7 and 15:

$$w(CASE(p, \acute{s_1}INST(\varepsilon x.(f(x), s_1), \acute{g}td(g, t)), \acute{s_2}INST(\varepsilon x.(f(x), s_2), \acute{g}te(g, t))))$$

Rewriting sequence 1:
$\rhd_\zeta CASE(p, \acute{s_1}w(INST(\varepsilon x.(f(x), s_1))), \acute{g}td(g, t)),$
$\qquad \acute{s_2}w(INST(\varepsilon x.(f(x), s_2), \acute{g}te(g, t))))$
$\rhd_\beta CASE(p, \acute{s_1}w(d(f, s_1)), \acute{s_2}w(INST(\varepsilon x.(f(x), s_2), \acute{g}te(g, t))))$
$\rhd_\beta CASE(p, \acute{s_1}w(d(f, s_1)), \acute{s_2}w(e(f, s_2)))$

Rewriting sequence 2:
$\rhd_\beta w(CASE(p, \acute{s_1}d(f, s_1), \acute{s_2}INST(\varepsilon x.(f(x), s_2), \acute{g}te(g, t))))$
$\rhd_\beta w(CASE(p, \acute{s_1}w(d(f, s_1)), \acute{s_2}e(f, s_2)))$
$\rhd_\zeta CASE(p, \acute{s_1}w(d(f, s_1)), \acute{s_2}w(e(f, s_2)))$

- Analyzing rules 8 and 15:

$$w(CASE(p, \acute{s_1}REWR(s_1(a, b), \acute{t}d(t)), \acute{s_2}REWR(s_2(a, b), \acute{r}g(r))))$$

Rewriting sequence 1:
$\rhd_\zeta CASE(p, \acute{s_1}w(REWR(s_1(a, b), \acute{t}d(t))), \acute{s_2}w(REWR(s_2(a, b), \acute{r}g(r))))$
$\rhd_\beta CASE(p, \acute{s_1}w(d(s_1)), \acute{s_2}w(REWR(s_2(a, b), \acute{r}g(r))))$
$\rhd_\beta CASE(p, \acute{s_1}w(d(s_1)), \acute{s_2}w(g(s_2)))$

Rewriting sequence 2:

$\triangleright_\beta w(CASE(p, \acute{s_1}d(s_1)), \acute{s_2}REWR(s_2(a, b), \acute{r}g(r))))$

$\triangleright_\beta w(CASE(p, \acute{s_1}w(d(s_1)), \acute{s_2}g(s_2))))$

$\triangleright_\zeta CASE(p, \acute{s_1}w(d(s_1)), \acute{s_2}w(g(s_2)))$

- Analyzing rules 10 and 15:

$w(CASE(p, \acute{s_1}CASE(s_1, \acute{a_1}inl(a_1), \acute{a_2}inr(a_2)),$
$$\acute{s_2}CASE(s_2, \acute{b_1}inl(b_1), \acute{b_2}inr(b_2)))$$

Rewriting sequence 1:

$\triangleright_\zeta CASE(p, \acute{s_1}w(CASE(s_1, \acute{a_1}inl(a_1), \acute{a_2}inr(a_2))),$
$$\acute{s_2}w(CASE(s_2, \acute{b_1}inl(b_1), \acute{b_2}inr(b_2)))$$

$\triangleright_\beta CASE(p, \acute{s_1}w(s_1), \acute{s_2}w(CASE(s_2, \acute{b_1}inl(b_1), \acute{b_2}inr(b_2)))$

$\triangleright_\beta CASE(p, \acute{s_1}w(s_1), \acute{s_2}w(s_2))$

Rewriting sequence 2:

$\triangleright_\beta w(CASE(p, \acute{s_1}s_1, \acute{s_2}CASE(s_2, \acute{b_1}inl(b_1), \acute{b_2}inr(b_2)))$

$\triangleright_\beta w(CASE(p, \acute{s_1}s_1, \acute{s_2}s_2))$

$\triangleright_\zeta CASE(p, \acute{s_1}w(s_1), \acute{s_2}w(s_2))$

- Analyzing rules 11 and 15:

$$w(CASE(p, \acute{s_1}\lambda x.APP(s_1, x), \acute{s_2}\lambda y.APP(s_2, y)))$$

Rewriting sequence 1:

$\triangleright_\zeta CASE(p, \acute{s_1}w(\lambda x.APP(s_1, x)), \acute{s_2}w(\lambda y.APP(s_2, y)))$

$\triangleright_\beta CASE(p, \acute{s_1}w(s_1), \acute{s_2}w(\lambda y.APP(s_2, y)))$

$\triangleright_\beta CASE(p, \acute{s_1}w(s_1), \acute{s_2}w(s_2))$

Rewriting sequence 2:

$\triangleright_\beta w(CASE(p, \acute{s_1}s_1, \acute{s_2}\lambda y.APP(s_2, y)))$

$\triangleright_\beta w(CASE(p, \acute{s_1}s_1, \acute{s_2}s_2))$

$\triangleright_\zeta CASE(p, \acute{s_1}w(s_1), \acute{s_2}w(s_2))$

The superposition of rules 12 and 15 is done in an analogous way, given that rules 11 and 12 are very similar.

- Analyzing rules 13 and 15:

$$w(CASE(p, \acute{s_1}INST(s_1, \acute{gt}\varepsilon y.(g(y), t)), \acute{s_2}INST(s_2, \acute{gt}\varepsilon x.(h(x), t))))$$

Rewriting sequence 1:

$\triangleright_\zeta CASE(p, \acute{s_1}w(INST(s_1, \acute{gt}\varepsilon y.(g(y), t))),$
$$\acute{s_2}w(INST(s_2, \acute{gt}\varepsilon x.(h(x), t))))$$

$\triangleright_\beta CASE(p, \acute{s_1}w(s_1), \acute{s_2}w(INST(s_2, \acute{gt}\varepsilon x.(h(x), t))))$

$\triangleright_\beta CASE(p, \acute{s_1}w(s_1), \acute{s_2}w(s_2))$

Rewriting sequence 2:

$\triangleright_\beta w(CASE(p, \acute{s_1}s_1, \acute{s_2}INST(s_2, \acute{gt}\varepsilon x.(h(x), t))))$

$\triangleright_\beta w(CASE(p, \acute{s_1}s_1, \acute{s_2}s_2))$

$\triangleright_\zeta CASE(p, \acute{s_1}w(s_1), \acute{s_2}w(s_2))$

- Analyzing rules 14 and 15:

$$w(CASE(p, \acute{s_1}REWR(s_1, \acute{t}t(a,b)), \acute{s_2}REWR(s_2, \acute{t}t(a,b))))$$

Rewriting sequence 1:
$\rhd_\zeta CASE(p, \acute{s_1}w(REWR(s_1, \acute{t}t(a,b))), \acute{s_2}w(REWR(s_2, \acute{t}t(a,b))))$
$\rhd_\beta CASE(p, \acute{s_1}w(s_1), \acute{s_2}w(REWR(s_2, \acute{t}t(a,b))))$
$\rhd_\beta CASE(p, \acute{s_1}w(s_1), \acute{s_2}w(s_2))$
Rewriting sequence 2:
$\rhd_\beta w(CASE(p, \acute{s_1}REWR(s_1, \acute{t}t(a,b)), \acute{s_2}REWR(s_2, \acute{t}t(a,b))))$
$\rhd_\beta w(CASE(p, \acute{s_1}s_1, \acute{s_2}s_2))$
$\rhd_\zeta CASE(p, \acute{s_1}w(s_1), \acute{s_2}w(s_2))$

- Analyzing rules 16 and 17:

  - Similarly to the case of rules 15 and 17.

- Analyzing rules 3 and 18:

$$CASE(inl(b_1), \acute{c_1}w(inl(c_1)), \acute{c_2}w(inr(c_2)))$$

Rewriting sequence 1:
$\rhd_\beta w(inl(b_1))$
Rewriting sequence 2:
$\rhd_\iota w(inl(b_1))$
Similarly, rules 4 and 18 do not generate any divergent critical pair.

- Analyzing rules 15 and 18:

$$w_1(CASE(p.\acute{s_1}w_2(inl(s_1)), \acute{s_2}w_2(inr(s_2))))$$

Rewriting sequence 1:
$\rhd_\zeta CASE(p.\acute{s_1}w_1(w_2(inl(s_1))), \acute{s_2}w_1(w_2(inr(s_2))))$
$\rhd_\iota w(z(p))$
Rewriting sequence 2:
$\rhd_\iota w((z(p))$

- Analyzing rules 7 and 19:

$$INST(\varepsilon x.(f(x),a), \acute{g}tw(\varepsilon y.(g(y),t)))$$

Rewriting sequence 1:
$\rhd_\beta w(\varepsilon y.(f(y),a))$
Rewriting sequence 2:
$\rhd_\iota w(\varepsilon x.(f(x),a))$

- Analyzing rules 16 and 19:

$$w_1(INST(c, \acute{g}tw_2(\varepsilon y.(g(y),t))))$$

Rewriting sequence 1:
$$\triangleright_\zeta INST(c, \acute{g}\acute{t}w_1(w_2(\varepsilon y.(g(y), t))))$$
$$\triangleright_\iota w_1(w_2(c))$$
Rewriting sequence 2:
$$\triangleright_\iota w_1(w_2(c))$$

- Analyzing rules 8 and 20:

$$REWR(s(a, b), \acute{t}w(t(a, b)))$$

Rewriting sequence 1:
$$\triangleright_\beta w(s(a, b))$$
Rewriting sequence 2:
$$\triangleright_\iota w(s(a, b))$$

- Analyzing rules 17 and 20:

$$w_1(REWR(c, \acute{t}w_2(t(a, b))))$$

Rewriting sequence 1:
$$\triangleright_\zeta REWR(c, \acute{t}w_1(w_2(t(a, b))))$$
$$\triangleright_\iota w_1(w_2(t(a, b)))$$
Rewriting sequence 2:
$$\triangleright_\iota w_1(w_2(t(a, b)))$$

Chapter 5

Natural Deduction for Equality

5.1 Introduction

The clarification of the notion of normal form for equality reasoning took an important step with the work of Statman in the late 1970s [Statman (1977, 1978)]. The concept of *direct computation* was instrumental in the development of Statman's approach. By way of motivation, let us take a simple example from the λ-calculus.

$$(\lambda x.(\lambda y.yx)(\lambda w.zw))v \vartriangleright_\eta (\lambda x.(\lambda y.yx)z)v \vartriangleright_\beta (\lambda y.yv)z \vartriangleright_\beta zv$$
$$(\lambda x.(\lambda y.yx)(\lambda w.zw))v \vartriangleright_\beta (\lambda x.(\lambda w.zw)x)v \vartriangleright_\eta (\lambda x.zx)v \vartriangleright_\beta zv.$$

There is at least a sequence of conversions from the initial term to the final term. (In this case we have given two!) Thus, in the formal theory of λ-calculus, the term $(\lambda x.(\lambda y.yx)(\lambda w.zw))v$ is declared to be *equal* to zv.

Now, some natural questions arise. Are the sequences themselves *normal*? Are there non-normal sequences? If yes, how are the latter to be identified and (possibly) normalized?

As rightly pointed out by Le Chenadec in [Chenadec (1989)], the notion of normal proof has been somewhat neglected by the systems of equational logic: "In proof-theory, since the original work of Gentzen (1969) on sequent calculus, much work has been devoted to the normalization process of various logics, Prawitz (1965), Girard (1988). Such an analysis was lacking in equational logic (the only exceptions we are aware of are Statman (1977), Kreisel and Tait (1961))." The works of Statman [Statman (1977, 1978)] and Le Chenadec [Chenadec (1989)] represent significant attempts to fill this gap. Statman studies proof transformations for the equational calculus E of Kreisel–Tait [Kreisel and Tait (1961)]. Le Chenadec defines an equational proof system (the *LE system*) and gives a normalization procedure.

The intention here is to show how the framework of labelled natural deduction can help us formulate a proof theory for the "logical connective" of propositional equality.[1] The connective is meant to be used in reasoning about equality between referents (i.e. the objects of the functional calculus), as well as with a general notion of substitution which is needed for the characterization of the so-called *term declaration logics* [Aczel (1991)]. The characterization of propositional equality may be useful for the establishment of a proof theory for 'descriptions'.

In order to account for the distinction between the equalities that are:

definitional, i.e. those equalities that are given as rewrite rules (equations), or else originate from general functional principles (e.g. β, η, etc.),

and those that are:

propositional, i.e. the equalities that are supported (or otherwise) by an evidence (a composition of rewrites),

we need to provide for an equality sign as a symbol for *rewrite* (i.e. as part of the functional calculus on the labels), and an equality sign as a symbol for a *relation* between referents (i.e. as part of the logical calculus on the formulas).

Definitional equalities.   Let us recall from the theory of λ-calculus, that:

Definition 5.1 ([Hindley and Seldin (2008)]). The formal theory of $\lambda\beta\eta$ equality has the following axioms:

$$(\alpha) \;\; \lambda x.M = \lambda y.[y/x]M \qquad (y \notin FV(M))$$
$$(\beta) \;\; (\lambda x.M)N = [N/x]M$$
$$(\eta) \;\; (\lambda x.Mx) = M \qquad\qquad (x \notin FV(M))$$
$$(\xi) \;\; \frac{M = M'}{\lambda x.M = \lambda x.M'}$$

$$(\mu) \;\; \frac{M = M'}{NM = NM'}$$

$$(\nu) \;\; \frac{M = M'}{MN = M'N}$$

---

[1] An old question is in order here: what is a logical connective? We shall take it that from the point of view of proof theory (natural deduction style) a logical connective is whatever logical symbol which is analysable into rules of *introduction* and *elimination*.

$$(\rho)\ M = M$$

$$(\sigma)\ \frac{M = N}{N = M}$$

$$(\tau)\ \frac{M = N \qquad N = P}{M = P}.$$

Propositional equality. Again, let us recall from the theory of λ-calculus, that:

Definition 5.2 ([Hindley and Seldin (2008)]). P is β-equal or β-convertible to Q (notation $P =_\beta Q$) iff Q is obtained from P by a finite (perhaps empty) series of β-contractions and reversed β-contractions and changes of bound variables. That is, $P =_\beta Q$ iff there exist $P_0, \ldots, P_n$ ($n \geq 0$) such that

$$P_0 \equiv P,\ P_n \equiv Q,$$
$$(\forall i \geq n - 1)(P_i \triangleright_{1\beta} P_{i+1}\ or\ P_{i+1} \triangleright_{1\beta} P_i\ or\ P_i \equiv_\alpha P_{i+1}).$$

NB: equality with an *existential* force.

Remark 5.1. In setting up a set of Gentzen's ND-style rules for equality we need to account for:
1. definitional versus propositional equality;
2. there may be more than one *normal* proof of a certain equality statement;
3. given a (possibly non-normal) proof, the process of bringing it to a normal form should be finite and confluent.

The missing entity. Within the framework of the functional interpretation (*à la* Curry–Howard [Howard (1980)]), the definitional equality is often considered by reference to a judgement of the form:

$$a = b : D$$

which says that a and b are equal elements from domain D. Notice that the 'reason' why they are equal does not play any part in the judgement. This aspect of 'forgetting contextual information' is, one might say, the first step towards 'extensionality' of equality, for whenever one wants to introduce intensionality into a logical system one invariably needs to introduce information of a 'contextual' nature, such as, where the identification of two terms (i.e. equation) comes from.

We feel that a first step towards finding an alternative formulation of the proof theory for propositional equality which takes care of the intensional aspect is to allow the 'reason' for the equality to play a more significant part in the form of judgement. We also believe that from the point of view of the logical calculus, if there is a 'reason' for two expressions to be considered equal, the proposition asserting their equality will be true, regardless of what particular composition of rewrites (definitional equalities) amounts to the evidence in support of the proposition concerned. Given these general guidelines, we shall provide what may be seen as a middle ground solution between the intensional [Martin-Löf (1975a,b)] and the extensional [Martin-Löf (1982)] accounts of Martin-Löf's propositional equality. The intensionality is taken care by the functional calculus on the labels, while the extensionality is catered by the logical calculus on the formulas. In order to account for the intensionality in the labels, we shall make the composition of rewrites (definitional equalities) appear as indexes of the equality sign in the judgement with a variable denoting a sequence of equality identifiers (we have seen that in the Curry–Howard functional interpretation there are at least four 'natural' equality identifiers: β, η, ξ and μ). So, instead of the form above, we shall have the following pattern for the equality judgement:

$$a =_s b : D,$$

where 's' is meant to be a sequence of equality identifiers.

In the sequel we shall be discussing in some detail the need to identify the kind of definitional equality, as well as the need to have a logical connective of 'propositional equality' in order to be able to reason about the functional objects (those to the left hand side of the ':' sign).

Term rewriting. Deductive systems based on the Curry–Howard isomorphism [Howard (1980)] have an interesting feature: normalization and strong normalization (Church–Rosser property) theorems can be proved by reductions on the terms of the functional calculus. Exploring this important characteristic, we have proved these theorems for the *labelled natural deduction* (*LND*) via a term rewriting system constructed from the *LND*-terms of the functional calculus [de Oliveira and de Queiroz (1999)]. Applying this same technique to the *LND* equational fragment, we obtain the normalization theorems for the equational logic of the *LND* system [de Oliveira (1995); de Oliveira and de Queiroz (1999)].

This technique is used given the possibility of defining two measures of redundancy for the *LND* system that can be dealt with in the object level: the terms on the functional calculus and the *rewrite reason* (composition of rewrites), the latter being indexes of the equations in the *LND* equational fragment.

In the *LND* equational logic [de Oliveira and de Queiroz (1999)], the equations have the following pattern:

$$a =_s b : D,$$

where one is to read: a is equal to b *because* of 's' ('s' being the *rewrite reason*); 's' is a term denoting a sequence of equality identifiers (β, η, α, etc.), i.e. a composition of rewrites. In other words, 's' is the *computational path* from a to b.

In this way, the *rewrite reason* (reason, for short) represents an *orthogonal measure of redundancy* for the *LND*, which makes the *LND* equational fragment an "enriched" system of equational logic. Unlike the traditional equational logic systems, in *LND* equational fragment there is a gain in local control by the use of *reason*. All the proof steps are recorded in the composition of rewrites (reasons). Thus, consulting the reasons, one should be able to see whether the proof has the normal form. We have then used this powerful mechanism of controlling proofs to present a precise normalization procedure for the *LND* equational fragment. Since the reasons can be dealt with in the object level, we can employ a computational method to prove the normalization theorems: we built a term rewriting system based on an algebraic calculus on the "*rewrite reasons*", which compute normal proofs. With this we believe we are making a step towards filling a gap in the literature on equational logic and on proof theory (natural deduction).

Kreisel–Tait's system. In [Kreisel and Tait (1961)] Kreisel and Tait define the system E for equality reasoning as consisting of axioms of the form $t = t$, and the following rules of inference:

$$(E1) \qquad \frac{E[t/x] \quad t = u}{E[u/x]}$$

$$(E2) \qquad \frac{s(t) = s(u)}{t = u}$$

$$(E3) \qquad \frac{0 = s(t)}{A} \quad \text{for any formula } A$$

$$(E4_n) \qquad \frac{t = s^n(t)}{A} \quad \text{for any formula } A,$$

where t and u are terms, '0' is the first natural number (zero), '$s(-)$' is the successor function.

Statman's normal form theorem. In order to prove the normalization results for the calculus E Statman defines two subsets of E: (i) a natural deduction based calculus for equality reasoning NE; (ii) a sequent style calculus SE.

The NE calculus is defined as having axioms of the form $a = a$, and the rule of substituting equals for equals:

$$(=) \quad \frac{E[a/u] \quad a \approx b}{E[b/u]},$$

where E is any set of equations, and $a \approx b$ is ambiguously $a = b$ and $b = a$.

Statman arrives at various important results on normal forms and bounds for proof search in NE. In this case, however, a rather different notion of normal form is being used: the 'cuts' do not arise out of an 'inversion principle', as it is the case for the logical connectives, but rather from a certain form of sequence of equations which Statman calls 'computation'. With the formulation of a proof theory for the 'logical connective' of propositional equality we wish to analyse equality reasoning into its basic components: rewrites, on the one hand, and statements about the existence of rewrites, on the other hand. This type of analysis came to the surface in the context of constructive type theory and the Curry–Howard functional interpretation.

Martin-Löf's equality type. There has been essentially two approaches to the problem of characterizing a proof theory for propositional equality, both of which originate in P. Martin-Löf's work on *intuitionistic type theory*: the intensional [Martin-Löf (1975b)] and the extensional [Martin-Löf (1982)] formulations.

The extensional version. In his [Martin-Löf (1982)] and [Martin-Löf (1984)] presentations of *intuitionistic type theory*, P. Martin-Löf defines the type of *extensional* propositional equality 'I' (here called 'I_{ext}') as:

I_{ext}-*formation*

$$\frac{D \; type \qquad a : D \qquad b : D}{I_{ext}(D, a, b) \; type}$$

I_{ext}-*introduction*

$$\frac{a = b : D}{r : I_{ext}(D, a, b)}$$

I_{ext}-*elimination*[2]

$$\frac{c : I_{ext}(D, a, b)}{a = b : D}$$

---

[2]The set of rules given in [Martin-Löf (1982)] contained the additional *elimination* rule:

$$\frac{c : I(D, a, b) \qquad d : C(r/z)}{J(c, d) : C(c/z)},$$

which may be seen as reminiscent of the previous *intensional* account of propositional equality [Martin-Löf (1975b)].

I_{ext}-*equality*

$$\frac{c : I_{ext}(D, a, b)}{c = r : I_{ext}(D, a, b)}.$$

Note that the above account of propositional equality does not 'keep track of all proof steps': both in the I_{ext}-*introduction* and in the I_{ext}-*elimination* rules there is a considerable loss of information concerning the deduction steps. While in the I_{ext}-*introduction* rule the 'a' and the 'b' do not appear in the 'trace' (the label/term alongside the logical formula), the latter containing only the canonical element 'r', in the rule of I_{ext}-*elimination* all the trace that might be recorded in the label 'c' simply disappears from label of the conclusion. If by 'intensionality' we understand a feature of a logical system which identifies as paramount the concern with issues of *context* and *provability*, then it is quite clear that any logical system containing I_{ext}-type can hardly be said to be 'intensional': as we have said above, neither its *introduction* rule nor its *elimination* rule carry the necessary *contextual* information from the premise to the conclusion.

And, indeed, the well-known statement of the extensionality of functions can be proved as a theorem of a logical system containing the I_{ext}-type such as Martin-Löf's *intuitionistic type theory* [Martin-Löf (1984)]. The statement says that if two functions return the same value in their common codomain when applied to each argument of their common domain (i.e. if they are equal pointwise), then they are said to be (extensionally) equal. Now, we can construct a derivation of the statement written in the formal language as:

$$\forall f, g^{A \to B}.(\forall x^A.I_{ext}(B, APP(f, x), APP(g, x)) \to I_{ext}(A \to B, f, g))$$

by using the rules of proof given for the I_{ext}, assuming we have the rules of proof given for the implication and the universal quantifier.

The intensional version.   Another version of the propositional equality, which has its origins in Martin-Löf's early accounts of *intuitionistic type theory* [Martin-Löf (1975a,b)], and is apparently in the most recent, as yet unpublished, versions of type theory, is defined in [Troelstra and van Dalen (1988)] and [Nordström *et al.* (1990)]. In a section dedicated to the *intensional vs. extensional* debate, [Troelstra and van Dalen (1988)] (p.633) says that:

> "Martin-Löf has returned to an intensional point of view, as in Martin-Löf (1975), that is to say, $t = t' \in A$ is understood as 't and t' are definitionally equal'. As a consequence the rules for identity types have to be adapted."

If we try to combine the existing accounts of the *intensional* equality type 'I' [Martin-Löf (1975b); Troelstra and van Dalen (1988); Nordström *et al.* (1990)], here denoted 'I_{int}', the rules will look like:

I_{int}-*formation*

$$\frac{D\ type \qquad a:D \qquad b:D}{I_{int}(D,a,b)\ type}$$

I_{int}-*introduction*

$$\frac{a:D}{e(a):I_{int}(D,a,a)} \qquad \frac{a=b:D}{e(a):I_{int}(D,a,b)}$$

I_{int}-*elimination*

$$\frac{a:D \quad b:D \quad c:I_{int}(D,a,b) \quad \overset{[x:D]}{d(x):C(x,x,e(x))} \quad \overset{[x:D,y:D,z:I_{int}(D,x,y)]}{C(x,y,z)\ type}}{J(c,d):C(a,b,c)}$$

I_{int}-*equality*

$$\frac{a:D \quad \overset{[x:D]}{d(x):C(x,x,e(x))} \quad \overset{[x:D,y:D,z:I_{int}(D,x,y)]}{C(x,y,z)\ type}}{J(e(a),d(x))=d(a/x):C(a,a,e(a))}.$$

With slight differences in notation, the 'adapted' rules for identity type given in [Troelstra and van Dalen (1988)] and [Nordström *et al.* (1990)] resemble the one given in [Martin-Löf (1975b)]. It is called *intensional* equality because there remains no direct connection between judgements like '$a = b : D$' and '$s : I_{int}(D,a,b)$'.

A labelled proof theory for propositional equality. Now, it seems that an alternative formulation of propositional equality within the functional interpretation, which will be a little more elaborate than the extensional I_{ext}-type, and simpler than the intensional I_{int}-type, could prove more convenient from the point of view of the 'logical interpretation'. It seems that whereas in the former we have a considerable loss of information in the I_{ext}-*elimination*, in the latter we have an I_{int}-*elimination* too heavily loaded with (perhaps unnecessary) information. If, on the one hand, there is an *over*explicitation of information in I_{int}, on the other hand, in I_{ext} we have a case of *under*explicitation. With the formulation of a proof theory for equality via labelled natural deduction we wish to find a middle ground solution between those two extremes.

5.2   Labelled deduction

The functional interpretation of logical connectives via deductive systems which use some sort of labelling mechanism [Martin-Löf (1984); Gabbay (1994)] can be seen as the basis for a general framework characterizing logics via a clear separation between a functional calculus on the *labels*, i.e. the referents (names of individuals, expressions denoting the record of proof steps used to arrive at a certain formula, names of 'worlds', etc.) and a logical calculus on the formulas. The key idea is to make these two dimensions as harmonious as possible, i.e. that the functional calculus on the labels matches the logical calculus on the formulas at least in the sense that to every abstraction on the variables of the functional calculus there corresponds a discharge of an assumption-formula of the logical calculus. One aspect of such interpretation which stirred much discussion in the literature of the past ten years or so, especially in connection with *intuitionistic type theory* [Martin-Löf (1984)] was that of whether the logical connective of propositional equality ought to be dealt with 'extensionally' or 'intensionally'.[3]

Here we attempt to formulate what appears to be a middle ground solution, in the sense that the intensional aspect is dealt with in the functional calculus on the labels, whereas the extensionality is kept to the logical calculus. We also intend to demonstrate that the connective of propositional equality (cf. Aczel's [Aczel (1980)] '$\doteq$') needs to be dealt with in a similar manner to 'Skolem-type' connectives (such as disjunction and existential quantification), where notions like *hiding*, *choice* and *dependent variables* play crucial rôles.

Our motivation: Where did it all start? The characterization of a proof theory for *labelled deductive systems* (LDS) has been the concern of some authors for some time now [Gabbay and de Queiroz (1991); de Queiroz and Gabbay (1995, 1997)]. Here we address two topics of special interest to logic and computation, namely, *substitution* and *unification*. As a starting point, we posed ourselves two interrelated questions: how could we incorporate the handling of rewrites and function symbols into the proof theory, and how could we 'give logical content', so to speak, to the procedures coming from unification algorithms?

For those not familiar with the *LDS* perspective, it suffices at this stage to say that the declarative unit of logical systems is seen as made up of two components: a formula and a label. The label is meant to carry information which may be of a less declarative nature than that carried by the formulas. The introduction of such an 'extra' dimension was motivated by the need to cope with the demands of computer science applications.

---

[3] A more recent approach to studying provability which comes close to the notion of labelled deduction is Sergei Artemov's "Logic of Proofs" [Artemov (1998)]. In fact, according to Artemov, the idea of replacing "implicit" provability (as in modal logic) by "explicit" provability (as in the bi-dimensional formal unit "$t : A$" read as "t is a proof of A") was put forward in Gödel's 1938 *Lecture at Zilsel's* [Gödel (1995)] which was published in S. Feferman *et al.* 1995 *Kurt Gödel Collected Works, Vol. III*, 1995.

Indeed, with the diversification of computer science applications to problems involving *reasoning*, there has been a proliferation of logics originated mainly from the need to tailor the logical system to the demands of the particular application area. If there were a number of 'logics' already developed and well established in the mathematical and philosophical logic literature (relevant, intuitionistic, minimal, etc.), the diversification was significantly increased with the contribution from computer science.

Gabbay observed that many of the distinctive features of most logics being studied by logicians and computer scientists alike, stemmed from 'meta-level' considerations: in order to consider a step to be a valid one, it was invariably the case that one had to take into account questions like: 'whether the assumptions have actually been used'; 'whether they have been used in a certain order'; 'whether the number of times an assumption was used has been in keeping with the need to take care of resources'; etc.

There are a number of inconveniences in having to cope with increasingly diverse logical systems, and Gabbay set out a research programme with at least the following desiderata:

- to find a unifying framework (sequent calculus by itself would not do, and we shall see why later on) *factoring out* meta- from object-level features;
- to keep the logic (and logical steps, for that matter) simple, handling meta-level features via a *separate*, yet *harmonious* calculus;
- to have means of structuring and combining logics;
- to make sure the relevant assumptions in a deduction are uncovered, paying more attention to the explicitation and use of resources.

The idea of *labelled deduction* seemed to be a natural evolution from the traditional logical systems. The development of a novel approach to logic, namely, *LDS*, where the meta-level features would be incorporated into the deductive calculus in an orderly manner, looked general enough to be an appropriate candidate for such an unifying framework.

In summary, it seems fair to say that *LDS* offer a new perspective on the discipline of *logic and computation*. Arising from computer science applications, it provides the essential ingredients for a framework whereby one can study

- meta-level features of logical systems, by 'knocking down' some of the elements of the meta-level reasoning to the object-level, and allowing each logical step to take care of *what has been done so far*;
- the 'logic' of Skolem functions and substitution (dependencies, term declaration).

Why sequent calculus by itself will not do. Boole did manage to formalize the algebra of logical connectives, with the aspect of *duality* coming out very neatly. The sequent calculus follows on this quest for duality:

$$\text{negative} \quad \vdash \text{positive}$$
$$\text{conjunctive} \vdash \text{disjunctive}$$

Nevertheless, since Frege, logic is *also* about quantifiers, predicates, functions, equality among referents, etc. In a few words, beyond *duality*, first-order logic also deals with *quantification* in a *direct* fashion, instead of via, say, Venn diagrams. Thus, a proof theory for first-order logic ought to account for the manipulation of function symbols, terms, dependencies and substitutions, as Herbrand already perceived.

We shall see a little more about this later on when we come to a brief discussion of the so-called 'sharpened Hauptsatz'. But it seems appropriate to add here that we are looking for strengthening the connections between Gentzen's and Herbrand's methods in proof theory. We believe that the two-dimensional approach of *LDS* is the right framework for the enterprise. This is because, if, on the one hand:

(+) Gentzen's methods come with a well defined mathematical theory of proofs,

and

(+) Herbrand's method show how to handle function symbols and terms in a direct fashion,

on the other hand:

(−) in Gentzen's calculi (plain natural deduction, sequent calculus) function symbols are not *citizens*,

and

(−) Herbrand's methods hardly give us means of looking at proofs (deductions) as the main objects of study.

By combining a functional calculus on the labels (which carry along referents, function symbols) with a logical calculus on the formulas, the *LDS* perspective can have the (+)'s without the (−)'s.

The generality of Herbrand base. Let us take the example which Leisenring uses to demonstrate the application of Herbrand's decision procedure to check the validity of the formula [Leisenring (1969)]:

$$\exists x.\forall y.(P(x) \rightarrow P(y)).$$

Herbrand's 'original' procedure.   The first step is to find the Herbrand resolution ($\exists$-prenex normal form), which can be done by introducing a new function symbol g, and obtaining:

$$\exists x.(P(x) \rightarrow P(g(x))).$$

As this would be equivalent to a disjunction of substitution instances like:

$$P(a) \rightarrow P(g(a)) \quad \vee \quad P(a') \rightarrow P(g(a')) \quad \vee \quad P(a'') \rightarrow P(g(a'')) \quad \vee \quad \cdots$$

the second step is to find a p-substitution instance (p finite) which is a tautology. For that, we take the 'Herbrand base' to be $\{a, g\}$, where a is an arbitrary individual from the domain, and g is an arbitrary function symbol which can construct, out of a, further elements of the domain. Thus, the 1-substitution instance is:

$$P(a) \rightarrow P(g(a)),$$

which is clearly not a tautology. Now, we can iterate the process, and find the 2-reduction as a disjunction of the 1-reduction and the formula made up with a 2-substitution (taking $a' = g(a)$), that is:

$$P(a) \rightarrow P(g(a)) \quad \vee \quad P(g(a)) \rightarrow P(g(g(a))),$$

which is a tautology.

In summary:

(1) $\exists x.\forall y.(P(x) \rightarrow P(y))$
(2) take g as a unary function
(3) $\exists x.(P(x) \rightarrow P(g(x)))$
(4) $P(a) \rightarrow P(g(a)) \quad \vee \quad P(a') \rightarrow P(g(a')) \quad \vee \quad P(a'') \rightarrow P(g(a'')) \quad \vee$
$\cdots$
(5) 1st substitution: $P(a) \rightarrow P(g(a))$
(6) take $a' = g(a)$
(7) 2nd substitution: $P(a) \rightarrow P(g(a)) \quad \vee \quad P(g(a)) \rightarrow P(g(g(a)))$ (tautology).

In checking the validity of $\exists x.\forall y.(P(x) \rightarrow P(y))$ we needed the following extra assumptions:

(1) the domain is non-empty (step 4).

(2) there is a way of identifying an arbitrary term with another one (step 6).

As we shall see below, the labelled deduction method will have helped us 'to bring up to the surface' those two (hidden) assumptions.

Now, how can we justify the *generality* of the 'base' $\{a, g\}$? Why is it that it does not matter which a and g we choose, the procedure always works? In other words, why is it that *for any* element a of the domain and *for any* 'function symbol' g, the procedure always works?

In a previous opportunity [de Queiroz and Gabbay (1995)] we have already demonstrated the *universal force* that is given to Skolem functions by the device of *abstraction* in the elimination of the existential quantifier. The point was that although there was no quantification over function symbols being made in the logic (the logical calculus on the formulas, that is), an abstraction in the name for the Skolem function was performed in the functional calculus on the labels. The observation suggested that, as in the statement of Skolem's theorem, *for any* (new) function symbol f we choose when Skolemizing $\forall x.\exists y.P(x, y)$ to $\forall x.P(x, f(x))$, if an arbitrary statement can be deduced from the latter then it can also be deduced from the former, regardless of the choice of f. We shall come back to this point later on when we will then demonstrate that in labelled natural deduction the Herbrand function gets abstracted away thus getting universal force.

Gentzen–Herbrand connections: the sharpened Hauptsatz. The connections between the proof theory as developed by Gentzen and the work on the proof theory of first order logic by Herbrand are usually seen through the so-called 'sharpened Hauptsatz'.

Theorem (Gentzen's sharpened Hauptsatz)
Given $\Gamma \vdash \Delta$ *(prenex formulae), if* $\Gamma \vdash \Delta$ *is provable then there is a cut-free, pure-variable proof which contains a sequent* $\Gamma' \vdash \Delta'$ *(the* midsequent*) with:*

(1) *Every formula in* $\Gamma' \vdash \Delta'$ *is quantifier free.*
(2) *No quantifier rule above* $\Gamma' \vdash \Delta'$.
(3) *Every rule below* $\Gamma' \vdash \Delta'$ *is either a quantifier rule, or a contraction or exchange structural rule (not a weakening).*

The theorem relies on the so-called *permutability* lemma, which states that quantifier rules can always be 'pushed down' in the deduction tree.

Lemma (Permutability [Kleene (1967)])
Let π *be a cut-free proof of* $\Gamma \vdash \Delta$ *(with only prenex formulas); then it is possible to construct another proof* π' *where:*

> *every quantifier rule can be permuted with a logical or structural rule applied below it (with some provisos).*

(The interested reader will find detailed expositions of the sharpened Hauptsatz in [Kleene (1967)] and [Girard (1987b)].)

Example.   For the sake of illustration, let us construct a sequent-calculus deduction of the formula used in our original example, i.e. let us build proof of the sequent

$$\vdash \exists x.\forall y.(P(x) \to P(y)),$$

and see what the *midsequent* means in this case:

$$
\frac{
\frac{
\frac{
\frac{
\frac{
\frac{
\frac{
\frac{
\frac{
\boxed{P(b), P(a) \vdash P(b), P(c)}^{\,-}
}{P(b) \vdash P(a) \to P(b), P(c)}
}{P(b) \vdash P(c), P(a) \to P(b)}
}{\boxed{\vdash P(b) \to P(c), P(a) \to P(b)}}
}{\vdash \forall y.(P(b) \to P(y)), P(a) \to P(b)}
}{\vdash \exists x.\forall y.(P(x) \to P(y)), P(a) \to P(b)}
}{\vdash P(a) \to P(b), \exists x.\forall y.(P(x) \to P(y))}
}{\vdash \forall y.(P(a) \to P(y)), \exists x.\forall y.(P(x) \to P(y))}
}{\vdash \exists x.\forall y.(P(x) \to P(y)), \exists x.\forall y.(P(x) \to P(y))}
}{\vdash \exists x.\forall y.(P(x) \to P(y))}.
$$

Note that every rule below the boxed sequent is either a quantifier rule, or a contraction or exchange rule (no weakenings). Due to the *eigenvariable* restrictions to the quantifiers rules ($\exists$ on the left, and $\forall$ on the right), we had to use contraction and choose the same b in different instantiations. (As we will see later on, when we use labelled natural deduction the 'assumption' that the firstly used b is the same as the other one is introduced as a logical formula using propositional equality.)

With such an example we intend to draw the attention to the fact that although the sharpened Hauptsatz brings Gentzen's methods closer to Herbrand's methods, it shows how less informative the former is with respect to the latter. The midsequent does not mention any function symbol, nor does it point to the inductive nature of the generation of the so-called Herbrand universe: the fact that the proof obtains is related to the 'meta-level' choice of the same instantiation constant b.

5.2.1   *Identifiers for (compositions of) equalities*

In the functional interpretation, where a functional calculus on the labels go hand in hand with a logical calculus on the formulas, we have a classification of equalities, whose identifications are carried along as part of the deduction: either β-, η-, ξ-, μ- or α- equality will have been part of an expression labelling a formula containing '$\doteq$'. There one finds the key to the idea of 'hiding' in the *introduction* rule, and opening local (Skolem-type) assumptions in the *elimination* rule. (Recall that in the case of disjunction we also have alternatives: either into the left disjunct, or into the right disjunct.) So, we believe that it is not unreasonable to start off the formalization of propositional equality with the parallel to the disjunction and existential cases in mind. Only, the witness of the type of propositional equality are not the 'a's and 'b's of '$a = b : D$', but the actual (sequence of) equalities (β-, η-, ξ-, α-) that might have been used to arrive at the judgement '$a =_s b : D$' (meaning '$a = b$' *because* of 's'), 's' being a sequence made up of β-, η-, ξ- and/or α-equalities, perhaps with some of the general equality rules of reflexivity, symmetry and transitivity. So, in the *introduction* rule of the type we need to form the canonical proof as if we were *hiding* the actual sequence. Also, in the rule of *elimination* we need to open a new local assumption introducing a new variable denoting a possible sequence as a (Skolem-type) new constant. That is, in order to eliminate the connective '$\doteq$' (i.e. to deduce something from a proposition like '$\doteq_D (a,b)$'), we start by choosing a new variable to denote the reason why the two terms are equal: 'let t be an expression (sequence of equalities) justifying the equality between the terms'. If we then arrive at an arbitrary formula 'C' labelled with an expression where the t still occurs free, then we can conclude that the same C can be obtained from the $\doteq$-formula regardless of the identity of the chosen t, meaning that the label alongside C in the conclusion will have been abstracted from the free occurrences of t.

Observe that now we are still able to 'keep track' of all proof steps (which does not happen with Martin-Löf's I_{ext}-type) [Martin-Löf (1982, 1984)], and we have an easier formulation (as compared with Martin-Löf's I_{int}-type) [Martin-Löf (1975a)] of how to perform the *elimination* step.

5.2.2 *The proof rules*

In formulating the propositional equality connective, which we shall identify by '$\doteq$', we shall keep the pattern of inference rules essentially the same as the one used for the other logical connectives (as in, e.g. [de Queiroz and Gabbay (1995)]), and we shall provide an alternative presentation of propositional equality as follows:

$\doteq$-*introduction*

$$\frac{a =_s b : D}{s(a,b) : \doteq_D (a,b)}$$

$\doteq$-*reduction*

$$\frac{\dfrac{a =_s b : D}{s(a,b) :\doteq_D (a,b)} \qquad \dfrac{[a =_t b : D]}{d(t) : C} \; \doteq\text{-}intr}{REW\,R(s(a,b), \acute{t}d(t)) : C} \; \doteq\text{-}elim \qquad \rhd_\beta \qquad \dfrac{[a =_s b : D]}{d(s/t) : C}$$

$\doteq$-*induction*

$$\frac{e :\doteq_D (a,b) \qquad \dfrac{\dfrac{[a =_t b : D]}{t(a,b) :\doteq_D (a,b)} \; \doteq\text{-}intr}{} \; \doteq\text{-}elim}{REW\,R(e, \acute{t}t(a,b)) :\doteq_D (a,b)} \qquad \rhd_\eta \qquad e :\doteq_D (a,b),$$

where '$\acute{}$' is an abstractor which binds the occurrences of the (new) variable 't' introduced with the local assumption '$[a =_t b : D]$' as a kind of 'Skolem'-type constant denoting the (presumed) 'reason' why 'a' was assumed to be equal to 'b'. (Recall the Skolem-type procedures of introducing new local assumptions in order to allow for the elimination of logical connectives where the notion of 'hiding' is crucial, e.g. disjunction and existential quantifier – in [de Queiroz and Gabbay (1995)].)

Now, having been defined as a 'Skolem'-type connective, '$\doteq$' needs to have a conversion stating the non-interference of the newly opened branch (the local assumption in the $\doteq$-*elimination* rule) with the main branch. Thus, we have:

$\doteq$-*(permutative) reduction*

$$\frac{\dfrac{e :\doteq_D (a,b) \qquad \dfrac{[a =_t b : D]}{d(t) : C}}{REW\,R(e, \acute{t}d(t)) : C}}{w(REW\,R(e, \acute{t}d(t))) : W} \; r \qquad \rhd_\zeta \qquad \frac{e :\doteq_D (a,b) \qquad \dfrac{\dfrac{[a =_t b : D]}{d(t) : C}}{w(d(t)) : W} \; r}{REW\,R(e, \acute{t}w(d(t))) : W},$$

provided w does not disturb the existing dependencies in the term e (the main branch), i.e. provided that rule 'r' does not discharge any assumption on which '$\doteq_D (a,b)$' depends. The corresponding ζ-equality is:

$$w(REW\,R(e, \acute{t}d(t))) =_\zeta REW\,R(e, \acute{t}w(d(t)))$$

The equality indicates that the operation w can be pushed inside the '$\acute{}$'-abstraction term, provided that it does not affect the dependencies of the term e.

Since we are defining the logical connective '$\doteq$' as a connective which deals with singular terms, where the 'witness' is supposed to be hidden, we shall not be using direct *elimination* like Martin-Löf's I_{ext}-*elimination*. Instead, we shall be using the following $\doteq$-*elimination*:

$$\frac{e :\doteq_D (a,b) \qquad \dfrac{[a =_t b : D]}{d(t) : C}}{REW\,R(e, \acute{t}d(t)) : C}.$$

The *elimination* rule involves the introduction of a new local assumption (and corresponding variable in the functional calculus), namely '$[a =_t b : D]$' (where 't' is the new variable) which is only discharged (and 't' bound) in the conclusion of the rule. The intuitive explanation would be given in the following lines. In order to eliminate the equality $\doteq$-connective, where one does not have access to the 'reason' (i.e. a sequence of 'β', 'η', 'ξ' or 'ζ' equalities) why the equality holds because '$\doteq$' is supposed to be a connective dealing with singular terms (as are '$\vee$' and '$\exists$'), in the first step one has to open a new local assumption supposing the equality holds because of, say 't' (a new variable). The new assumption then stands for 'let t be the unknown equality'. If a third (arbitrary) statement can be obtained from this new local assumption via an unspecified number of steps which does not involve any binding of the new variable 't', then one discharges the newly introduced local assumption binding the free occurrences of the new variable in the label alongside the statement obtained, and concludes that statement is to be labelled by the term '$REWR(e, \acute{t}d(t))$' where the new variable (i.e. t) is bound by the '´'-abstractor.

Another feature of the $\doteq$-connective which is worth noticing at this stage is the equality under 'ξ' of all its elements (see second *introduction* rule). This does not mean that the labels serving as evidences for the $\doteq$-statement are all identical to a constant (cf. constant 'r' in Martin-Löf's I_{ext}-type), but simply that if two (sequences of) equality are obtained as witnesses of the equality between, say 'a' and 'b' of domain D, then they are taken to be equal under ξ-equality. It would not seem unreasonable to think of the $\doteq$-connective of propositional equality as expressing the proposition which, whenever true, indicates that the two elements of the domain concerned are equal under some (unspecified, *hidden*) composition of definitional equalities. It is as if the proposition points to the existence of a term (witness) which depends on both elements and on the kind of equality judgements used to arrive at its proof. So, in the logical side, one forgets about what was the actual witness. Cf. the existential generalization:

$$\frac{F(t)}{\exists x.F(x)},$$

where the actual witness is in fact 'abandoned'. Obviously, as we are interested in keeping track of relevant information introduced by each proof step, in labelled natural deduction system the witness is not abandoned, but is carried over as an unbounded name in the label of the corresponding conclusion formula.

$$\frac{t : D \qquad f(t) : F(t)}{\varepsilon x(f(x), t) : \exists x^D.F(x)}.$$

Note, however, that it is carried along *only* in the functional side, the logical side not keeping any trace of it at all.

Now, notice that if the functional calculus on the labels is to match the logical calculus on the formulas, than we must have the resulting label on the left of the '$\triangleright_\beta$' as β-convertible to the concluding label on the right. So, we must have the convertibility equality:

$$REWR(s(a,b), \acute{t}d(t)) =_\beta d(s/t) : C.$$

The same holds for the η-equality:

$$REWR(e, \acute{t}t(a,b)) =_\eta e \mathbin{\dot{:}=_D} (a,b).$$

Parallel to the case of disjunction, where two different constructors distinguish the two alternatives, namely 'inl' and 'inr', we here have any (sequence of) equality identifiers ('β', 'η', 'μ', 'ξ', etc.) as constructors of proofs for the $\dot{=}$-connective. They are meant to denote the alternatives available.

General rules of equality. Apart from the already mentioned 'constants' (identifiers) which compose the reasons for equality (i.e. the indexes to the equality on the functional calculus), it is reasonable to expect that the following rules are taken for granted: *reflexivity*, *symmetry* and *transitivity*.

Substitution without involving quantifiers. We know from logic programming, i.e. from the theory of unification, that substitution can take place even when no quantifier is involved. This is justified when, for some reason a certain referent can replace another under some condition for identifying the one with the other.

Now, what would be counterpart to such a 'quantifier-less' notion of substitution in a labelled natural deduction system. Without the appropriate means of handling equality (definitional and propositional) we would hardly be capable of finding such a counterpart. Having said all that, let us think of what we ought to do at a certain stage in a proof (deduction) where the following two premises would be at hand:

$$a =_g y : D \qquad \text{and} \qquad f(a) : P(a).$$

We have that a and y are equal ('identifiable') under some arbitrary sequence of equalities (rewrites) which we name g. We also have that the predicate formula $P(a)$ is labelled by a certain functional expression f which depends on a. Clearly, if a and y are 'identifiable', we would like to infer that P, being true of a, will also be true of y. So, we shall be happy in inferring (on the logical calculus) the formula $P(y)$. Now, given that we ought to compose the label of the conclusion out of a composition of the labels of the premises, what label should we insert along-side $P(y)$? Perhaps various good answers could be given here, but we shall choose one which is in line with our 'keeping record of what (relevant) data was used in a deduction'. We have already stated how much importance we attach to names of individuals, names of formula instances, and of course, what kind of deduction was performed (i.e. what kind of connective was introduced or eliminated). In this section we have also insisted on the importance of, not only 'classifying' the equalities, but also having variables for the kinds of equalities that may be used in a deduction. Let us then formulate our rule of 'quantifier-less' substitution as:

$$\frac{a =_g y : D \qquad f(a) : P(a)}{g(a, y) \cdot f(a) : P(y)},$$

which could be explained in words as follows: if a and y are 'identifiable' due to a certain g, and $f(a)$ is the evidence for $P(a)$, then let the composition of $g(a, y)$ (the label for the propositional equality between a and y) with $f(a)$ (the evidence for $P(a)$) be the evidence for $P(y)$.

By having this extra rule of substitution added to the system of rules of infer-ence, we are able to validate one half of the so-called 'Leibniz's law', namely,

$$\forall x^D.\forall y^D.(\doteq_D (x, y) \to (P(x) \to P(y))).$$

The *LND* equational fragment. As we already mentioned, in the *LND* equa-tional logic, the equations have an index (the *reason*) which keeps all proof steps. The *reasons* is defined by the kind of rule used in the proof and the equational axioms (*definitional equalities*) of the system. The rules are divided into the fol-lowing classes: (i) general rules; (ii) subterm substitution rule; (iii) ξ- and μ-rules.

Since the *LND* system is based on the Curry–Howard isomorphism [Howard (1980)], terms represent proof constructions, thus proof transformations corre-spond to equalities between terms. In this way, the *LND* equational logic can deal with equalities between *LND* proofs. The proofs in the *LND* equational fragment which deals with equalities between deductions are built from the basic proof transformations for the *LND* system. These basic proof transformations form an equational system, composed by *definitional equalities* (β, η and ζ).

General rules.

Definition 5.3 (equation). An equation in $\mathrm{LND_{EQ}}$ is of the form:

$$s =_r t : D,$$

where s and t are terms, r is the identifier for the rewrite reason, and D is the type (*formula*).

Definition 5.4 (system of equations). A system of equations S is a set of equations:

$$\{s_1 =_{r_1} t_1 : D_1, \ldots, s_n =_{r_n} t_n : D_n\},$$

where r_i is the *rewrite reason* identifier for the ith equation in S.

Definition 5.5 (rewrite reason). Given a system of equations S and an equation $s =_r t : D$, if $S \vdash s =_r t : D$, i.e. there is a deduction/computation of the equation starting from the equations in S, then the rewrite reason r is built up from:

(i) the constants for rewrite reasons: $\{\rho, \beta, \eta, \zeta\}$;
(ii) the r_i's;

using the substitution operations:

(iii) $\mathsf{sub_L}$;
(iv) $\mathsf{sub_R}$;

and the operations for building new rewrite reasons:

(v) σ, τ, ξ, μ.

Definition 5.6 (general rules of equality). The general rules for equality (reflexivity, symmetry and transitivity) are defined as follows:

$$\begin{array}{ccc} \textit{reflexivity} & \textit{symmetry} & \textit{transitivity} \\[2mm] \dfrac{x : D}{x =_\rho x : D} & \dfrac{x =_t y : D}{y =_{\sigma(t)} x : D} & \dfrac{x =_t y : D \qquad y =_u z : D}{x =_{\tau(t,u)} z : D}. \end{array}$$

The "subterm substitution" rule. Equational logic as usually presented has the following inference rule of substitution:

$$\frac{s = t}{s\theta = t\theta},$$

where θ is a substitution.

Note that the substitution θ "appeared" in the conclusion of the rule. As rightly pointed out by Le Chenadec in [Chenadec (1989)], from the view point of the subformula property (objects in the conclusion of some inference should be sub-objects of the premises), this rule is unsatisfactory. He then defines two rules:

$$IL\frac{M = N \quad C[N] = O}{C[M] = O} \qquad IR\frac{M = C[N] \quad N = O}{M = C[O]},$$

where M, N and O are terms and the context $C[\_]$ is adopted in order to distinguish subterms.

In [de Oliveira and de Queiroz (1999)] we have formulated an inference rule called "subterm substitution" which deals in a *explicit way* [4] with substitutions. In fact, the *LND* system can be seen as an enriched system which brings to the object language terms, and now substitutions.

Definition 5.7 (subterm substitution). The rule of "subterm substitution" is framed as follows:

$$\frac{x =_r C \mid y \mid: D \quad y =_s u : D'}{x =_{\text{sub}_L(r,s)} C \mid u \mid: D} \qquad \frac{x =_r w : D' \quad C \mid w \mid =_s u : D}{C \mid x \mid =_{\text{sub}_R(r,s)} u : D},$$

where C is the context in which the subterm detached by '$\mid \ \mid$' appears and D' could be a subdomain of D, equal to D or disjoint to D.

The symbols sub_L and sub_R denote in which side (L – *left* or R – *right*) is the premise that contains the subterm to be substituted.

Note that the transitivity rule previously defined can be seen as a special case for this rule when $D' = D$ and the context C is empty.

The ξ- and μ-rules. In the Curry–Howard "formulae-as-types" interpretation [Howard (1980)], the ξ-rule[5] states when two *canonical* elements are equal,

---

[4]Let us recall Girard, who describes the intimate connections between constructivity and explicitation, and claim that "...one of the aims of inserting a label alongside formulas (accounting for the steps made to arrive at each particular point in the deduction) is exactly that of making *explicit* the use of formulas (and instances of formulas and individuals) throughout a deduction ..."

[5]The ξ-rule is the formal counterpart to Bishop's constructive principle of definition of a set [Bishop (1967)] (p. 2) which says: "To define a set we prescribe, at least implicitly, what we have (the constructing intelligence) must do in order to construct an element of the set, and what we must do to show that two elements of the set are equal." Cf. also [Bishop (1967)] (p. 12) Bishop defines a product of set as "The *cartesian product*, or simply *product*, $X \equiv X_1 \times \ldots \times X_n$ of sets $X_1, X_2, \ldots, X_n$ is defined to be the set of all ordered n-tuples $(x_1, \ldots, x_n)$ and $(y_1, \ldots, y_n)$ of X are equal if the coordinates x_i and y_i are equal for each i." See also [Martin-Löf (1984)] (p.8): "... a set A is defined by prescribing how a canonical element of A is formed as well as how two equal canonical elements of A are formed." We also know from the theory of Lambda Calculus the definition of ξ-rule, see e.g. [Barendregt (1981)] (pp. 23 and 78): "ξ : $M = N \Rightarrow \lambda x.M = \lambda x.N$"

168   *The Functional Interpretation of Logical Deduction*

and the μ-rule[6] states when two *noncanonical* elements are equal. So, each introduction rule for the *LND* system has associated to it a ξ-rule and each elimination rule has a related μ-rule. For instance, the ξ-rule and μ-rule for the connective ∧ are defined as follows:

$$\frac{x =_u y : A \qquad s =_v t : B}{\langle x, s \rangle =_{\xi(u,v)} \langle y, t \rangle : A \wedge B}$$

$$\frac{x =_r y : A \wedge B}{FST(x) =_{\mu(r)} FST(y) : A} \qquad \frac{x =_r y : A \wedge B}{SND(x) =_{\mu(r)} SND(y) : B}.$$

Term rewriting system for *LND* with equality. In [de Oliveira and de Queiroz (1999)] we have proved termination and confluence for the rewriting system arising out of the proof rules given for the proposed natural deduction system for equality.

Back to our example. Having defined the proof rules for our formulation of propositional equality, the proof of the logically valid formula used in our example related to the connections between Gentzen's and Herbrand's method would be as follows:

$$\frac{[a : D] \quad \dfrac{\dfrac{[a =_g y : D] \qquad [f(a) : P(a)]}{g(a,y) \cdot f(a) : P(y)}}{\dfrac{\lambda f.g(a,y) \cdot f(a) : P(a) \rightarrow P(y)}{\Lambda y.\lambda f.g(a,y) \cdot f(a) : \forall y^D.(P(a) \rightarrow P(y))}}}{\varepsilon x.((\Lambda y.\lambda f.g(x,y) \cdot f(x)), a) : \exists x^D.\forall y^D.(P(x) \rightarrow P(y))}.$$

Notice, however, that the proof term '$\varepsilon x.((\Lambda y.\lambda f.g(x,y) \cdot f(x)), a)$' contains two free variables, namely a and g, associated to, respectively, two assumptions:

(1) the domain is nonempty: $[a : D]$
(2) there is a way of identifying a term with another one: $[a =_g y : D]$

The variable g, which denotes a sequence of rewrites taking from a to y, has to be dealt with before the quantification of y. And, from our proof rules defined above,

---

[6]The μ-rule is also defined in the theory of Lambda Calculus, see e.g. [Mitchell and Scedrov (1993)]: "The equational axioms and inference rules are as follows, where $[N/x]M$ denotes substitution of N for x in M. ....

$$(\mu) \quad \frac{\Gamma \triangleright M_1 = M_2 : \sigma \Rightarrow \tau \quad \Gamma \triangleright N_1 = N_2 : \sigma}{\Gamma \triangleright M_1 N_1 = M_2 N_2 : \tau},,$$

and is divided into two equalities μ and ν in [Hindley and Seldin (2008)]:

$$(\mu) \frac{M = M'}{NM = NM'} \qquad (\nu) \frac{M = M'}{MN = M'N}$$

the assumption '$[a =_g y : D]$' appears in the application of an elimination rule to a statement like '$\doteq_D (a, y)$'. Thus, our formal derivation takes the following form:

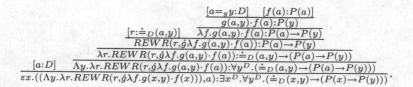

5.3 Finale

The conception of the very first decision procedures for first-order sentences in the 1920s brought about the need for giving 'logical' citizenship to function symbols (e.g. Skolem functions). We have taken the view that a closer look at proof procedures for first-order sentences with equality brings about the need for introducing what we have called the "missing entity": (function) symbols for rewrites. This, we have argued, is appropriately done via the framework of labelled natural deduction which allows to formulate a proof theory for the "logical connective" of propositional equality. The basic idea is that when analysing an equality sentence into (i) proof conditions (*introduction*) and (ii) immediate consequences (*elimination*), it becomes clear that we need to bring in identifiers (i.e. function symbols) for sequences of rewrites, and this is what we have claimed should be the missing entity in P. Martin-Löf's equality types, both intensional and extensional. What we end up with is a formulation of what appears to be a middle ground solution to the 'intensional' versus 'extensional' dichotomy that permeates most of the work on characterising propositional equality in natural deduction style.

Chapter 6

Normalisation for the Equality Fragment

Preamble

As we already seen in the previous chapter, in the *LND* equational logic, the equations have an index (the *reason*) which keeps all proof steps. The *reasons* is defined by the kind of rule used in the proof and the equational axioms (*definitional equalities*) of the system. The rules are divided into the following classes:

- General rules.
- Subterm substitution rule.
- ξ- and μ-rules.

Since the *LND* system is based on the Curry–Howard isomorphism [Howard (1980)], terms represent proof constructions, thus proof transformations correspond to equalities between terms. In this way, the *LND* equational logic can deal with equalities between *LND* proofs. The proofs in the *LND* equational fragment which deals with equalities between deductions are built from the basic proof transformations for the *LND* system, summarized in Section 6.9. These basic proof transformations form an equational system, composed by *definitional equalities* (β, η and ζ), shown in Appendix 6.9.

6.1 General rules

Definition 6.1 (equation). An equation in our LND_{EQ} is of the form:

$$s =_r t : D$$

where s and t are terms, r is the identifier for the rewrite reason, and D is the type (formula).

Definition 6.2 (system of equations). A system of equations S is a set of equations:

$$\{s_1 =_{r_1} t_1 : D_1, \ldots, s_n =_{r_n} t_n : D_n\}$$

where r_i is the rewrite reason identifier for the ith equation in S.

Definition 6.3 (rewrite reason). Given a system of equations S and an equation $s =_r t : D$, if $S \vdash s =_r t : D$, i.e. there is a deduction/computation of the equation starting from the equations in S, then the rewrite reason r is built up from:

 (i) the constants for rewrite reasons: $\{\beta, \eta, \zeta, \rho\}$;
 (ii) the r_i's;

using the substitution operations:

 (iii) $\mathsf{sub_L}$;
 (iv) $\mathsf{sub_R}$;

and the operations for building new rewrite reasons:

 (v) $\sigma, \tau, \xi, \mu, \nu$.

Definition 6.4 (general rules of equality). The general rules for equality (reflexivity, symmetry and transitivity) are defined as follows:

$$\begin{array}{ccc} \textit{reflexivity} & \textit{symmetry} & \textit{transitivity} \\[2mm] \dfrac{x : D}{x =_\rho x : D} & \dfrac{x =_t y : D}{y =_{\sigma(t)} x : D} & \dfrac{x =_t y : D \qquad y =_u z : D}{x =_{\tau(t,u)} z : D} \end{array}.$$

6.2 The 'subterm substitution' rule

Equational logic as usually presented has the following inference rule of substitution:

$$\frac{s = t}{s\theta = t\theta},$$

where θ is a substitution.

Note that the substitution θ 'appeared' in the conclusion of the rule. As rightly pointed out by Le Chenadec in [Chenadec (1989)], from the view point of the subformula property (objects in the conclusion of some inference should be subobjects of the premises), this rule is unsatisfactory. He then defines two rules:

$$IL \frac{M = N \quad C[N] = O}{C[M] = O} \qquad IR \frac{M = C[N] \quad N = O}{M = C[O]},$$

where M, N and O are terms and the context $C[\_]$ is adopted in order to distinguish subterms.

In [de Oliveira and de Queiroz (1999)] we have formulated an inference rule called "subterm substitution" which deals in a *explicit way*[1] with substitutions. In fact, the *LND* can be seen as an enriched system which brings to the object language terms, and now substitutions.

Definition 6.5 (subterm substitution). The rule of "subterm substitution" is framed as follows:

$$\frac{x =_r C \mid y \mid : D \quad y =_s u : D'}{x =_{\text{sub}_L(r,s)} C \mid u \mid : D} \qquad \frac{x =_r w : D' \quad C \mid w \mid =_s u : D}{C \mid x \mid =_{\text{sub}_R(r,s)} u : D},$$

where C is the context in which the subterm detached by '$\mid \ \mid$' appears and D' could be a subdomain of D, equal to D or disjoint to D.

The symbols sub_L and sub_R denote in which side (L – *left* or R – *right*) is the premiss that contains the subterm to be substituted.

Note that the transitivity rule previously defined can be seen as a special case for this rule when $D' = D$ and the context C is empty.

Example 6.1. The following subdeduction illustrates this rule:

$$\frac{\dfrac{z =_r \langle FST(z), SND(z)\rangle : A \wedge B \quad FST(z) =_e a : A}{z =_{\text{sub}_L(r,e)} \langle a, SND(z)\rangle : A \wedge B} \quad SND(z) =_w b : B}{z =_{\text{sub}_L(\text{sub}_L(r,e),w)} \langle a, b\rangle : A \wedge B}.$$

Note that A is a subformula of $A \wedge B$ while the term $FST(z)$ is a subterm of $\langle FST(z), SND(z)\rangle$.

---

[1] In [de Queiroz and Gabbay (199?)] de Queiroz and Gabbay recall Girard, who describes the intimate connections between constructivity and explicitation, and claim:

"...one of the aims of inserting a label alongside formulas (accounting for the steps made to arrive at each particular point in the deduction) is exactly that of making *explicit* the use of formulas (and instances of formulas and individuals) throughout a deduction ..."

6.3 The ξ- and μ-rules

In the Curry–Howard "formulae-as-types" interpretation [Howard (1980)], the ξ-rule[2] states when two *canonical* elements are equal, and the μ-rule[3] states when two *noncanonical* elements are equal. So, each introduction rule for the *LND* system has associated to it a ξ-rule and each elimination rule has a related μ-rule. For instance, the ξ-rule and μ-rule for the connective ∧ are defined as follows:

$$\frac{x =_u y : A \qquad s =_v t : B}{\langle x, s \rangle =_{\xi(u,v)} \langle y, t \rangle : A \wedge B}$$

$$\frac{x =_r y : A \wedge B}{FST(x) =_{\mu(r)} FST(y) : A} \qquad \frac{x =_r y : A \wedge B}{SND(x) =_{\mu(r)} SND(y) : B} .$$

6.4 Term rewriting systems

In this section we give a brief overview of the theory of term rewriting systems, which we employ in the proofs of the normalization theorems for the *LND* equational logic.

The theory of term rewriting systems provides efficient means to solve the famous "word problem", i.e. the problem of deciding if an equation can be derived from a set of equations of a given equational theory; the problem of deciding the equivalence between terms in an algebraic system, defined by equational axioms; etc. [Buchberger (1987)]. The idea of term rewriting systems (TRSs) is quite simple: instead of having equations like $r = s$, equational systems are replaced by term rewriting systems which use rules such as $r \longrightarrow s$, stating that r may be replaced by s but not vice-versa. The equations are oriented in such a way that terms are replaced by simpler terms, thereby reducing the complexity of the terms being

---

[2]The ξ-rule is the formal counterpart to Bishop's constructive principle of definition of a set [Bishop (1967)] (p. 2) which says: "To define a set we prescribe, at least implicitly, what we have (the constructing intelligence) must to do in order to construct an element of the set, and what we must do to show that two elements of the set are equal." Cf. also [Bishop (1967)] (p. 12) Bishop defines a product of set as "The *cartesian product*, or simply *product*, $X \equiv X_1 \times \ldots \times X_n$ of sets $X_1, X_2, \ldots, X_n$ is defined to be the set of all ordered n-tuples $(x_1, \ldots, x_n)$ and $(y_1, \ldots, y_n)$ of X are equal if the coordinates x_i and y_i are equal for each i." See also [Martin-Löf (1984)] (p. 8): "... a set A is defined by prescribing how a canonical element of A is formed as well as how two equal canonical elements of A are formed." We also know from the theory of Lambda Calculus the definition of ξ-rule, see e.g. [Barendregt (1981)] (pp. 23 and 78): "ξ : $M = N \Rightarrow \lambda x.M = \lambda x.N$"

[3]The μ-rule is also defined in the theory of Lambda Calculus, see e.g. [Mitchell and Scedrov (1993)]: "The equational axioms and inference rules are as follows, where $[N/x]M$ denotes substitution of N for x in M. ...

$$(\mu) \quad \frac{\Gamma \triangleright M_1 = M_2 : \sigma \Rightarrow \tau \quad \Gamma \triangleright N_1 = N_2 : \sigma}{\Gamma \triangleright M_1 N_1 = M_2 N_2 : \tau} \text{,,}$$

rewritten at each stage of the proof. It is obtained by imposing a *well-founded ordering* on the set of terms.

The "word problem" when studied in the theory of term rewriting systems can be restricted to the problem of reducing the terms to a *normal form* and to compare them. This procedure is possible if the TRS has two fundamental properties: *termination* and *confluence*. Whilst the termination property guarantees that the rewriting process is finite, the confluence assures the unicity of the normal form. In addition, if the TRS has these two properties, it is called *convergent*, or *canonic*, or *complete* [Dershowitz and Jouannaud (1990); Plaisted (1994)].

6.4.1 *Termination property*

The importance of study such property is that if a system R is terminating, we can always find a normal form of a term by any rewrite sequence.

Definition 6.6 (terminating system). A term rewriting system is terminating if it has no infinite rewrite sequences. For instance, the system $R = \{x \rightarrow f(x)\}$ is not a terminating system.

In order to prove this property it is necessary to choose an appropriate order between the terms, in such a way that in each rewrite stage the complexity of the term being rewritten is reduced and the process stops with an irreducible term achieved. Several orderings were defined to prove such property, and in [Dershowitz (1987)] the reader can find an excellent survey in this subject.

6.4.2 *Confluence property*

While the terminating property guarantees that the rewriting process applied to a term will stop and will lead to a normal form of a term, the confluence property states that there will be only one normal form for each term, if it exists.

The intuitive idea of the *confluence* notion is that any two rewrite sequences from a given term can always be "brought together" [Plaisted (1994)]. Formally, a system R is confluent if for all terms s and t, $s \uparrow t$ implies $s \downarrow t$. Where $s \downarrow t$ means that there is a term v such that $s \rightarrow^* v$ and $t \rightarrow^* v$. And, we say that $s \uparrow t$ if there is a term r such that $r \rightarrow^* s$ and $r \rightarrow^* t$.

The test for confluence is based on two theorems: one given by Newman [Newman (1942)] and the other by Knuth-Bendix [Knuth and Bendix (1970)].

Newman, in [Newman (1942)], showed that in order to determine the confluence property for terminating TRSs it is sufficient to identify the *local confluence condition*. Knuth and Bendix, in [Knuth and Bendix (1970)], showed that the *local confluence condition* can be characterized by the convergence of the *critical pairs*. Thus, a terminating TRS is confluent if and only if its critical pairs converge.

Definition 6.7 (local confluence condition). A TRS is locally confluent when for all terms t, z and s, if $z \to t$ and $z \to s$ then $t \downarrow s$.

The test for confluence is established by the *superposition algorithm*, given in [Knuth and Bendix (1970)], which checks overlaps between the left-hand side of the rules of the system. The term resulting from the overlaps can be rewritten to two other terms, and these two terms are called *critical pair*. If the critical pair is convergent, the system is confluent.

6.4.3 *The completion procedure*

When a term rewriting system is not confluent it is possible to add rules in such a way that the resulting system becomes confluent. The so-called *completion* procedure, introduced by Knuth and Bendix in [Knuth and Bendix (1970)][4], gives a way in which to perform this transformation.

6.5 The transformations between proofs in the equational fragment of the *LND*

The *LND* equational logic has a powerful mechanism of controlling proofs: the *rewrite reasons*, which appears as indexes of the equations of the system. All rewrite steps from one term to another are reflected in the *reason*, on which we define an orthogonal measure of redundancy to the *LND* system. Since reasons represent proof constructions, proof transformations correspond to equalities between reasons. Thus, the normalization procedure becomes easier: the procedure is based on the normality of the *rewrite reasons*. As we will see in the next section, from the equational system formed by the basic transformations between proofs in the *LND* equational logic, we construct a term rewriting system to compute the normal form of proofs [de Oliveira (1995)].

The basic transformations between proofs to the equational logic of *LND* are grouped as follows:

---

[4]After the work of Knuth and Bendix, several refinements were proposed to improve the *completion* procedure. In [Buchberger (1987)], the author presents a historical survey about the development of the *completion* procedure. In [Dershowitz and Jouannaud (1990); Snyder (1991)] a non-deterministic version of the algorithm is shown. This non-deterministic version is based on the works of Bachmair (*Proof Methods for Equational Theories*, Ph.D thesis, University of Illinois, Urbana Champaign, Illinois (1987)).

- Reductions on the rewrite reasons. They capture the redundancies on the equational proofs.
- Transformations on the rewrite reasons. They are given by the way in which the proof is built up. For instance, the order of the premises and rules.

6.5.1 *Reductions on the rewrite sequence*

6.5.1.1 *Symmetry*

The rule of *symmetry* is the only rule which changes the direction of an equation. So, its use must be controlled. Here we group two reductions related to this rule.

Definition 6.8 (reductions involving ρ and σ).

$$\frac{x =_\rho x : D}{x =_{\sigma(\rho)} x : D} \quad \triangleright \quad x =_\rho x : D$$

$$\frac{\dfrac{x =_r y : D}{y =_{\sigma(r)} x : D}}{x =_{\sigma(\sigma(r))} y : D} \quad \triangleright \quad x =_r y : D$$

Associated rewritings:
- $\sigma(\rho) \triangleright \rho$
- $\sigma(\sigma(r)) \triangleright r$.

6.5.1.2 *Transitivity*

The *transitivity* rule when combined with *symmetry* and *reflexivity* causes redundancy on the proof. Thus the following reductions are defined.

Definition 6.9 (reductions involving τ).

$$\frac{x =_r y : D \quad y =_{\sigma(r)} x : D}{x =_{\tau(r,\sigma(r))} x : D} \quad \triangleright \quad x =_\rho x : D$$

$$\frac{y =_{\sigma(r)} x : D \quad x =_r y : D}{y =_{\tau(\sigma(r),r)} y : D} \quad \triangleright \quad y =_\rho y : D$$

$$\frac{u =_r v : D \quad v =_\rho v : D}{u =_{\tau(r,\rho)} v : D} \quad \triangleright \quad u =_r v : D$$

$$\frac{u =_\rho u : D \quad u =_r v : D}{u =_{\tau(\rho,r)} v : D} \quad \triangleright \quad u =_r v : D$$

Note that the first two reductions identify the case in which a *reason* which is part of a rewrite sequence meets its inverse.

These reductions can be generalized to transformations where the reasons r and $\sigma(r)$ (transf. 1 and 2) and r and ρ (transf. 3 and 4) appear in some context, as illustrated by the following example:

Example 6.2.

$$\dfrac{\dfrac{x =_r y : A}{inl(r) =_{\xi(r)} inl(y) : A \vee B} \quad \dfrac{\begin{array}{c} x =_r y : A \\ y =_{\sigma(r)} x : A \end{array}}{inl(y) =_{\xi(\sigma(r))} inl(x) : A \vee B}}{inl(x) =_{\tau(\xi(r), \xi(\sigma(r)))} inl(x) : A \vee B} \quad \rhd \quad \dfrac{\dfrac{x : A}{x =_\rho x : A}}{inl(x) =_{\xi(\rho)} inl(x) : A \vee B}$$

Associated rewritings:

- $\tau(\mathcal{C}[r], \mathcal{C}[\sigma(r)]) \rhd \mathcal{C}[\rho]$
- $\tau(\mathcal{C}[\sigma(r)], \mathcal{C}[r]) \rhd \mathcal{C}[\rho]$
- $\tau(\mathcal{C}[r], \mathcal{C}[\rho]) \rhd \mathcal{C}[r]$
- $\tau(\mathcal{C}[\rho], \mathcal{C}[r]) \rhd \mathcal{C}[r]$.

6.5.1.3 *Substitution*

Definition 6.10 (substitution rules).

$$\dfrac{u =_r \mathcal{C}[x] : D \quad x =_\rho x : D'}{u =_{\mathrm{sub_L}(r, \rho)} \mathcal{C}[x] : D} \quad \rhd \quad u =_r \mathcal{C}[x] : D$$

$$\dfrac{x =_\rho x : D' \quad \mathcal{C}[x] =_r z : D}{\mathcal{C}[x] =_{\mathrm{sub_R}(\rho, r)} z : D} \quad \rhd \quad \mathcal{C}[x] =_r z : D$$

$$\dfrac{\dfrac{z =_s \mathcal{C}[y] : D \quad y =_r w : D'}{z =_{\mathrm{sub_L}(s, r)} \mathcal{C}[w] : D} \quad \dfrac{y =_r w : D'}{w =_{\sigma(r)} y : D'}}{z =_{\mathrm{sub_L}(\mathrm{sub_L}(s, r), \sigma(r))} \mathcal{C}[y] : D} \quad \rhd \quad z =_s \mathcal{C}[y] : D$$

$$\dfrac{\dfrac{z =_s \mathcal{C}[y] : D \quad y =_r w : D'}{z =_{\mathrm{sub_L}(s, r)} \mathcal{C}[w] : D} \quad \dfrac{y =_r w : D'}{w =_{\sigma(r)} y : D'}}{z =_{\mathrm{sub_L}(\mathrm{sub_L}(s, r), \sigma(r))} \mathcal{C}[y] : D} \quad \rhd \quad z =_s \mathcal{C}[y] : D$$

$$\dfrac{\dfrac{\dfrac{x =_s w : D'}{w =_{\sigma(s)} x : D'} \quad \mathcal{C}[x] =_r z : D}{\mathcal{C}[w] =_{\mathrm{sub_R}(\sigma(s), r)} z : D}}{\mathcal{C}[x] =_{\mathrm{sub_R}(s, \mathrm{sub_R}(\sigma(s), r))} z : D} \quad \rhd \quad \mathcal{C}[x] =_r z : D$$

$$\dfrac{\dfrac{x =_s w : D'}{w =_{\sigma(s)} x : D'} \quad \dfrac{x =_s w : D' \quad \mathcal{C}[w] =_r z : D}{\mathcal{C}[x] =_{\mathrm{sub_R}(s, r)} z : D}}{\mathcal{C}[w] =_{\mathrm{sub_R}(\sigma(s), \mathrm{sub_R}(s, r))} z : D} \quad \rhd \quad \mathcal{C}[w] =_r z : D$$

Associated rewritings:

- $\mathrm{sub_L}(\mathcal{C}[r], \mathcal{C}[\rho]) \rhd \mathcal{C}[r]$
- $\mathrm{sub_R}(\mathcal{C}[\rho], \mathcal{C}[r]) \rhd \mathcal{C}[r]$

- $\mathrm{sub_L(sub_L}(s, \mathcal{C}[r]), \mathcal{C}[\sigma(r)]) \rhd s$
- $\mathrm{sub_L(sub_L}(s, \mathcal{C}[\sigma(r)]), \mathcal{C}[r]) \rhd s$
- $\mathrm{sub_R}(s, \mathrm{sub_R}(\mathcal{C}[\sigma(s)], r)) \rhd r$
- $\mathrm{sub_R}(\mathcal{C}[\sigma(s)], \mathrm{sub_R}(\mathcal{C}[s], r)) \rhd r.$

6.5.1.4 *The β_{rewr} reduction*

When a μ operator has as argument a ξ operator we have a redundant step called β_{rewr} reduction, which reflects a redundancy on the logical calculus of the *LND* removed by a β-reduction. The β-type reductions have the role of spelling out the effect of an elimination inference on the result of introduction steps [de Queiroz and Gabbay (1995)].

The following example illustrates the β_{rewr} reduction:

Example 6.3.

$$\cfrac{\cfrac{x =_r y : A \qquad z =_s w : B}{\langle x, z \rangle =_{\xi(r,s)} \langle y, w \rangle : A \wedge B} \wedge \text{-}intr}{FST(\langle x, z \rangle) =_{\mu(\xi(r,s))} FST(\langle y, w \rangle) : A} \wedge \text{-}elim$$

$$\cfrac{\cfrac{x =_r y : A \qquad z =_s w : B}{\langle x, z \rangle =_{\xi(r,s)} \langle y, w \rangle : A \wedge B} \wedge \text{-}intr}{SND(\langle x, z \rangle) =_{\mu(\xi(r,s))} SND(\langle y, w \rangle) : B} \wedge \text{-}elim$$

Note that this subrewrite captures the redundancy on the terms. In this example, the terms "$FST(\langle x, y \rangle)$", "$FST(\langle y, w \rangle)$", "$SND(\langle x, y \rangle)$" and "$SND(\langle y, w \rangle)$" are β-redexes. Thus, the β_{rewr} reductions is defined by the following procedure:

- Firstly a β-reduction [de Queiroz and Gabbay (1995)] must be applied on the terms.
- After, a reduction is applied to the rewrite sequence in the following way:
  - If the reduction is applied to terms related with the connective $\wedge$ the rewrite operators μ and ξ are dropped and the argument of ξ is selected depending on the 'DESTRUCTOR'. If the 'DESTRUCTOR' is FST then the first argument is chosen to form the remained reason, otherwise if the 'DESTRUCTOR' is SND then the second argument is chosen.
  - If the reduction is applied to terms related to the connective $\vee$, the resulting reason is:

       * the second argument of μ, if the "constructor" of the β-redex is *inl*;

       * the third argument of μ, if the "constructor" is *inr*.

    – If the reduction is applied to the other connectives then we just select the second argument of μ. If the external symbol of the second argument of μ is ξ, the argument of ξ is chosen.

Applying this procedure on labels of the above example we have the following reduction:

$$FST(\langle x, z \rangle) =_{\mu(\xi(r,s))} FST(\langle y, w \rangle) : A \quad \triangleright_{\beta_{rewr}} \quad x =_r y : A$$

$$SND(\langle x, z \rangle) =_{\mu(\xi(r,s))} SND(\langle y, w \rangle) : B \quad \triangleright_{\beta_{rewr}} \quad z =_s w : B.$$

Definition 6.11 (β_{rewr}).

β_{rewr}-$\wedge$-*reduction*

$$\dfrac{\dfrac{x =_r y : A \qquad z =_s w : B}{\langle x, z \rangle =_{\xi(r,s)} \langle y, w \rangle : A \wedge B} \wedge\text{-}intr}{FST(\langle x, z \rangle) =_{\mu(\xi(r,s))} FST(\langle y, w \rangle) : A} \wedge\text{-}elim$$

$$\triangleright_{\beta_{rewr}} \quad x =_r y : A$$

$$\dfrac{\dfrac{x =_r y : A \qquad z =_s w : B}{\langle x, z \rangle =_{\xi(r,s)} \langle y, w \rangle : A \wedge B} \wedge\text{-}intr}{SND(\langle x, z \rangle) =_{\mu(\xi(r,s))} SND(\langle y, w \rangle) : B} \wedge\text{-}elim$$

$$\triangleright_{\beta_{rewr}} \quad z =_s w : B$$

Associated rewritings:

- (FST) $\mu(\xi(r, s)) \triangleright r$
- (SND) $\mu(\xi(r, s)) \triangleright s$

β_{rewr}-$\vee$-*reduction*

$$\dfrac{\dfrac{a =_r a' : A}{inl(a) =_{\xi(r)} inl(a') : A \vee B} \vee\text{-}intr \quad \dfrac{\overset{[x:A]}{f(x) =_s k(x):C} \quad \overset{[y:B]}{g(y) =_u h(y):C}}{}}{CASE(inl(a), \hat{x}f(x), \hat{y}g(y)) =_{\mu(\xi(r),s,u)} CASE(inl(a'), \hat{x}k(x), \hat{y}h(y)):C} \vee\text{-}elim$$

$$\triangleright_{\beta_{rewr}} \quad \begin{array}{c} a =_r a' : A \\ f(a/x) =_s k(a'/x) : C \end{array}$$

$$\dfrac{\dfrac{b =_r b' : B}{inr(b) =_{\xi(r)} inr(b') : A \vee B} \vee\text{-}intr \quad \dfrac{\overset{[x:A]}{f(x) =_s k(x):C} \quad \overset{[y:B]}{g(y) =_u h(y):C}}{}}{CASE(inr(b), \hat{x}f(x), \hat{y}g(y)) =_{\mu(\xi(r),s,u)} CASE(inr(b'), \hat{x}k(x), \hat{y}h(y)):C} \vee\text{-}elim$$

$$\triangleright_{\beta_{rewr}} \quad \begin{array}{c} [b =_s b' : B] \\ g(b/y) =_u h(b'/y) : C \end{array}$$

Associated rewritings:

- (inl) $\mu(\xi(r), s, u) \triangleright s$
- (inr) $\mu(\xi(r), s, u) \triangleright u$

β_{rewr}-$\rightarrow$-*reduction*

$$[x{:}A]$$

$$\vdots$$

$$\frac{a=_s a'{:}A \quad \dfrac{b(x)=_r g(x){:}B}{\lambda x.b(x)=_{\xi(r)}\lambda x.g(x){:}A{\rightarrow}B}\rightarrow\text{-}intr}{APP(\lambda x.b(x),a)=_{\mu(s,\xi(r))}APP(\lambda x.g(x),a'){:}B}\rightarrow\text{-}elim$$

$$\triangleright_{\beta_{rewr}}\quad b(a/x)=_r g(a'/x):B$$

Associated rewriting:

- $\mu(s,\xi(r))\triangleright r$

β_{rewr}-$\forall$-*reduction*

$$[x{:}D]$$

$$\frac{a{:}D\quad \dfrac{f(x)=_r g(x){:}P(x)}{\Lambda x.f(x)=_{\xi(s)}\Lambda x.g(x){:}\forall x^D.P(x)}}{EXTR(\Lambda x.f(x),a)=_{\mu(s,\xi(r))}EXTR(\Lambda x.g(x),a){:}\forall x^D.P(x)}$$

$$\triangleright_{\beta_{rewr}}\quad \begin{array}{c}[a:D]\\ f(a/x)=_r g(a/x):P(a)\end{array}$$

Associated rewriting:

- $\mu(s,\xi(r))\triangleright r$

β_{rewr}-$\exists$-*reduction*

$$\frac{\dfrac{a{:}D\quad f(a)=_r i(a){:}P(a)}{\varepsilon.x(f(x),a)=_{\xi(r)}\varepsilon x.(i(x),a){:}\exists x^D.P(x)}\quad \dfrac{[t{:}D,g(t){:}P(t)]}{d(g,t)=_s h(g,t){:}C}}{INST(\varepsilon x.(f(x),a),\hat{g}\hat{t}d(g,t))=_{\mu(\xi(r),s)}INST(\varepsilon x.(i(x),a),\hat{g}\hat{t}h(g,t)){:}C}$$

$$\triangleright_{\beta_{rewr}}\quad \begin{array}{c}[a:D\quad f(a)=_r g(a):P(a)]\\ d(f/g,a/t)=_s h(i/g,a/t):C\end{array}$$

Associated rewriting:

- $\mu(\xi(r),s)\triangleright s$.

6.5.1.5 *The η_{rewr} reduction*

When a ξ operator has as argument a μ operator we have an η_{rewr} reduction. As the β_{rewr}-reduction, the η_{rewr}-reduction captures the redundancy on the *LND* terms. It occurs when an η-reduction must be applied on the logical calculus, i.e. when an elimination inference is followed by an introduction rule of the same connective [de Queiroz and Gabbay (199?, 1995)].

The η_{rewr} is defined by the procedure described below:

- An η-reduction [de Queiroz and Gabbay (199?, 1995)] is applied on the terms.
- The rewrite operators ξ and μ are deleted from the rewrite sequence and the argument related to the major premiss of the corresponding elimination rule is selected.

Definition 6.12 (η_{rewr}).

η_{rewr}- $\wedge$-*reduction*

$$\dfrac{\dfrac{x=_r y:A\wedge B}{FST(x)=_{\mu(r)}FST(y):A}\wedge\text{-}elim \quad \dfrac{x=_r y:A\wedge B}{SND(x)=_{\mu(r)}SND(y):B}\wedge\text{-}elim}{\langle FST(x),SND(x)\rangle=_{\xi(\mu(r))}\langle FST(y),SND(y)\rangle:A\wedge B}\wedge\text{-}intr$$

$$\rhd_{\eta_{rewr}}\ x=_r y:A\wedge B$$

η_{rewr}- $\vee$-*reduction*

$$\dfrac{c=_t d:A\vee B \quad \dfrac{[a_1=_r a_2:A]}{inl(a_1)=_{\xi(r)}inl(a_2):A\vee B}\vee i \quad \dfrac{[b_1=_s b_2:B]}{inr(b_1)=_{\xi(s)}inr(b_2):A\vee B}\vee\text{-}intr}{CASE(c,a_1'inl(a_1),b_1'inr(b_1))=_{\mu(t,\xi(r),\xi(s))}CASE(d,a_2'inl(a_2),b_2'inr(b_2))}\vee\text{-}elim$$

$$\rhd_{\eta_{rewr}}\ c=_t d:A\vee B$$

η_{rewr}-$\rightarrow$-*reduction*

$$\dfrac{\dfrac{[x=_r y:A] \quad c=_s d:A\rightarrow B}{APP(c,x)=_{\mu(r,s)}APP(d,y):B}\rightarrow\text{-}elim}{\lambda x.APP(c,x)=_{\xi(\mu(r,s))}\lambda y.APP(d,y):A\rightarrow B}\rightarrow\text{-}intr$$

$$\rhd_{\eta_{rewr}}\ c=_s d:A\rightarrow B$$

x does not occur free in c or d.

$\forall$-η_{rewr}-*reduction*

$$\dfrac{\dfrac{[t:D]\quad c=_r d:\forall x^D.P(x)}{EXTR(c,t)=_{\mu(r)}EXTR(d,t):P(t)}\forall\text{-}elim}{\Lambda t.EXTR(c,t)=_{\xi(\mu(r))}\Lambda t.EXTR(d,t):\forall t^D.P(t)}\forall\text{-}intr$$

$$\rhd_{\eta_{rewr}}\ c=_r d:\forall x^D.P(x)$$

where c and d do not depend on x.

$\exists$-η_{rewr}-*reduction*

$$\dfrac{c:\exists x^D.P(x)\quad \dfrac{[t:D]\quad [g(t)=_r h(t):P(t)]}{\varepsilon y.(g(y),t)=_{\xi(r)}\varepsilon y.(h(y),t):\exists y^D.P(y)}\exists\text{-}intr}{INST(c,g't\varepsilon y.(g(y),t))=_{\mu(s,\xi(r))}INST(b,h't\varepsilon y.(h(y),t)):\exists y^D.P(y)}\exists\text{-}elim$$

$$\rhd_{\eta_{rewr}}\ c=_s b:\exists x^D.P(x)$$

Associated rewritings:

- $\xi(\mu(r))\rhd r$
- $\mu(t,\xi(r),\xi(s))\rhd t$
- $\xi(\mu(r,s))\rhd s$
- $\mu(s,\xi(r))\rhd s$.

6.5.2 *Transformations on the rewrite reasons*

Most of the transformations defined here are permutations. They express equality between proofs and can make evident a hidden redundancy.

6.5.2.1 *Symmetry*

Here we group all possible transformations which could be made when the rewrite operator σ is combined with τ, sub_L, sub_R, ξ and μ.

Definition 6.13 (σ and τ).

$$\cfrac{\cfrac{x=_r y:D \qquad y=_s w:D}{x=_{\tau(r,s)} w:D}}{w=_{\sigma(\tau(r,s))} x:D} \qquad \triangleright \qquad \cfrac{\cfrac{y=_s w:D}{w=_{\sigma(s)} y:D} \qquad \cfrac{x=_r y:D}{y=_{\sigma(r)} x:D}}{w=_{\tau(\sigma(s),\sigma(r))} x:D}$$

Associated rewriting:

• $\sigma(\tau(r,s)) \triangleright \tau(\sigma(s),\sigma(r))$

Definition 6.14 (σ and sub).

$$\cfrac{\cfrac{x=_r \mathcal{C}[y]:D \qquad y=_s w:D'}{x=_{\mathrm{sub_L}(r,s)} \mathcal{C}[w]:D}}{\mathcal{C}[w]=_{\sigma(\mathrm{sub_L}(r,s))} x:D} \qquad \triangleright \qquad \cfrac{\cfrac{y=_s w:D'}{w=_{\sigma(s)} y:D'} \qquad \cfrac{x=_r \mathcal{C}[y]:D}{\mathcal{C}[y]=_{\sigma(r)} x:D}}{\mathcal{C}[w]=_{\mathrm{sub_R}(\sigma(s),\sigma(r))} x:D}$$

$$\cfrac{\cfrac{x=_r y:D' \qquad \mathcal{C}[y]=_s w:D}{\mathcal{C}[x]=_{\mathrm{sub_R}(r,s)} w:D}}{w=_{\sigma(\mathrm{sub_R}(r,s))} \mathcal{C}[x]:D} \qquad \triangleright \qquad \cfrac{\cfrac{\mathcal{C}[y]=_s w:D}{w=_{\sigma(s)} \mathcal{C}[y]:D} \qquad \cfrac{x=_r y:D'}{y=_{\sigma(r)} x:D'}}{w=_{\mathrm{sub_L}(\sigma(s),\sigma(r))} \mathcal{C}[x]:D}$$

Associated rewritings:

• $\sigma(\mathrm{sub_L}(r,s)) \triangleright \mathrm{sub_R}(\sigma(s),\sigma(r))$

• $\sigma(\mathrm{sub_R}(r,s)) \triangleright \mathrm{sub_L}(\sigma(s),\sigma(r))$.

Definition 6.15 (σ and ξ).

$$\cfrac{\cfrac{x=_r y:A}{inl(x)=_{\xi(r)} inl(y):A\vee B}}{inl(y)=_{\sigma(\xi(r))} inl(x):A\vee B} \qquad \triangleright \qquad \cfrac{\cfrac{x=_r y:A}{y=_{\sigma(r)} x:A}}{inl(y)=_{\xi(\sigma(r))} inl(x):A\vee B}$$

$$\cfrac{\cfrac{x=_r y:A \qquad z=_s w:B}{\langle x,z\rangle=_{\xi(r,s)} \langle y,w\rangle:A\wedge B}}{\langle y,w\rangle=_{\sigma(\xi(r,s))} \langle x,z\rangle:A\wedge B} \qquad \triangleright \qquad \cfrac{\cfrac{x=_r y:A}{y=_{\sigma(r)} x:A} \qquad \cfrac{z=_s w:B}{w=_{\sigma(s)} z:B}}{\langle y,w\rangle=_{\xi(\sigma(r),\sigma(s))} \langle x,z\rangle:A\wedge B}$$

$$\cfrac{\cfrac{\overset{[x:A]}{\vdots}}{\cfrac{f(x)=_r g(x):B}{\lambda x.f(x)=_{\xi(r)} \lambda x.g(x):A\to B}}}{\lambda x.g(x)=_{\sigma(\xi(r))} \lambda x.f(x):A\to B} \qquad \triangleright \qquad \cfrac{\cfrac{\overset{[x:A]}{\vdots}}{\cfrac{f(x)=_r g(x):B}{g(x)=_{\sigma(r)} f(x):B}}}{\lambda x.g(x)=_{\xi(\sigma(r))} \lambda x.f(x):A\to B}$$

$$\cfrac{\cfrac{\overset{[x:D]}{\vdots}}{\cfrac{f(x)=_s g(x):P(x)}{\Lambda x.f(x)=_{\xi(s)} \Lambda x.g(x):\forall x^D.P(x)}}}{\Lambda x.g(x)=_{\sigma(\xi(s))} \Lambda x.f(x):\forall x^D.P(x)} \qquad \triangleright \qquad \cfrac{\cfrac{\overset{[x:D]}{\vdots}}{\cfrac{f(x)=_s g(x):P(x)}{g(x)=_{\sigma(s)} f(x):P(x)}}}{\Lambda x.g(x)=_{\xi(\sigma(s))} \Lambda x.f(x):\forall x^D.P(x)}$$

Associated rewritings:

• $\sigma(\xi(r)) \triangleright \xi(\sigma(r))$

• $\sigma(\xi(r,s)) \triangleright \xi(\sigma(r),\sigma(s))$.

Definition 6.16 (σ and μ).

$$\cfrac{\cfrac{x=_r y:A\wedge B}{FST(x)=_{\mu(r)} FST(y):A}}{FST(y)=_{\sigma(\mu(r))} FST(x):A} \qquad \triangleright \qquad \cfrac{\cfrac{x=_r y:A\wedge B}{y=_{\sigma(r)} x:A\wedge B}}{FSt(y)=_{\mu(\sigma(r))} FST(x):A}$$

$$\cfrac{\cfrac{x=_s y:A \qquad f=_r g:A\to B}{APP(f,x)=_{\mu(s,r)} APP(g,y):B}}{APP(g,y)=_{\sigma(\mu(s,r))} APP(f,x):B}$$

$$\triangleright \qquad \cfrac{\cfrac{x=_s y:A}{y=_{\sigma(s)} x:A} \qquad \cfrac{f=_r g:A\to B}{g=_{\sigma(r)} f:A\to B}}{APP(g,y)=_{\mu(\sigma(s),\sigma(r))} APP(f,x):B}$$

$$\dfrac{x=_ry{:}A\vee B \quad \dfrac{[s{:}A]}{\vdots\ \ d(s)=_uf(s){:}C} \quad \dfrac{[t{:}B]}{\vdots\ \ e(t)=_vg(t){:}C}}{\dfrac{CASE(x,\acute{s}d(s),\acute{t}e(t))=_{\mu(r,u,v)}CASE(y,\acute{s}f(s),\acute{t}g(t)){:}C}{CASE(y,\acute{s}f(s),\acute{t}g(t)){:}C=_{\sigma(\mu(r,u,v))}CASE(x,\acute{s}d(s),\acute{t}e(t)){:}C}}$$

$$\rhd\quad \dfrac{\dfrac{x=_ry{:}A\vee B}{y=_{\sigma(r)}x{:}A\vee B} \quad \dfrac{[s{:}A]\quad d(s)=_uf(s){:}C}{f(s)=_{\sigma(u)}d(s){:}C} \quad \dfrac{[t{:}B]\quad e(t)=_vg(t){:}C}{g(t)=_{\sigma(v)}e(t){:}C}}{CASE(y,\acute{s}f(s),\acute{t}g(t))=_{\mu(\sigma(r),\sigma(u),\sigma(v))}CASE(x,\acute{s}d(s),\acute{t}e(t)){:}C}$$

$$\dfrac{e=_sb{:}\exists x^D.P(x) \quad \dfrac{[t{:}D,\ g(t){:}P(t)]}{d(g,t)=_rf(g,t){:}C}}{\dfrac{INST(e,\acute{g}\acute{t}d(g,t))=_{\mu(s,r)}INST(b,\acute{g}\acute{t}f(g,t)){:}C}{INST(b,\acute{g}\acute{t}f(g,t))=_{\sigma(\mu(s,r))}INST(e,\acute{g}\acute{t}d(g,t)){:}C}}$$

$$\rhd\quad \dfrac{\dfrac{e=_sb{:}\exists x^D.P(x)}{b=_{\sigma(s)}e{:}\exists x^D.P(x)} \quad \dfrac{[t{:}D,\ g(t){:}P(t)]\quad d(g,t)=_rf(g,t){:}C}{f(g,t)=_{\sigma(r)}d(g,t){:}C}}{INST(b,\acute{g}\acute{t}f(g,t))=_{\mu(\sigma(s),\sigma(r))}INST(e,\acute{g}\acute{t}d(g,t)){:}C}$$

Associated rewritings:

- $\sigma(\mu(r)) \rhd \mu(\sigma(r))$
- $\sigma(\mu(s,r)) \rhd \mu(\sigma(s),\sigma(r))$
- $\sigma(\mu(r,u,v)) \rhd \mu(\sigma(r),\sigma(u),\sigma(v))$.

6.5.2.2   *Transitivity*

Definition 6.17 (τ and sub).

$$\dfrac{\dfrac{x=_r\mathcal{C}[y]:D \quad y=_sw:D'}{x=_{\mathsf{sub_L}(r,s)}\mathcal{C}[w]:D} \quad \mathcal{C}[w]=_tz:D}{x=_{\tau(\mathsf{sub_L}(r,s),t)}z:D}$$

$$\rhd\quad \dfrac{x=_r\mathcal{C}[y]:D \quad \dfrac{y=_sw:D' \quad \mathcal{C}[w]=_tz:D}{\mathcal{C}[y]=_{\mathsf{sub_R}(s,t)}z:D}}{x=_{\tau(r,\mathsf{sub_R}(s,t))}z:D}$$

$$\dfrac{\dfrac{y=_sw:D \quad \mathcal{C}[w]=_tz:D}{\mathcal{C}[y]=_{\mathsf{sub_R}(s,t)}z:D} \quad z=_uv:D}{\mathcal{C}[y]=_{\tau(\mathsf{sub_R}(s,t),u)}v:D}$$

$$\rhd\quad \dfrac{y=_sw:D' \quad \dfrac{\mathcal{C}[w]=_tz:D \quad z=_uv:D}{\mathcal{C}[w]=_{\tau(t,u)}v:D}}{\mathcal{C}[y]=_{\mathsf{sub_R}(s,\tau(t,u))}v:D}$$

$$\dfrac{x=_r\mathcal{C}[z]:D \quad \dfrac{\mathcal{C}[z]=_\rho\mathcal{C}[z]:D \quad z=_sw:D'}{\mathcal{C}[z]=_{\mathsf{sub_L}(\rho,s)}\mathcal{C}[w]:D}}{x=_{\tau(r,\mathsf{sub_L}(\rho,s))}\mathcal{C}[w]:D}$$

$$\rhd\quad \dfrac{x=_r\mathcal{C}[z]:D \quad z=_sw:D'}{x=_{\mathsf{sub_L}(r,s)}\mathcal{C}[w]:D}$$

$$x =_r \mathcal{C}[w] : D \quad \dfrac{w =_s z : D' \quad \mathcal{C}[z] =_\rho \mathcal{C}[z] : D}{\mathcal{C}[w] =_{\text{sub}_R(s,\rho)} \mathcal{C}[z] : D}$$
$$\overline{\qquad\qquad x =_{\tau(r,\text{sub}_R(s,\rho))} \mathcal{C}[z] : D \qquad\qquad}$$

$$\triangleright \quad \dfrac{x =_r \mathcal{C}[w] : D \quad w =_s z : D'}{x =_{\text{sub}_L(r,s)} \mathcal{C}[z] : D}$$

Definition 6.18 (τ and τ).

$$\dfrac{\dfrac{x =_t y : D \quad y =_r w : D}{x =_{\tau(t,r)} w : D} \quad w =_s z : D}{x =_{\tau(\tau(t,r),s)} z : D}$$

$$\triangleright \quad \dfrac{x =_t y : D \quad \dfrac{y =_r w : D \quad w =_s z : D}{y =_{\tau(r,s)} z : D}}{x =_{\tau(t,\tau(r,s))} z : D}$$

Associated rewritings:

- $\tau(\text{sub}_L(r,s),t) \triangleright \tau(r, \text{sub}_R(s,t))$
- $\tau(\text{sub}_R(s,t),u)) \triangleright \text{sub}_R(s, \tau(t,u))$
- $\tau(r, \text{sub}_L(\tau,s)) \triangleright \text{sub}_L(r,s)$
- $\tau(r, \text{sub}_R(s,\tau)) \triangleright \text{sub}_L(r,s)$
- $\tau(\tau(t,r),s) \triangleright \tau(t, \tau(r,s))$.

6.6 The rewriting system for the *LND* equational logic

The reductions and transformations between proofs shown in the previous section, define a rewriting system for the equational fragment of *LND* based on the *rewrite reasons*. This system is built from the equations between the rewrite reasons [de Oliveira (1995)].

The equations are oriented according to the direction of the reductions and transformations. Thus, the rewriting system defined here computes the normal form of proofs for the equational logic of the *LND*. It is defined as follows:

Definition 6.19 (LND_{EQ}-TRS). The term rewriting system for the equational fragment of *LND* (LND_{EQ}-TRS) computes the normal form of proofs for this fragment. It is composed by the following rules:

1. $\sigma(\rho) \triangleright \rho$
2. $\sigma(\sigma(r)) \triangleright r$
3. $\tau(\mathcal{C}[r], \mathcal{C}[\sigma(r)]) \triangleright \mathcal{C}[\rho]$
4. $\tau(\mathcal{C}[\sigma(r)], \mathcal{C}[r]) \triangleright \mathcal{C}[\rho]$
5. $\tau(\mathcal{C}[r], \mathcal{C}[\rho]) \triangleright \mathcal{C}[r]$

6. $\tau(\mathcal{C}[\rho], \mathcal{C}[r]) \rhd \mathcal{C}[r]$
7. $\text{sub}_L(\mathcal{C}[r], \mathcal{C}[\rho]) \rhd \mathcal{C}[r]$
8. $\text{sub}_R(\mathcal{C}[\rho], \mathcal{C}[r]) \rhd \mathcal{C}[r]$
9. $\text{sub}_L(\text{sub}_L(s, \mathcal{C}[r]), \mathcal{C}[\sigma(r)]) \rhd s$
10. $\text{sub}_L(\text{sub}_L(s, \mathcal{C}[\sigma(r)]), \mathcal{C}[r]) \rhd s$
11. $\text{sub}_R(\mathcal{C}[s], \text{sub}_R(\mathcal{C}[\sigma(s)], r)) \rhd r$
12. $\text{sub}_R(\mathcal{C}[\sigma(s)], \text{sub}_R(\mathcal{C}[s], r)) \rhd r$
13. $(FST)\ \mu(\xi(r, s)) \rhd r$
14. $(SND)\ \mu(\xi(r, s)) \rhd s$
15. $(inl)\ \mu(\xi(r), s, u) \rhd s$
16. $(inr)\ \mu(\xi(r), s, u) \rhd u$
17. $(\beta_{rewr}\text{-}\{\rightarrow, \forall\})\ \mu(s, \xi(r)) \rhd r$
18. $(\beta_{rewr}\text{-}\exists)\ \mu(\xi(r), s) \rhd s$
19. $(\eta_{rewr}\text{-}\wedge)\ \xi(\mu(r)) \rhd r$
20. $(\eta_{rewr}\text{-}\vee)\ \mu(t, \xi(r), \xi(s)) \rhd t$
21. $(\eta_{rewr}\text{:}\{\rightarrow, \forall\})\ \xi(\mu(r, s)) \rhd s$
22. $(\eta_{rewr}\text{-}\exists)\ \mu(s, \xi(r)) \rhd s$
23. $\sigma(\tau(r, s)) \rhd \tau(\sigma(s), \sigma(r))$
24. $\sigma(\text{sub}_L(r, s)) \rhd \text{sub}_R(\sigma(s), \sigma(r))$
25. $\sigma(\text{sub}_R(r, s)) \rhd \text{sub}_L(\sigma(s), \sigma(r))$
26. $\sigma(\xi(r)) \rhd \xi(\sigma(r))$
27. $\sigma(\xi(s, r)) \rhd \xi(\sigma(s), \sigma(r))$
28. $\sigma(\mu(r)) \rhd \mu(\sigma(r))$
29. $\sigma(\mu(s, r)) \rhd \mu(\sigma(s), \sigma(r))$
30. $\sigma(\mu(r, u, v)) \rhd \mu(\sigma(r), \sigma(u), \sigma(v))$
31. $\tau(r, \text{sub}_L(\rho, s)) \rhd \text{sub}_L(r, s)$
32. $\tau(r, \text{sub}_R(s, \rho)) \rhd \text{sub}_L(r, s)$
33. $\tau(\text{sub}_L(r, s), t) \rhd \tau(r, \text{sub}_R(s, t))$
34. $\tau(\text{sub}_R(s, t), u) \rhd \text{sub}_R(s, \tau(t, u))$
35. $\tau(\tau(t, r), s) \rhd \tau(t, \tau(r, s))$.

The $LND_{EQ}\text{-}TRS$, shown above, contains two more rules, which will be added to the system by the completion procedure, applied to the set of equations between rewrite reasons. These new rules represent more transformations between equational proofs.

As we will see, the proofs of the *termination* and *confluence* properties for the $LND_{EQ}\text{-}TRS$ are the same as proof of the *normalization* and *strong normalization* theorems for the *LND* equational fragment, respectively.

6.6.1 Termination property for the LND_{EQ}-TRS

Theorem 6.1 (Termination property for LND_{EQ}-TRS). LND_{EQ}-TRS is terminating.

The proof of the termination property for LND_{EQ}-TRS is obtained by using a special kind of ordering: *recursive path ordering*, proposed by Dershowitz in [Dershowitz (1982)]:

Definition 6.20 (recursive path ordering). Let $>$ be a partial ordering on a set of operators F. The recursive path ordering $>^*$ on the set T(F) of terms over F is defined recursively as follows:

$$s = f(s_1, \ldots, s_m) >^* g(t_1, \ldots, t_n) = t,$$

if and only if

(1) $f = g$ and $\{s_1, \ldots, s_m\} \gg^* \{t_1, \ldots, t_n\}$, or
(2) $f > g$ and $\{s\} \gg^* \{t_1, \ldots, t_n\}$, or
(3) $f \not\geq g$ and $\{s_1, \ldots, s_m\} \gg^*$ or $= \{t\}$,

where $\gg^*$ is the extension of $>^*$ to multisets.

Note that this definition uses the notion of ordering on multisets. A given partial ordering $>$ on a set S may be extended to a partial ordering $\gg$ on finite multisets of elements of S, wherein a multiset is reduced by removing one or more elements and replacing them with any finite number of elements, each of which is smaller than one of the elements removed [Dershowitz (1982)].

The proof of termination property via a recursive path ordering is made by showing that for all rules $e \rightarrow d$ of the system, $e >^* d$.

The *recursive path ordering* can be extended in order to allow some function of a term $f(t_1, \ldots, t_n)$ to play the role of the operator f. As explained in [Dershowitz (1982)], we can consider the k-th operand t_k to be the operator, and compare two terms by first recursively comparing their k-th operands.

In the proof of termination property for the LND_{EQ}-TRS, we use the precedence ordering on the rewrite operators for the rules from 1 to 32 defined as follows:

$$\sigma > \tau > \rho,$$
$$\sigma > \xi,$$
$$\sigma > \mu,$$
$$\sigma > \text{sub}_\text{L},$$
$$\sigma > \text{sub}_\text{R},$$
$$\tau > \text{sub}_\text{L}$$

We can combine the recursive path idea used for the rules 1–32 with an extension of recursive path ordering for rules from 33 to 35, where the first operand is used as operator. When comparing two "τ" and "τ with sub$_\text{R}$", we use the first operand as operator. This proof is similar to Example (H) given in pp. 299–300 of [Dershowitz (1982)]. The full proof is shown in detail in the Appendix 6.7.[5]

6.6.2 *Confluence property for the LND_{EQ}-TRS*

Unlike standard TRSs, the LND_{EQ}-TRS contains some rules which are similar to the rules of *conditional rewriting systems*. In conditional rewriting systems, the rules have associated conditions. Thus, during the computation of the critical pairs, these conditions must be considered. The definition of a conditional TRS, given in [Plaisted (1994)], is as follows:

Definition 6.21 (conditional term rewriting systems). In conditional term rewriting systems, the rules have conditions attached, which must be true for the rewrite occur. For example, a rewrite rule $e \rightarrow d$ with condition C is expressed as

$$C \mid e \rightarrow d$$

The condition C may be of several forms; for example they may be logical formulas, equations, inequations, etc.

The only rules of the LND_{EQ}-TRS which have associated conditions are those which come from the β_{rewr} and η_{rewr} reductions. This is because these reductions reflect the redundancies on the logical calculus. For example, for the rule (FST) $\mu(\xi(r,s)) \rhd r$, the rewrite is performed only if the terms of the associated equation is a β-redex like "$FST\langle x,y\rangle$". Thus, the confluence proof must consider these conditions.

---

[5]We are grateful for one referee who noticed that our previous explanation of the proof was incomplete. It was necessary to make explicit our use of the *recursive path ordering*, such as it was done in Dershowitz's example.

The confluence proof is made by the superposition algorithm [Knuth and Bendix (1970)] applied to the rules of the system. For example:

- The rules 1 and 2 causes the following superposition: $\sigma(\sigma(\rho))$. This term rewrites in two different ways:

  (1) $\rhd_1 \sigma(\rho) \rhd_1 \rho$
  (2) $\rhd_2 \rho$

No divergent critical pair was generated.

The proof proceeds in this way, each rule is compared to the others in order to find a critical pair. If a critical pair is found, the completion procedure [Knuth and Bendix (1970)] shows how to add new rules to the system in such a way that it becomes confluent. In the case of $LND_{EQ}\text{-}TRS$ it was necessary to include two more rules:

36. $\tau(\mathcal{C}[u], \tau(\mathcal{C}[\sigma(u)], v)) \rhd v$
37. $\tau(\mathcal{C}[\sigma(u)], \tau(\mathcal{C}[u], v)) \rhd u$.

These rules define a new basic set of transformations between proofs:

$$\dfrac{x =_s u : D \qquad \dfrac{\dfrac{x =_s u : D}{u =_{\sigma(s)} x : D} \quad x =_v w : D}{u =_{\tau(\sigma(s),v)} w : D}}{x =_{\tau(s,\tau(\sigma(s),v))} w : D} \quad \rhd \quad x =_v w : D$$

$$\dfrac{\dfrac{x =_s w : D}{w =_{\sigma(s)} x : D} \quad \dfrac{x =_s w : D \quad w =_v z : D}{x =_{\tau(s,v)} z :}}{w =_{\tau(\sigma(s),\tau(s,v))} z : D} \quad \rhd \quad w =_v z : D.$$

6.6.3 *The normalization theorems*

The normalization theorems are obtained as a consequence of the termination and confluence properties of the $LND_{EQ}\text{-}TRS$.

Theorem 6.2 (normalization). *Every derivation in the LND system converts to a normal form.*

Proof. Since the *rewrite reasons* represent an orthogonal measure of redundancy associated to the *LND* equational logic, thus being a "record" of all proof steps, then when a TRS is built from the reductions on the *rewrite reasons*, the proof of its termination implies that all proofs in the *LND* equational fragment have a normal form. ∎

Theorem 6.3 (strong normalization). *Every derivation in the LND system converts to a unique normal form.*

Proof.   This theorem is proved by using the same argument as in the previous proof. The *confluence* proof of the LND_{EQ}-TRS implies the unicity of normal form proofs in the *LND* equational fragment.   ∎

6.7   The normalization procedure: some examples

In this section, we illustrate the potential of the normalization procedure of *LND* equational fragment through two examples of proof transformations. By just manipulating the composition of rewrites with the rules of the LND_{EQ}-TRS one can identify redundancies on the proofs and transform them into its normal form.

Example 6.4. The following deduction is not in normal form:

$$\cfrac{\cfrac{\cfrac{\cfrac{f(x,z) =_s f(w,y) : D}{f(w,y) =_{\sigma(s)} f(x,z) : D} \quad x =_r c : D}{f(w,y) =_{\mathrm{sub_L}(\sigma(s),r)} f(c,z) : D}}{f(c,z) =_{\sigma(\mathrm{sub_L}(\sigma(s),r))} f(w,y) : D \quad y =_t b : D}}{f(c,z) =_{\mathrm{sub_L}(\sigma(\mathrm{sub_L}(\sigma(s),r)),t)} f(w,b) : D}$$

The resulting composition of rewrites '$\mathrm{sub_L}(\sigma(\mathrm{sub_L}(\sigma(s),r)),t)$' is reduced by the LND_{EQ}-TRS to its normal form:

$$\mathrm{sub_L}(\sigma(\mathrm{sub_L}(\sigma(s),r)),t) \triangleright_{24} \mathrm{sub_L}(\mathrm{sub_R}(\sigma(r),\sigma(\sigma(s))),t)$$
$$\triangleright_2 \ \mathrm{sub_L}(\mathrm{sub_R}(\sigma(r),s),t)$$

From the resulting reason '$(\mathrm{sub_L}(\mathrm{sub_R}(\sigma(r),s),t))$' obtained by the rewriting process, an equivalent normal proof is built up:

$$\cfrac{\cfrac{\cfrac{x =_r c : D}{c =_{\sigma(r)} x : D} \quad f(x,z) =_s f(w,y) : D}{f(c,z) =_{\mathrm{sub_R}(\sigma(r),s)} f(w,y) : D} \quad y =_t b : D}{f(c,z) =_{\mathrm{sub_L}(\mathrm{sub_R}(\sigma(r),s),t)} f(w,b) : D}$$

Example 6.5.

$$\cfrac{\cfrac{x =_r y : A}{inl(x) =_{\xi(r)} inl(y) : A \vee B} \quad \cfrac{\cfrac{x =_r y : A}{inl(x) =_{\xi(r)} inl(y) : A \vee B}}{inl(y) =_{\sigma(\xi(r))} inl(x) : A \vee B}}{inl(x) =_{\tau(\xi(r),\sigma(\xi(r)))} inl(x) : A \vee B}$$

This proof is reduced to its normal form by the following rewrites applied to '$\tau(\xi(r), \sigma(\xi(r)))$':

(1) $\triangleright_3 \xi(\rho)$

(2) $\triangleright_{26} \tau(\xi(r), \xi(\sigma(r)))\ \triangleright_3\ \xi(\rho)$

Note that two differents rewrites were performed. However, since the LND_{EQ}-TRS is a confluent system, only one reason was found.

The equivalent normal proof is the following:

$$\frac{\dfrac{\dfrac{x : A}{x =_\rho x : A}}{inl(x) =_\rho inl(x) : A \vee B}}{}$$

6.8   Final remarks

We have shown that starting from the basic transformations between proofs in the *LND* equational fragment, one can define a set of rules, based on the 'rewrite *reasons*', which forms a *term rewriting system (TRS)*: the LND_{EQ}-TRS. Proving the *termination* and *confluence* properties for the LND_{EQ}-TRS, we have in fact proved the normalization and strong normalization theorems for the *LND* equational fragment, respectively.

Applying the so-called *completion procedure*, proposed by Knuth and Bendix in [Knuth and Bendix (1970)], to the LND_{EQ}-TRS, a rather wide and complete set of proof transformations was generated: a total of 37 rewrite rules, each one corresponding to a proof transformation.

6.9   Appendix: The β- and η-reductions for the *LND* system

6.9.1   β-*type reductions*

The explanation of the normalization of noncanonical proofs, i.e. those which contain 'redundant' steps identified by an *introduction* inference immediately followed by an *elimination* inference, are framed in the following way (where '$\triangleright_\beta$' represents 'β-converts/normalizes to'):

β-∧-*reduction*

$$\frac{\dfrac{a : A \qquad b : B}{\langle a, b \rangle : A \wedge B}\ \wedge\text{-intr}}{FST(\langle a, b \rangle) : A}\ \wedge\text{-elim} \qquad \triangleright_\beta \qquad a : A$$

$$\frac{\dfrac{a:A \qquad b:B}{\langle a,b \rangle : A \wedge B} \wedge \text{-}intr}{SND(\langle a,b \rangle) : B} \wedge \text{-}elim \qquad\qquad \triangleright_\beta \qquad\qquad b:B$$

$\beta\text{-}\vee\text{-reduction}$

$$\frac{\dfrac{a:A}{inl(a):A \vee B} \vee \text{-}intr \qquad \dfrac{[x:A] \qquad [y:B]}{d(x):C \qquad e(y):C}}{CASE(inl(a), \acute{x}d(x), \acute{y}e(y)) : C} \vee \text{-}elim \quad \triangleright_\beta \qquad \begin{array}{c} a:A \\ d(a/x):C \end{array}$$

$$\frac{\dfrac{b:B}{inr(b):A \vee B} \vee \text{-}intr \qquad \dfrac{[x:A] \qquad [y:B]}{d(x):C \qquad e(y):C}}{CASE(inr(b), \acute{x}d(x), \acute{y}e(y)) : C} \vee \text{-}elim \quad \triangleright_\beta \qquad \begin{array}{c} b:B \\ e(b/y):C \end{array}$$

where '´' is an abstractor which forms value-range terms such as '$\acute{x}d(x)$' where 'x' is bound, discharging the corresponding assumption labelled by x.

$\beta\text{-}\rightarrow\text{-reduction}$

$$\frac{a:A \qquad \dfrac{\begin{array}{c}[x:A]\\ b(x):B\end{array}}{\lambda x.b(x):A \rightarrow B} \rightarrow \text{-}intr}{APP(\lambda x.b(x),a):B} \rightarrow \text{-}elim \qquad \triangleright_\beta \qquad \begin{array}{c} a:A \\ b(a/x):B \end{array}$$

$\beta\text{-}\forall\text{-reduction}$

$$\frac{a:D \qquad \dfrac{\begin{array}{c}[x:D]\\ f(x):P(x)\end{array}}{\Lambda x.f(x):\forall x^D.P(x)} \forall\text{-}intr}{EXTR(\Lambda x.f(x),a):P(a)} \forall\text{-}elim \qquad \triangleright_\beta \qquad \begin{array}{c} a:D \\ f(a/x):P(a) \end{array}$$

$\beta\text{-}\exists\text{-reduction}$

$$\frac{\dfrac{a:D \qquad f(a):P(a)}{\varepsilon x.(f(x),a):\exists x^D.P(x)} \exists\text{-}intr \qquad \dfrac{[t:D,g(t):P(t)]}{d(g,t):C}}{INST(\varepsilon x.(f(x),a), \acute{g}\acute{t}d(g,t)):C} \exists\text{-}elim \quad \triangleright_\beta \qquad \begin{array}{c} a:D, f(a):P(a) \\ d(f/g,a/t):C \end{array}$$

where '´' is an abstractor which binds the free variables of the label, discharging the corresponding assumptions made in eliminating the existential quantifier, namely the 'Skolem'-type assumptions '$[t:D]$' and '$[g(t):P(t)]$', forming the value-range term '$\acute{g}\acute{t}d(g,t)$' where both the Skolem-constant 't', and the Skolem-function 'g', are bound. In the $\exists$-*elimination* the variables 't' and 'g' must occur free at least once in the term alongside the formula 'C' in the premise, and will be bound alongside the same formula in the conclusion of the rule.

6.9.1.1 β-*equality*

To each β-*reduction* rule there corresponds an equation which is induced by the so-called Curry–Howard correspondence.

$$FST(\langle a, b \rangle) =_\beta a$$
$$SND(\langle a, b \rangle) =_\beta b$$

$$CASE(inl(a), \acute{x}d(x), \acute{y}e(y)) =_\beta d(a/x)$$
$$CASE(inr(b), \acute{x}d(x), \acute{y}e(y)) =_\beta e(b/y)$$

$$APP(\lambda x.b(x), a) =_\beta b(a/x)$$

$$EXTR(\Lambda x.f(x), a) =_\beta f(a/x)$$

$$INST(\varepsilon x.(f(x), a), \acute{g}\acute{t}d(g, t)) =_\beta d(f/g, a/t)$$

6.9.2 η-*type reductions*

In a natural deduction proof system there is another way of making 'redundant' steps that one can make, apart from the above '*introduction* followed by *elimination*'. It is the exact inverse of this previous way of introducing redundancies: an *elimination* step is followed by an *introduction* step.

η-∧-*reduction*

$$\cfrac{\cfrac{c : A \wedge B}{FST(c) : A} \wedge \text{-}elim \quad \cfrac{c : A \wedge B}{SND(c) : B} \wedge \text{-}elim}{\langle FST(c), SND(c) \rangle : A \wedge B} \wedge \text{-}intr \qquad \rhd_\eta \qquad c : A \wedge B$$

η-∨-*reduction*

$$\cfrac{c : A \vee B \quad \cfrac{[x : A]}{inl(x) : A \vee B} \vee \text{-}intr \quad \cfrac{[y : B]}{inr(y) : A \vee B} \vee \text{-}intr}{CASE(c, \acute{x}inl(x), \acute{y}inr(y)) : A \vee B} \vee \text{-}elim$$

$$\rhd_\eta c : A \vee B$$

η-→-*reduction*

$$\cfrac{\cfrac{[x : A] \quad c : A \rightarrow B}{APP(c, x) : B} \rightarrow \text{-}elim}{\lambda x.APP(c, x) : A \rightarrow B} \rightarrow \text{-}intr \qquad \rhd_\eta \qquad c : A \rightarrow B$$

where c does not depend on x.

η-∀-reduction

$$\cfrac{\cfrac{[t:D] \qquad c:\forall x^D.P(x)}{EXTR(c,t):P(t)}\forall\text{-}elim}{\Lambda t.EXTR(c,t):\forall t^D.P(t)}\forall\text{-}intr \qquad\qquad \rhd_\eta \qquad\qquad c:\forall x^D.P(x)$$

where x does not occur free in c.

η-∃-reduction

$$\cfrac{c:\exists x^D.P(x) \qquad \cfrac{[t:D] \qquad [g(t):P(t)]}{\varepsilon y.(g(y),t):\exists y^D.P(y)}\exists\text{-}intr}{INST(c,\acute{g}t\varepsilon y.(g(y),t)):\exists y^D.P(y)}\exists\text{-}elim \quad \rhd_\eta \quad c:\exists x^D.P(x)$$

6.9.2.1   *η-equality*

The η-rules of proof transformations induce the following equations (according to the Curry–Howard correspondence):

$$\langle FST(c), SND(c)\rangle =_\eta c$$

$$CASE(c, \acute{a}inl(a), \acute{b}inr(b)) =_\eta c$$

$$\lambda x.APP(c,x) =_\eta c \quad (x \text{ is not free in } c)$$

$$\Lambda t.EXTR(c,t) =_\eta c \quad (t \text{ is not free in } c)$$

$$INST(c, \acute{g}t\varepsilon y.(g(y),t)) =_\eta c$$

6.9.3   *ζ-type transformations: The* **permutative** *transformations*

For the connectives that make use of 'Skolem'-type procedures of opening local branches with new assumptions, locally introducing new names and making them 'disappear' (or lose their identity via an abstraction) just before coming out of the local context or scope, there is another way of transforming proofs, which goes hand-in-hand with the properties of 'value-range' terms resulting from abstractions.

In natural deduction terminology, these proof transformations are called 'permutative' transformations.

ζ-∨-transformation

$$\cfrac{\cfrac{p:A\vee B \quad \cfrac{[x:A]}{d(x):C} \quad \cfrac{[y:B]}{e(y):C}}{CASE(p,\acute{x}d(x),\acute{y}e(y)):C}}{w(CASE(p,\acute{x}d(x),\acute{y}e(y))):W}r \quad \rhd_\zeta \quad \cfrac{p:A\vee B \quad \cfrac{\cfrac{[x:A]}{d(x):C}}{w(d(x)):W}r \quad \cfrac{\cfrac{[y:B]}{e(y):C}}{w(e(y)):W}r}{CASE(p,\acute{x}w(d(x)),\acute{y}w(e(y))):W}$$

provided 'r' does not interfere with the dependencies in the main branch, i.e. that 'w' does not bind any variable in p.

ζ-$\exists$-*transformation*

$$\frac{e : \exists x^D.P(x) \qquad \begin{array}{c}[t : D, g(t) : P(t)]\\ d(g,t) : C\end{array}}{INST(e, \acute{g}\acute{t}d(g,t)) : W}r \quad \triangleright_\zeta \quad \frac{e : \exists x^D.P(x) \qquad \begin{array}{c}[t : D, g(t) : P(t)]\\ \dfrac{d(g,t) : C}{w(d(g,t)) : W}r\end{array}}{INST(e, \acute{g}\acute{t}w(d(g,t))) : W}$$

provided 'r' does not interfere with the dependencies in the main branch, i.e. that 'w' does not bind any variable in e.

6.9.3.1 ζ-equality

Now, if the functional calculus on the labels is to match the logical calculus on the formulas, we must have the following ζ-equality (read 'zeta'-equality) between terms:

$$w(CASE(p, \acute{x}d(x), \acute{y}e(y))) =_\zeta CASE(p, \acute{x}w(d(x)), \acute{y}w(e(y)))$$

$$w(INST(e, \acute{g}\acute{t}d(g,t))) =_\zeta INST(e, \acute{g}\acute{t}w(d(g,t)))$$

Note that both in the case of '$\vee$' and '$\exists$' the operator 'w' could be 'pushed inside' the value-range abstraction terms. In the case of disjunction, the operator could be pushed inside the ''-abstraction terms, and in the $\exists$-case, the 'w' could be pushed inside the 'σ'-abstraction term.

6.10 Termination property for LND_{EQ}-TRS

The proof of the termination property for LND_{EQ}-TRS uses the *subterm condition* of recursive path orderings, which says that $f(\ldots t \ldots) \geq t$ [Dershowitz (1982)].

- The proof for the rules 1 and 2 and from 5 to 22 follows from the subterm condition of recursive path oroderings.
- (rule 3) $\tau(C \mid r \mid, C \mid \sigma(r) \mid) >^* \rho$:

  - $\tau > \rho$ from the precedence ordering on the rewrite operators.
  - $\{\tau(C \mid r \mid, C \mid \sigma(r) \mid)\} \gg^* \emptyset$.

- (rule 4) $\tau(C \mid \sigma(r) \mid, C \mid r \mid) >^* \rho$:

  - $\tau > \rho$ from the precedence ordering on the rewrite operators.

- $\{\tau(C \mid \sigma(r) \mid, \mid r \mid)\} \gg^* \emptyset$.

- (rule 23) $\sigma(\tau(r, s)) >^* \tau(\sigma(s), \sigma(r))$:

  - $\sigma > \tau$ from the precedence ordering on the rewrite operators.
  - $\{\sigma(\tau(r, s))\} \gg^* \{\sigma(s), \sigma(r)\}$:

    - $\sigma(\tau(r, s)) >^* \sigma(s)$: $\sigma = \sigma$ and $\{\tau(r, s)\} \gg^* \{s\}$ from the subterm condition.
    - $\sigma(\tau(r, s)) >^* \sigma(r)$: $\sigma = \sigma$ and $\{\tau(r, s)\} \gg^* \{r\}$ from the subterm condition.

- (rule 24) $\sigma(\mathrm{sub_L}(r, s)) >^* \mathrm{sub_R}(\sigma(s), \sigma(r))$:

  - $\sigma > \mathrm{sub_R}$ from the precedence ordering on the rewrite operators.
  - $\{\sigma(\mathrm{sub_L}(r, s))\} \gg^* \{\sigma(s), \sigma(r)\}$:

    - $\sigma(\mathrm{sub_L}(r, s)) >^* \sigma(s)$ e $\sigma(\mathrm{sub_L}(r, s)) >^* \sigma(r)$:

      - $\sigma = \sigma$
      - $\{\mathrm{sub}(r, s)\} \gg^* \{s\}$ from the subterm condition.
      - $\{\mathrm{sub}(r, s)\} \gg^* \{r\}$ from the subterm condition.

- (rule 25) $\sigma(\mathrm{sub_R}(r, s)) \rhd \mathrm{sub_L}(\sigma(s), \sigma(r))$:

  - $\sigma > \mathrm{sub_L}$ from the precedence ordering on the rewrite operators.
  - $\{\sigma(\mathrm{sub_R}(r, s))\} \gg^* \{\sigma(s), \sigma(r)\}$:

    - $\sigma = \sigma$
    - $\{\mathrm{sub_R}(r, s)\} \gg^* \{s\}$ from the subterm condition.
    - $\{\mathrm{sub_R}(r, s)\} \gg^* \{r\}$ from the subterm condition.

- (rule 26) $\sigma(\xi(r_1)) >^* \xi(\sigma(r_1))$:

  - $\sigma > \xi$ from the precedence ordering on the rewrite operators.
  - $\{\sigma(\xi(r_1))\} \gg^* \{\sigma(r_1)\}$:

    - $\sigma = \sigma$ and $\{(\xi(r_1)\} \gg^* \{r_1\}$ from the subterm condition.

- (rule 27) $\sigma(\xi(r_1, r_2)) >^* \xi(\sigma(r_1), \sigma(r_2))$:

  - $\sigma > \xi$ from the precedence ordering on the rewrite operators.
  - $\{\sigma(\xi(r_1, r_2))\} \gg^* \{\sigma(r_1), \sigma(r_2)\}$:

    - $\sigma = \sigma$
    - $\{\xi(r_1, r_2)\} \gg^* \{r_1\}$ from the subterm condition.
    - $\{\xi(r_1, r_2)\} \gg^* \{r_2\}$ from the subterm condition.

- (rule 28) $\sigma(\mu(r_1)) >^* \mu(\sigma(r_1))$:

  - $\sigma > \mu$ from the precedence ordering on the rewrite operators.

- $\{\sigma(\mu(r_1))\} \gg^* \{\sigma(r_1)\}$:

  - $\sigma = \sigma$
  - $\{\mu(r_1)\} \gg^* \{r_1\}$ from the subterm condition.

- (rule 29) $\sigma(\mu(r_1, r_2)) >^* \mu(\sigma(r_1), \sigma(r_2))$:

  - $\sigma > \mu$ from the precedence ordering on the rewrite operators.
  - $\{\sigma(\mu(r_1, r_2))\} \gg^* \{\sigma(r_1), \texttt{symm}(r_2)\}$:

    - $\sigma = \sigma$
    - $\{\mu(r_1, r_2)\} \gg^* \{r_1\}$ from the subterm condition.
    - $\{\mu(r_1, r_2)\} \gg^* \{r_2\}$ from the subterm condition.

- (rule 30) $\sigma(\mu(r_1, r_2, r_3)) >^* \mu(\sigma(r_1), \sigma(r_2), \sigma(r_3))$:

  - $\sigma > \mu$ from the precedence ordering on the rewrite operators.
  - $\{\sigma(\mu(r_1, r_2, r_3))\} \gg^* \{\sigma(r_1), \sigma(r_2), \sigma(r_3)\}$:

    - $\sigma = \sigma$
    - $\{\mu(r_1, r_2, r_3)\} \gg^* \{r_1\}$, $\{\mu(r_1, r_2, r_3)\} \gg^* \{r_2\}$ and $\{\mu(r_1, r_2, r_3)\} \gg^* \{r_3\}$ from the subterm condition.

- (rule 33) $\tau(\texttt{sub}_\texttt{L}(r, s), t) >^* \tau(r, \texttt{sub}_\texttt{R}(s, t))$:

  - $\texttt{sub}_\texttt{L}(r, s) >^* r$ from the subterm condition.
  - $\{\tau(\texttt{sub}_\texttt{L}(r, s), t)\} \gg^* \{\texttt{sub}_\texttt{R}(s, t)\}$:

    - $\texttt{sub}_\texttt{L}(r, s) >^* s$ from the subterm condition.
    - $\{\tau(\texttt{sub}_\texttt{L}(r, s), t)\} \gg^* \{t\}$ from the subterm condition.

- (rule 34) $\tau(\texttt{sub}_\texttt{R}(s, t), u)) >^* \texttt{sub}_\texttt{R}(s, \texttt{trans}(t, u))$:

  - $\texttt{sub}_\texttt{R}(s, t) >^* s$
  - $\{\tau(\texttt{sub}_\texttt{R}(s, t), u))\} \gg^* \{\tau(t, u)\}$:

    - $\texttt{sub}_\texttt{R}(s, t) >^* t$ from the subterm condition.
    - $\{\tau(\texttt{sub}_\texttt{R}(s, t), u))\} \gg^* \{u\}$ from the subterm condition.

- (rule 35) $\tau(\tau(t, r), s) >^* \tau(t, \tau(r, s))$:

  - $\tau(t, r) >^* t$ from the subterm condition.
  - $\{\tau(\tau(t, r), s)\} \gg^* \{\tau(r, s)\}$:

    - $\tau(t, r) >^* r$ from the subterm condition.
    - $\{\tau(\tau(t, r), s)\} \gg^* \{s\}$ from the subterm condition.

Chapter 7

Modal Logics

Preamble

Since the early days of Kripke-style possible-worlds semantics for modalities, there has been a significant amount of research into the development of mechanisms for handling and characterising modal logics by means of 'naming' possible worlds, either directly by introducing identifiers, or indirectly by some other means (such as, for example, using formulas to identify the possible world). To list a few, we have: Gabbay's [Gabbay (1994)] 'labelling' formulas with names for worlds in the framework of *labelled deductive systems*; Fitch's [Fitch (1966b)] tree-proof deduction procedures for modal logics; Thomason and Stalnaker's [Thomason and Stalnaker (1968)] device of *predicate abstraction* introduced to handle skolemisation across worlds and the problem of characterising non-rigid designators; Fitting's [Fitting (1972, 1975)] ε-calculus-based axiom systems for modal logics, as well as his tableau systems for modal logics [Fitting (1981)] with *predicate abstraction* [Fitting (1989)]; the *irreflexivity rule* of Gabbay and Hodinson's [Gabbay and Hodkinson (1990)] axiomatic systems of temporal logic; the explicit reference to possible worlds in the deterministic modal logics of Fariñas and Herzig [Fariñas del Cerro and Herzig (1990)]; Ohlbach's [Ohlbach (1991)] semantics-based translation methods for modal logics and its functional representation of possible worlds structures; etc. In first-order predicate logics the individuals over which one quantifies (either universally or existentially) are rather naturally assumed to have names. The main connectives of the so-called modal logics are such that they quantify over (higher-order) objects which, unlike the first-order case, are not usually given names, in some cases for methodological reasons such as that which says that the main paradigm of modal logics is exactly that of variable-freeness (name-freeness).

The so-called Curry–Howard functional interpretation of logics [Curry (1934);

Howard (1980)] appears to be particularly suitable to deal with the handling of 'arbitrary' objects. Once names for arbitrary objects are introduced, the mechanism of *abstraction*, which goes back at least to Frege's [Frege (1893)] *Grundgesetze* mechanism of binding free-variables via 'course-of-values' function-terms, serves to 'discharge' assumptions such as that the particular name denoted an arbitrary object.

7.1 The functional interpretation

In the functional interpretation (with a system of 'labelled natural deduction', as described in Chapter 1) the rule of universal generalisation — $\forall$-*introduction* —, discharges the assumption that an arbitrary element inhabits the domain over which one wants to quantify by Λ-abstracting over the particular variable in the term that shows why that arbitrary individual does satisfy the property, which is being asserted for all individuals from that particular domain:

$\forall$-*introduction*

$$\frac{\begin{array}{c}[x:D]\\ f(x):F(x)\end{array}}{\Lambda x.f(x):\forall x^D.F(x)}$$

Note that the assumption '$x : D$', literally 'x' inhabits 'D', is discharged by the *introduction* of the universal quantifier, the free occurrences of 'x' in '$f(x)$' being bound by the Λ-abstraction. Observe also that '$f(x)$' may actually contain none, one, many, etc., occurrences of 'x' and so, according to recipe as to how assumptions are to be discharged, the abstraction might be actually binding none, exactly one, many, etc., free occurrences of the variable 'x'. This element of vagueness, as we shall see below, will serve to help us determine what kind of universal quantifier we want to have.[1] The characterisation of the connective in question as indeed a universal quantifier will come as its functionality is explained, which is not yet made clear by the mere explanation of its assertability conditions formalised by the *introduction* rule.

The key semantical notion in the functional interpretation is that of *normalisation* of proofs, which is a proof transformation step demonstrating how the *elimination* rules operate on the result of *introduction* inferences. The characterisation of the universal quantifier is then established by the *normalisation* of $\forall$

---

[1] The various kinds of universal quantifier can be analysed via the framework using a similar technique to control abstractions as described in Chapter 2. By an explicit control of the rule of universal quantification one can study how this will have an effect on the kind of universal quantifier one is dealing with. For example in certain areas one might be dealing with branching quantifiers.

proofs: given an arbitrary element 'a' from the domain 'D', and given a proof-construction of 'forall x in D, $F(x)$ is verified', then the $EXTR$action of the proof-construction for $F(a)$ is made by applying the Λ-term to the argument 'a' eliminating the universal quantifier.

Thus, by fixing the rule of β-normalisation on proofs,[2] namely,

∀-reduction

$$\frac{a : D \quad \dfrac{\begin{array}{c}[x : D]\\ f(x) : F(x)\end{array}}{\Lambda x. f(x) : \forall x :^D . F(x)} \text{∀-intr}}{EXTR(\Lambda x. f(x), a) : F(a)} \text{∀-elim} \quad \rightsquigarrow_\beta \quad \begin{array}{c}[a : D]\\ f(a/x) : F(a),\end{array}$$

which means that β-equality on terms is fixed by the rule of computation (cf. [Tait (1965)], where the semantics of convertibility is advocated):

$$EXTR(\Lambda x. f(x), a) =_\beta f(a/x) : F(a)$$

the distinct kinds of universal quantifiers (linear, relevant, etc.) can be obtained by imposing a corresponding Λ-abstraction discipline on the terms allowing some free occurrences of the variable to be bound (therefore allowing the corresponding assumption to be discharged), and disallowing others to become bound at a single step. It seems natural to take account of this extra-condition imposed on the Λ-abstraction discipline, given that the natural deduction *introduction* rule for conditionals ($\rightarrow$, $\forall$, $\Box$) does leave room for manoeuvre: in the rule of ∀-*introduction* above, which is to be read as 'given the assumption that 'x' is an arbitrary element from the domain 'D', and *from this assumption one obtains* an element '$f(x)$' of (dependent) type '$F(x)$', one can conclude that '$\Lambda x. f(x)$' is an element of type '$\forall x^D . F(x)$'.' Note that the italicized part of the reading does not contain any specific recipe as to if, how or when the assumption '$x : D$' was actually used to obtain '$f(x) : F(x)$'. In some natural deduction presentations such an element of looseness is made explicit by the inclusion of three vertical dots meant to be understood as 'after a certain number of proof steps', as in:

$$[x : D]$$
$$\vdots$$
$$\frac{f(x) : F(x)}{\Lambda x. f(x) : \forall x^D . F(x)}.$$

One is then supposed to read the rule as saying that, assuming x inhabits D and, after a certain number of proof steps, we obtain that x satisfies $F(x)$, we can

---

[2] See [Prawitz (1965)] (Chapter II, §2) for an elegant explanation of the notion of proof reduction.

conclude that for all elements in D, $F(x)$ is verified. And the room for manoeuvre, so to speak, as to how to obtain different universal quantifiers through making the three dots a more specific sequence of proof steps, is clearly seen to be given by the (somewhat vague) character of the rule of $\forall$-*introduction*.

7.1.1 *Handling assumptions with 'world' variables*

It so happens that essentially the same procedure of capitalising on this loose aspect of the natural deduction *introduction* rules for conditionals was used in [Gabbay and de Queiroz (1992)] to obtain the functional interpretation of various notions of implication below and above intuitionistic implication. By fixing the normalisation rule on terms of $\rightarrow$-type, and parameterising the rule that naturally leaves room for manoeuvre, namely that which involves a λ-abstraction procedure, one can make use of various λ-calculi to obtain the correspondence between normalisation of proofs in these various implicational calculi and the convertibility of terms in the corresponding term-calculus.

Now, we can look at the modality of necessity, namely '$\Box$', as a second-order universal quantification, in a way such that:

$$\Box A \equiv \forall W^{\mathcal{U}}.A(W),$$

reading 'for all worlds W from the (structured) collection of worlds $\mathcal{U}$, A is true at (demonstrable at, forced by, etc.) W', with corresponding *introduction* rule:

$\Box$-*introduction*:

$$\frac{\begin{array}{c}[W:\mathcal{U}]\\ F(W):A(W)\end{array}}{\underbrace{\Lambda W.f(W):\forall W^{\mathcal{U}}.A(W)}_{\Box A}} \qquad \text{or} \qquad \frac{\begin{array}{c}[W:\mathcal{U}]\\ F(W):A(W)\end{array}}{\Lambda W.f(W):\Box A},$$

where '$\mathcal{U}$' would be a collection of 'worlds' (where a world can be taken to be, e.g. structured collections (lists, bags, trees, etc.) of labelled formulas) and '$F(W)$' is an expression which may depend on the world-variable 'W'.

The *reduction* rule would be framed in a similar way to the rule of *reducion* of the implication and of the universal quantifier:

$\Box$-*reduction*

$$\frac{S:\mathcal{U} \quad \dfrac{\dfrac{[W:\mathcal{U}]}{F(W):A(W)}}{\Lambda W.f(W):\Box A}\Box\text{-}intr}{\mathcal{EXTR}(\Lambda W.f(W),S):A(S)}\Box\text{-}elim \quad \leadsto_{\beta} \quad \frac{[S:\mathcal{U}]}{F(S/W):A(S)},$$

which in terms of equality would look like:

$$\frac{[\mathbb{W} : \mathcal{U}]}{\mathbb{S} : \mathcal{U} \qquad F(\mathbb{W}) : A(\mathbb{W})}{\mathcal{EXTR}(\Lambda\mathbb{W}.f(\mathbb{W}), \mathbb{S}) =_\beta F(\mathbb{S}/\mathbb{W}) : A(\mathbb{S})}.$$

Notice that also here — as in the propositional and the first-order cases — it appears to make good sense to follow the 'semantics of convertibility' mentioned above: the rule of *normalisation*, which shows the effect of the *elimination* operator (here '$\mathcal{EXTR}$') on a term resulting from an *introduction* inference (here '$\Lambda\mathbb{W}.f(\mathbb{W})$') is the key semantical device used to provide the functional interpretation of '$\Box$'.[3]

So, our $\Box$-*elimination* also follows the pattern established for the other conditionals ($\rightarrow$, $\forall$), and is presented similarly to *modus ponens* and *universal extraction*:

---

[3] As we have previously mentioned in Chapter 1, there is an interesting connection between the semantics of convertibility and the so-called 'dialogue-game' semantics of Lorenzen's [Lorenzen (1961, 1969)] and Hintikka's [Hintikka (1983)]. It is enough to look at the canonical operator (coming from *introduction* steps) as the marks for the 'assertion' (resp. 'myself') moves, and the non-canonical operator(s) (coming from *elimination* inferences) as the representatives of the 'attack' (resp. 'nature') moves ([Lorenzen (1969)] (p. 25ff); resp. [Hintikka (1983)] (p. 3)). The rules of *reduction* then become explanations of how the 'attack' (resp. 'nature') moves operate on 'assertion' (resp. 'myself') moves.

On a section dedicated to modal logic (Ibid., pp. 61ff), Lorenzen explains the dialogical game for the modal necessity by an explanation of how the connective is to be *eliminated*. The main limitation of Lorenzen's method for modal logic is similar to many other methods, as we mentioned in the beginning of this chapter, namely the lack of devices to 'name' worlds: the 'assertion' being 'necessarily *A*' (notated by Lorenzen as 'ΔA'), the 'attack' operates by attempting to eliminate the 'necessary' connective, but it does not give a name to the world, as it happens with the dialogue rule for the first order universal quantifier where an arbitrary individual is presented by the 'attack'.

"For the all-quantifier we may now propose the following attack-defense-rule:

Assertion	Attack	Defense
$\bigwedge_x A(x)$	S?	$A(S)$

(Ibid., p. 25.)

whereas for the universal quantification over 'worlds' (Lorenzen prefers to think in terms of 'systems of sentences') the rule is given as:

Assertion	Attack	Defense
ΔA	?	A

(Ibid., p. 64.)

□-elimination[4]

$$\frac{\mathbb{S}:\mathcal{U} \qquad l:\Box A}{\mathcal{EXTR}(l,\mathbb{S}):A(\mathbb{S})}$$

which can be read as 'given that for all possible worlds A is true (where 'l' is the corresponding proof-construction), and $\mathbb{S}$ is a possible world, then A is true at $\mathbb{S}$'.

Observe that, similarly to the $\forall$ case aforementioned, even if we fix the $\Box$-*reduction* rule, the *introduction* rule for the modal conditional '$\Box$' still leaves a 'natural' gap to be explored, namely the three dots below:

$$[\mathbb{W}:\mathcal{U}]$$

$$\vdots$$

$$\frac{F(\mathbb{W}):A(\mathbb{W})}{\Lambda \mathbb{W}.f(\mathbb{W}):\Box A}$$

And, indeed, this loose aspect of the *introduction* rule makes room for the characterisation of distinct kinds of 'necessity' (K, T, D, S4, S5, etc.) simply by making more explicit the nature of the three dots (e.g. by saying that the world variable '$\mathbb{W}$' must occur exactly once in the expression '$F(\mathbb{W})$', at least once, etc.), and yet retaining the same general pattern of the *introduction* rule.[5]

Thus, it should be easy to see that the same reasoning made for implication and first-order quantification can be carried through for the case of modal logics:

---

[4]Cf. Prawitz' [Prawitz (1965)] natural deduction rules for Necessity:

$$\text{NI)} \quad \frac{A}{NA} \qquad \text{NE)} \quad \frac{NA}{A}$$

where the difference is only that we have in our functional interpretation a procedure to introduce names of worlds, and a procedure to get rid of them by Λ-abstraction on world-variables. Without having some special provisos, such as e.g. the classification of the rule of NI as *improper inference rule* whose assumptions can only be modalised (and negation of modalised) formulas in the case of S4 (resp. S5), Prawitz's NE-rule might look like a trivialisation of modal logics.

Perhaps our formulation of Necessity *introduction* is better at least in the sense that it shows rather explicitly the fact that the rule must indeed be seen as an *improper* rule (following the terminology of Prawitz' [Prawitz (1965)]), similar to the *introduction* rule for the other conditionals, namely propositional implication and first-order universal quantifier.

[5]Note that here, where we are trying to explore the 'natural' gaps which exist in the general framework of natural deduction, there is no place for ad hoc treatments of modal logics with natural deduction where the distinction from one kind of necessity to another is the presence or absence of certain rules of inference, such as, e.g. in [Fitch (1966a)] and Fitting's [Fitting (1993)] expositions of how one can handle modal logics with natural deduction with a so-called 'strict (re-)iteration rule':

"The main difference of the systems S4 and S5 from the systems M and B is that $\ulcorner\Box\Box\phi\urcorner$ is derivable from $\ulcorner\Box\phi\urcorner$ in the former two systems but not in the latter two systems, while the main difference of the systems B and S5 from M and S4 is that $\ulcorner\Box \sim \Box\phi\urcorner$ is derivable from $\ulcorner\sim \phi\urcorner$ in B and S5, but not in M or S4. Also, of these four systems, S5 is the only one in which $\ulcorner\Box \sim \Box\phi\urcorner$ is derivable from $\ulcorner\sim \Box\phi\urcorner$."   [Fitch (1966a)] (p. 33)

"In addition [to the 'strict iteration rule'] there are special rules that vary from logic to logic. These are given below, with the understanding that any logic not mentioned does not need any special rules:

we can fix the β-convertibility on terms of □-type, and allow for variations on the Λ-abstraction discipline over terms containing variables naming worlds. So, in the case of this higher-order quantification connective we would be distinguishing the kinds of '□' by the distinct ways of handling assumptions of the form '$\mathbb{W} : \mathcal{U}$', in a similar way that we distinguished the various notions of first-order quantification by varying the handling of assumptions of the form '$x : D$' ('x is an element from the domain D').[6]

Being aware of the 'vagueness' of the three dots in the □-*introduction* rule, we hope that by making it more specific we can establish

(1) conditions for world-quantification and
(2) ways of distinguishing one kind of □ from another.

Now, recall:

(1) using these guidelines we were able to handle the '→' case (relevant, linear, intuitionistic, etc.) in [Gabbay and de Queiroz (1992)].
(2) the origin of '□' is attributed to Lewis' alternative to classical (material) implication [Lewis and Langford (1932)].

So, to summarise, by using labels/terms alongside the formulas, we can

(1) keep track of proof steps (giving local control) and
(2) handle 'arbitrary' names (via variable abstraction operators).

B, T, S4 and S5 $\dfrac{\Box X}{X}$

D, DB and D4 $\dfrac{\Box X}{\neg\Box\neg X}$ ".

[Fitting (1993)] (p. 28)

Here we are attempting to explore the 'natural' gap: the rule of □-*introduction* allows room for manoeuvre in the same way that the rules of →-*introduction* and ∀-*introduction* do. So, □-*introduction* and □-*elimination* are the only inference rules for all notions of necessity here, with the single exception of a kind of *modal reductio ad absurdum* needed for the modal counterpart of classical propositional logic shown below.

[6]Note, however, that we are not trying to find correspondences between first-order formulas and modal formulas, such as in the well-known Tarski's translation of S4 into intuitionistic logic, in Sahlqvist's [Sahlqvist (1975)] correspondence and completeness theorems, and many other well-established 'translation'-based approaches. The objects we are dealing with in our functional interpretation of modal logics are meant to be of a higher-order nature, and we do not intend to treat them as (first-order) individuals. Furthermore, we do not wish to represent the 'accessibility relation' explicitly as most translations do. Rather than relying on the device of introducing accessibility relations, we want to find disciplines of Λ-abstraction over 'world variables' that would allow the distinction of the various kinds of *necessity* on the basis of how one handles (withdraws) assumptions of the sort '$\mathbb{S}$ is an arbitrary world'.

and our labelled natural deduction system gives us at least two advantages over the usual plain natural deduction systems:

(1) It matches
    - the functional calculus on terms with
    - the logical calculus on formulas.

(2) It takes care of 'contexts' and 'scopes' in a more explicit fashion.

A question that may naturally arise is 'what are the labels?', and we recall that in the framework of the functional interpretation labels/terms 'name':

(1) Instances of formulas (propositional case)

$$a : A$$

(2) Individuals (first-order case)

$$t : D$$

and also instances of formulas as in the case above, possibly with open first-order formulas such as

$$f(a) : F(a)$$

(3) Structured databases (higher-order, modal)

$$\mathbb{W} : \mathcal{U}$$

and, of course, also individuals and instances of formulas, as in the previous cases.

7.1.2 *Natural deduction with an extra parameter*

One aspect of our interpretation of '$F(\mathbb{W}) : A(\mathbb{W})$' as 'A is true at $\mathbb{W}$ because $F(\mathbb{W})$ shows how $A(\mathbb{W})$ was obtained starting from the assumption that $\mathbb{W}$ was an arbitrary world from the universe of worlds (i.e. '$\mathbb{W} : \mathcal{U}$')', is that the usual rules of deduction (i.e. *introduction* and *elimination*) for the propositional and first-order connectives still apply, for example:

$\wedge$-*introduction*

$$\frac{a_1 : A_1(\mathbb{W}) \qquad a_2 : A_2(\mathbb{W})}{\langle a_1, a_2 \rangle : (A_1 \wedge A_2)(\mathbb{W})}$$

$\wedge$-*elimination*

$$\frac{c : (A_1 \wedge A_2)(\mathbb{W})}{FST(c) : A_1(\mathbb{W})} \qquad \frac{c : (A_1 \wedge A_2)(\mathbb{W})}{SND(c) : A_2(\mathbb{W})}$$

$$\vee\text{-}introduction$$

$$\frac{a_1 : A_1(\mathbb{W})}{inl(a_1) : (A_1 \vee A_2)(\mathbb{W})} \qquad \frac{a_2 : A_2(\mathbb{W})}{inr(a_2) : (A_1 \vee A_2)(\mathbb{W})}$$

$$\vee\text{-}elimination$$

$$\frac{c : (A_1 \vee A_2)(\mathbb{W}) \qquad \overset{[s_1 : A_1(\mathbb{W})]}{d(s_1) : C(\mathbb{W})} \qquad \overset{[s_2 : A_2(\mathbb{W})]}{e(s_2) : C(\mathbb{W})}}{CASE(c, \acute{s_1}d(s_1), \acute{s_2}e(s_2)) : C(\mathbb{W})}$$

$$\rightarrow\text{-}introduction \qquad\qquad \rightarrow\text{-}elimination$$

$$\frac{\overset{[x : A(\mathbb{W})]}{b(x) : B(\mathbb{W})}}{\lambda x b(x) : (A \rightarrow B)(\mathbb{W})} \qquad \frac{a : A(\mathbb{W}) \qquad y : (A \rightarrow B)(\mathbb{W})}{APP(y, a) : B(\mathbb{W})}$$

$$\forall\text{-}introduction \qquad\qquad \forall\text{-}elimination$$

$$\frac{\overset{[x : D(\mathbb{W})]}{f(x) : F(x)(\mathbb{W})}}{\Lambda x f(x) : (\forall x^D F(x))(\mathbb{W})} \qquad \frac{a : D(\mathbb{W}) \qquad y : (\forall x^D F(x))(\mathbb{W})}{EXTR(y, a) : F(a)(\mathbb{W})}$$

$$\exists\text{-}introduction \qquad\qquad \exists\text{-}elimination$$

$$\frac{a : D(\mathbb{W}) \qquad f(a) : F(a)(\mathbb{W})}{\varepsilon x(f(x), a) : (\exists x^D F(x))(\mathbb{W})} \qquad \frac{e : (\exists x^D F(x))(\mathbb{W}) \qquad \overset{[t : D(\mathbb{W}), g(t) : F(t)(\mathbb{W})]}{d(g, t) : C(\mathbb{W})}}{INST(e, \acute{g}\acute{t}d(g, t)) : C(\mathbb{W})}.$$

By using the inference rules above we can show that a (logic-dependent) tautology is 'necessarily' true, or true in 'all possible worlds'. For example, the intuitionistic tautology '$A \rightarrow (B \rightarrow A)$' is proved to be *necessarily* true in the following way:

$$\frac{\frac{\frac{\frac{[\mathbb{W} : \mathcal{U}]}{[y : B(\mathbb{W})]}}{\frac{[x : A(\mathbb{W})]}{\lambda y x : (B \rightarrow A)(\mathbb{W})}(*)}}{\lambda x \lambda y x : (A \rightarrow (B \rightarrow A))(\mathbb{W})}}{\Lambda \mathbb{W} \lambda x \lambda y x : \Box(A \rightarrow (B \rightarrow A))}.$$

Notice that this is 'necessarily' true if we take the implication to be non-relevant: in the step marked '$(*)$' a non-relevant discharge of assumptions at the propositional level was made.

Strict Implication.   Using the device of world variables for possible worlds the definition of strict implication in terms of necessity, namely,

$$A \prec B \equiv \Box(A \to B)$$

('$A \prec B$' to be read as 'A strictly implies B') comes out naturally from the rule of $\to$-*introduction* augmented with the extra parameter indicating the use of world variables:

$$\frac{\dfrac{[\mathbb{W} : \mathcal{U}]}{\dfrac{[x : A(\mathbb{W})]}{\dfrac{b(x) : B(\mathbb{W})}{\dfrac{\lambda x.b(x) : (A \to B)(\mathbb{W})}{\Lambda\mathbb{W}\lambda x b(x) : \underbrace{\Box(A \to B)}_{A \prec B}}}}}}{}$$

The $\to$-*introduction* now says that '$A \prec B$' if for any arbitrary world '$\mathbb{W}$', if A is true at (demonstrable at, forced by, etc.) $\mathbb{W}$ then B is also true at (resp. demonstrable at, forced by, etc.) $\mathbb{W}$.

7.1.2.1   *Montague's types, and world-variables as senses*

Let us for a moment think of a connection to another system of language analysis which would seem to have some similarity in the underlying ontological assumption, with respect to the idea of dividing the logical calculus into two dimensions, i.e. functional versus logical. The semantical framework defined in Montague's [Montague (1970)] intensional logic defines the semantic *types* of objects dealt with are given as e, t and s, which in words could be described as: *entities*, *truth-values*, and *senses*. The idea was that logic (language) was supposed to deal with objects of three kinds: names of entities, formulas denoting truth-values, and possible-worlds/contexts of use. Now, here when we say that we wish to have the bi-dimensional calculus, we are saying that the entities which are *namable* (i.e. individuals, possible-worlds, etc.) ought to be dealt with separately from (yet harmoniously) the logical calculus on the formulas, by a calculus of functional expressions. Whereas the variables for individuals are handled 'naturally' in the interpretation of first-order logic with our labelled natural deduction, the intro-duction of variables to denote *contexts*, or *possible-worlds* (structured collection of labelled formulas), as in our labelled natural deduction with an extra parameter, is supposed to account for Montague's *senses*.

7.1.2.2 *Open assumptions and the extra parameter*

The axiom of implication which corresponds to *permutation* (taking us from con-catenation implication to linear implication), namely,

$$(A \to (B \to C)) \to (B \to (A \to C))$$

would seem to find a counterpart in an inference rule of the following form:

$$\frac{\vdash \Box(B \to C)}{B \to \Box C}$$

whose proof-tree would look like:

$$\frac{[x : B(\mathbb{S})] \quad \dfrac{[\mathbb{S} : \mathcal{U}] \quad l : \Box(B \to C)}{\mathcal{EXTR}(l, \mathbb{S}) : (B \to C)(\mathbb{S})}}{\dfrac{APP(\mathcal{EXTR}(l, \mathbb{S}), x) : C(\mathbb{S})}{\dfrac{\Lambda \mathbb{S} APP(\mathcal{EXTR}(l, \mathbb{S}), x) : \Box C}{\lambda x \Lambda \mathbb{S} APP(\mathcal{EXTR}(l, \mathbb{S}), x) : B \to \Box C}} (*)$$

Notice, however, that the discharge of the assumption '$[\mathbb{S} : \mathcal{U}]$' was made at a point where there was an open assumption '$x : B(\mathbb{S})$' with '$\mathbb{S}$' as an extra parameter, namely step '$(*)$'. The point is that at that step '$\mathbb{S}$' could not be taken as arbitrary, therefore be universally quantified over, because it had '$x : B$' in it.

This is an example of how the scope of the connective of 'necessity' might be confused with the scope of 'implication', as it is rightly remarked in [Hughes and Cresswell (1968)].[7]

7.1.3 *Connections with Kripke's truth definition*

In order to generalise the notion of 'possible world' as defined in an early account of frame-based semantics [Kripke (1959)] where there could be no two worlds in which the same truth-value would be assigned to each atomic formula, Kripke [Kripke (1963)] introduced the notion of *accessibility relation*: in addition to a set of possible worlds $\mathcal{K}$ and one distinguished element $\mathbb{G}$ (the 'real world'), the generalised definition of a model would also involve the notion of 'accessibility

---

[7]When explaining the equivalence between a definition of strict implication ('$A \prec B$') using 'possi-bility' ('$\Diamond$') and another using 'necessity' ('$\Box$'), they say:

"It is important not to confuse $L(p \supset q)$, which means that the whole hypothetical 'if p then q' is a necessary truth, or that q follows logically from p, with $p \supset Lq$, which means that if p is true then q is a necessary truth. Unhappily, these are often confused in ordinary discourse, sometimes with disastrous results; and neglect of the distinction is made all the easier by the ambiguity of such common idioms as 'If ... then it must be (*or* is bound to be) the case that—'. To make things worse, the structure of such sentences is more closely analogous to that of $p \supset Lq$, but one suspects that most frequently what the speaker intends to assert (or at least all he is entitled to assert) is something of the form $L(p \supset q)$." [Hughes and Cresswell (1968)] (footnote 17 of p. 27).

relation' between worlds. Given two worlds $\mathbb{W}_1$ and $\mathbb{W}_2$, $\mathbb{W}_2$ is *accessible* (relative to $\mathbb{W}_1$) if every proposition true in $\mathbb{W}_2$ is to be possible in $\mathbb{W}_1$. Thus, the 'absolute' notion of possible world (as in [Kripke (1959)]) gives way to a relative notion, of one world being possible relative to another.

Truth definition.   Using the notation $\mathbb{W}_1 R \mathbb{W}_2$ to denote '$\mathbb{W}_2$ is accessible from $\mathbb{W}_1$', the definition of truth in the general model is given in a way such that:

$\Box A$ is true at $\mathbb{W}_1$ iff A is true at $\mathbb{W}_2$ for each $\mathbb{W}_2$ s.t. $\mathbb{W}_1 R \mathbb{W}_2$.

$\Diamond A$ is true at $\mathbb{W}_1$ iff there exists $\mathbb{W}_2$, s.t. $\mathbb{W}_1 R \mathbb{W}_2$, and A is true at $\mathbb{W}_2$.

As pointed out in [Kripke (1963)], the 'absolute' notion of possible world would be convenient for S5, but would hardly allow room for the treatment of normal propositional calculi in general. For the semantics to be general, there had to be ways of distinguishing various kinds of 'necessity' within the same framework.

In setting out a unified framework for the treatment of necessity via the functional interpretation where the distinction of various kinds of necessities is made on the basis of the discipline of handling 'world variables', we are replacing the notion of *accessibility relation* by the notion of 'assumption handling discipline' in much the same way we have done for the propositional implication [Gabbay and de Queiroz (1992)].

7.1.4   *Connections with Gentzen's sequent calculus*

As the more perspicuous reader will have noticed, our notation for 'A is true at (demonstrable at, forced by, etc.) $\mathbb{W}$', namely:

$$A(\mathbb{W}),$$

where '$\mathbb{W}$' is a structured collection (list, bag, tree, etc.) of formulas (together with their labels), is another way of writing (using Gentzen's [Gentzen (1935)] style sequent calculus):

$$\Gamma \vdash A,$$

where Gentzen's 'Γ' is our '$\mathbb{W}$'. There is an important difference, though: while the operations on the sets (or sequences) of formulas (such as 'Γ') are restricted to the so-called *structural* rules (namely, *weakening*, *contraction* and *exchange*), our structured collections of formulas will be handled according to their own structural properties. They could be sets, multisets, bags, lists, queues, trees, etc., and in each case we shall be able to prescribe particular ways of handling the Γ's (or $\mathbb{W}$'s in our notation). We shall not be restricted to any specific structure (sequence of formulas in Gentzen's sequent calculus, commutative sequence of formulas in Girard's [Girard (1987a)] linear logic, non-commutative string of formulas in Lambek's [Lambek (1958)] calculus, etc.), but we shall simply have a general framework where those collections will have names (identifiers) within the calculus, so that we can refer to them when performing the appropriate operations on them (e.g. removing or inserting an element in a certain Γ which is a tree of formulas, etc.).

The deduction theorem and the cut rule. Now, two questions seem to arise out of this attempted explanation of the consequence relation between a structured collection of formulas and a formula without the use of the turnstile sign ('$\vdash$'), namely:

(1) What is the form (and meaning) of the Deduction Theorem in this context?
(2) How is the cut rule framed, and what does it represent?

The answers seem to be rather easy to give. First of all, the Deduction Theorem, which is usually presented as:

$$\frac{\mathbb{W}, A \vdash B}{\mathbb{W} \vdash A \to B},$$

and proved to justify the soundness of the rule of $\to$-*introduction*, is now seen as the $\to$-*introduction* rule itself augmented with the extra parameter denoting the structured collection of assumptions:

$$\frac{[x : A(\mathbb{W})]}{\lambda x b(x) : (A \to B)(\mathbb{W})},$$

Girard's one-sided sequent calculus for linear logic. Here we shall demonstrate how Girard's one-sided sequent calculus for linear logic [Girard (1987a)] (p. 22) can be represented in our labelled natural deduction with an extra parameter denoting a structured collection of assumptions. (We abbreviate linear logic as 'L.L.', and labelled natural deduction as 'L.N.D.')

& (with)

L.L. $\quad \dfrac{\vdash A, \mathbb{C} \qquad \vdash B, \mathbb{C}}{\vdash A\&B, \mathbb{C}}$ L.N.D. $\quad \dfrac{a : A(\mathbb{C}) \qquad b : B(\mathbb{C})}{\langle a,b \rangle : (A \wedge B)(\mathbb{C})},$

and here we have the exact presentation of our (labelled natural deduction with an extra-parameter) rule of $\wedge$-*introduction*. When the 'world' variable for the first premise 'A' is different from the one for the second premise 'B' we will have to do something about joining not only the propositions, but the (structured) collections of labelled formulas denoted by the world-variables. (See the '$\otimes$' below.)

$\otimes$ (times)

$$\text{L.L.} \quad \frac{\vdash A, \mathbb{C} \qquad \vdash B, \mathbb{D}}{\vdash A \otimes B, \mathbb{C}, \mathbb{D}} \qquad \text{L.N.D.} \quad \frac{a : A(\mathbb{C}) \qquad b : B(\mathbb{D})}{\langle a, b \rangle : (A \otimes B)(\mathbb{C} \odot \mathbb{D})},$$

where '$\odot$' would be interpreted as multiset union.

$\oplus$ (plus)

$$\text{L.L.} \quad \frac{\vdash A, \mathbb{C}}{\vdash A \oplus B, \mathbb{C}} \qquad \frac{\vdash B, \mathbb{C}}{\vdash A \oplus B, \mathbb{C}}$$

$$\text{L.N.D.} \quad \frac{a : A(\mathbb{C})}{inl(a) : (A \vee B)(\mathbb{C})} \qquad \frac{b : B(\mathbb{C})}{inr(b) : (A \vee B)(\mathbb{C})}$$

par (par)

$$\text{L.L.} \quad \frac{\vdash A, B, \mathbb{C}}{\vdash A\mathrm{par}B, \mathbb{C}} \qquad \text{L.N.D.} \quad \frac{a : A(\mathbb{C}) \qquad b : B(\mathbb{C})}{inl(a) : (A \vee B)(\mathbb{C})},$$

where we have an extra premise for our rule of $\vee$-*introduction*.

7.2   Modal logics and the functional interpretation

Now, within this framework the distinct kinds of '$\Box$' will be characterised by the discipline of Λ-abstraction over world-variables, in a similar way to the propositional and first-order conditionals. So, we identify which disciplines of Λ-abstraction on world-variables such that we can prove the axioms of well-known modal logics, namely:

K. $\Box(A \to B) \to (\Box A \to \Box B)$
D. $\Box A \to \Diamond A$
T. $\Box A \to A$
B. $A \to \Box \Diamond A$
4. $\Box A \to \Box \Box A$
5. $\Diamond A \to \Box \Diamond A$

using the specific discipline of assumption discharge.

7.2.1 *Chellas' RM rule for standard normative logics*

As an example of how appropriate the treatment of 'necessity' via an extension of the treatment of propositional implication may prove to be, we can show that the weakest rule characterising the so-called standard normative logics according to Chellas [Chellas (1980)], namely:

$$\text{RM} \quad \frac{\vdash A \to B}{\Box A \to \Box B}$$

(sometimes called 'regularity' principle) mirrors the axiom schema characterising the weakest implication, namely 'concatenation' implication [Gabbay (1994)]. Similarly to our previous characterisation of the propositional implication via the framework of the functional interpretation, in our calculus here the proof of derivability of Chellas' RM uses the most 'well-behaved' discipline of Λ-abstraction over world-variables. We can see this from the following proof tree:

$$\frac{\frac{[\mathbb{S}:\mathcal{U}] \quad [w:\Box A]}{\mathcal{EXTR}(w,\mathbb{S}):A(\mathbb{S})} \quad p:(A \to B)(\mathbb{S})}{\frac{APP(p,\mathcal{EXTR}(w,\mathbb{S})):B(\mathbb{S})}{\frac{\Lambda\mathbb{S}APP(p,\mathcal{EXTR}(w,\mathbb{S})):\Box B}{\lambda w\Lambda\mathbb{S}APP(p,\mathcal{EXTR}(w,\mathbb{S})):\Box A \to \Box B}}}$$

where '$p:(A \to B)(\mathbb{S})$' indicates that '$A \to B$' is true in the 'actual' world we here name '$\mathbb{S}$', either because is an 'axiom' labelled 'p', or because it is a tautology (according to the notion of implication one is using) and so 'p' represents a closed λ-term, or even because the formula is derivable (with label 'p') by means of the rules of inference of the particular logic.

Note that in the proof-tree mirroring 'RM' the order of discharge of assumptions (and therefore lambda-abstractions) is not violated: the discharge of '$[\mathbb{S}:\mathcal{U}]$' and corresponding Λ-abstraction on '$\mathbb{S}$' is being made before the discharge of '$[w:\Box A]$' and corresponding λ-abstraction on 'w'. The order of abstraction over free-variables in the 'ordered' term '$APP(p,\mathcal{EXTR}(w,\mathbb{S}))$' is not violated. (We say that the term is 'ordered' because we take 'APP' and '$\mathcal{EXTR}$' as functional application operators which take a course-of-values term and apply it to an argument, the course-of-values term understood to be of a 'higher' type than the argument.)

7.2.2 *Modal logic K*

The axiom for the modal logic K finds a parallel in the axioms for implication and first-order universal quantification that correspond to the axiom introduced to the axiomatics of linear implication to obtain relevant implication, needs a discipline of Λ-abstraction over world-variables such that it binds more than one occurrence of the variable, similarly to the λ-abstraction discipline for the relevant implication:

$$\textbf{K.} \quad \Box(A \to B) \to (\Box A \to \Box B),$$

which could be rewritten as

$$\forall \mathbb{W}^{\mathcal{U}}((A \to B)(\mathbb{W})) \to (\forall \mathbb{W}^{\mathcal{U}} A(\mathbb{W}) \to \forall \mathbb{S}^{\mathcal{U}} B(\mathbb{S})),$$

where the proof-tree would look like

$$\cfrac{\cfrac{\cfrac{[\mathbb{S}:\mathcal{U}] \qquad [z:\Box A]}{\mathcal{EXTR}(z,\mathbb{S}):A(\mathbb{S})} \qquad \cfrac{[\mathbb{S}:\mathcal{U}] \qquad [w:\Box(A \to B)]}{\mathcal{EXTR}(w,\mathbb{S}):(A \to B)(\mathbb{S})}}{\cfrac{APP(\mathcal{EXTR}(w,\mathbb{S}),\mathcal{EXTR}(z,\mathbb{S})):B(\mathbb{S})}{\cfrac{\Lambda\mathbb{S}APP(\mathcal{EXTR}(w,\mathbb{S}),\mathcal{EXTR}(z,\mathbb{S})):\Box B}{\lambda z\Lambda\mathbb{S}APP(\mathcal{EXTR}(w,\mathbb{S}),\mathcal{EXTR}(z,\mathbb{S})):\Box A \to \Box B}}(*)}}{\lambda w\lambda z\Lambda\mathbb{S}APP(\mathcal{EXTR}(w,\mathbb{S}),\mathcal{EXTR}(z,\mathbb{S})):\Box(A \to B) \to (\Box A \to \Box B)},$$

and the $\Lambda\mathbb{S}$-abstraction at step marked '$(*)$' binds two free-ocurrences of the world-variable '$\mathbb{S}$' in the term '$APP(\mathcal{EXTR}(w,\mathbb{S}),\mathcal{EXTR}(z,\mathbb{S}))$', in the increasing order: first the 'lower' variable '$\mathbb{S}$', then the one a step higher ('z'), and finally the higher 'w'.

7.2.3 Modal logic D

The modal logic D is characterised by the existence of an accessible possible world, wherever world one is supposed to be in. In axiomatic presentation, it becomes

$$\textbf{D.} \quad \Box A \to \Diamond A,$$

which reads 'if for all possible worlds we have A then there exists a possible world where we have A'. Now, we draw your attention to the structural similarity of such an axiom to the axiom which distinguishes first-order logic from *inclusive* logics (i.e. those logics which deal with non-empty domains)[8]:

---

[8]Cf. [Fine (1985)] (Chapter 21): "An *inclusive* logic is one that is meant to be correct for both empty and non-empty domains. There are certain standard difficulties in formulating a system of inclusive logic. If, for example, we have the usual rules of UI, EG and conditional proof, then the following derivation of the theorem $\forall x F x \supset \exists x F x$ goes through (...) But the formula $\forall x F x \supset \exists x F x$ is not valid in the empty domain; the antecedent is true; while the consequent is false." [Fine (1985)] (p. 205).

$\forall x.F(x) \rightarrow \exists x.F(x)$

$$\cfrac{\cfrac{\cfrac{[\forall x F(x)]}{F(t)}(*)}{\exists x F(x)}}{\forall x F(x) \rightarrow \exists x F(x)}.$$

Notice that at the step marked '$(*)$' an assumption was made, namely that the domain over which the '$\forall$' is quantifying is a non-empty domain containing at least an element which is being named 't'. This assumption does not get recorded in the system of plain natural deduction.

Using the functional interpretation, where the presence of terms, and the expliciting of domains of quantification make the framework a much richer instrument for deduction calculi, we have:

$$\cfrac{\cfrac{\cfrac{[t:D] \qquad [z:\forall x^D F(x)]}{EXTR(z,t):F(t)}}{\varepsilon x(EXTR(z,x),t):\exists x^D.F(x)}}{\lambda z \varepsilon x(EXTR(z,x),\boxed{t}):\forall x^D F(x) \rightarrow \exists x^D F(x)}.$$

Here the existence of a free variable (namely 't') indicates that the assumption '$[t : D]$' remains to be discharged. By making the domain of quantification explicit one does not have the antecedent (vacuously) true and the consequent trivially false in the case of empty domain: the proof of the proposition is still depending on the assumption '$[t : D]$', i.e. that the type 'D' is non-empty. To be categorical the above proof would still have to proceed one step, as in:

$$\cfrac{\cfrac{\cfrac{\cfrac{[t:D] \qquad [z:\forall x^D F(x)]}{EXTR(z,t):F(t)}}{\varepsilon x(EXTR(z,x),t):\exists x^D F(x)}}{\lambda z \varepsilon x(EXTR(z,x),\boxed{t}):\forall x^D F(x) \rightarrow \exists x^D F(x)}}{\underbrace{\lambda t \lambda z \varepsilon x(EXTR(z,x),t)}:D \rightarrow (\forall x^D F(x) \rightarrow \exists x^D F(x))}.$$

no free variable

At this point we look at the proof-construction ('$\lambda t \lambda z \varepsilon x(EXTR(z,x),t)$') and we can see no free variables, thus the corresponding proof is categorical, i.e. does not depend on any assumption. And the result is that for the proof to be categorical, the domain of individuals must be non-empty.

Now, by considering variables (such as 'x') as ranging over possible worlds, and the domain (such as 'D') as being the universe of possible worlds, we immediately see the parallel: in order for the axiom to be true the domain 'D', now seen as the universe of possible worlds, must be non-empty. In other words, there is at least one possible world accessible from wherever one is, which in standard modal Kripke-style semantics represents the seriality of the accessibility relation.

7.2.4   Modal logic T

The modal logic T is characterised axiomatically by the schema:

$$\text{T.} \quad \Box P \to P,$$

but because we are working with 'closed' formulas (sentences)[9] and the above axiom implicitly leaves the 'actual' ('current') world free (non-quantified), let us look at the proof construction for the axiom schema '$\Box(\Box P \to P)$' which is the universal closure of the T-axiom schema. The proof tree is constructed in the following way:

$$\frac{\displaystyle \frac{[\mathbb{S}:\mathcal{U}] \qquad \frac{[\mathbb{S}:\mathcal{U}]}{[l:\Box P(\mathbb{S})]}}{\mathcal{EXTR}(l,\mathbb{S}):P(\mathbb{S})}}{\displaystyle \frac{\lambda l \mathcal{EXTR}(l,\mathbb{S}):(\Box P \to P)(\mathbb{S})}{\Lambda\mathbb{S}\lambda l \mathcal{EXTR}(l,\mathbb{S}):\Box(\Box P \to P)}},$$

and the distinguishing characteristic of the Λ-abstraction over the world-variable '$\mathbb{S}$' is that it breaks the order of Λ-abstraction in the sense that it is first abstracting over the variable at the 'function' place, namely 'l', when the variable at the 'argument' place, namely '$\mathbb{S}$' is still free. Recall that the weak implication called

---

[9]Note that the axiom schema '$\Box P \to P$', when translated to a formula using universal quantification over world-variables becomes:

$$\forall \mathbb{W}^{\mathcal{U}} P(\mathbb{W}) \to P(\mathbb{S}),$$

where '$\mathbb{S}$' is supposed to name the 'current' world. To obtain a closed formula, which would not name the 'current' world, but rather make use of a bound variable, we would be able to quantify over '$\mathbb{S}$':

$$\forall \mathbb{S}^{\mathcal{U}} (\forall \mathbb{W}^{\mathcal{U}} P(\mathbb{W}) \to P(\mathbb{S})),$$

which when translated back gives '$\Box(\Box P \to P)$'. The reading would then be 'for any possible world (including the 'current' one), if P is true in all possible worlds, then P is true in 'this' world.' Obviously, if '$\Box P$' is interpreted as 'P is provable', then we may start to encounter incongruencies if we are not careful with the notion of provability we actually mean. Gödel [Gödel (1933)] remarked that

$$\Box(\Box P \to P),$$

where '$\Box$' is taken to be 'provable in an axiomatic theory', making it possible to read the axiom as 'it is provable that (if P is provable then P is true)', could not be part of the axiomatic of the notion of provability itself because it would be contradicting his Second Incompleteness Theorem. The axiom schema would be stating the consistency of the axiomatic theory of provability, which goes against the 'unprovability of consistency of any axiomatic system at least as powerful as Peano arithmetic'.

Thus, it seems worth recalling here the distinguishing characteristic of the so-called GL-logic of *provability* ('G' for Gödel [Gödel (1933)] and 'L' for Löb [Löb (1955)]), which is syntactically characterised in the literature by the axiom schema (first presented in [Löb (1955)]):

$$\text{GL.} \quad \Box(\Box P \to P) \to \Box P$$

saying that '$\Box P$' is obtainable when '$\Box(\Box P \to P)$' is. It is then clearly 'irreflexive' given that the 'current' world is not left 'unbound'.

'concatenation' [Gabbay (1994)] admits the withdrawing of assumptions only in the increasing order, which is being obeyed by the Λ-abstraction discipline for K.

7.2.5 *Modal logic B*

The 'Brouwerian' modal logic is characterised as the modal logic of reflexive and symmetric accessibility relation. It is often characterised by the axiom schema:

$$\textbf{B.} \quad A \to \Box\Diamond A$$

for any formula A.

The proof of this axiom can be constructed in the following way:

$$\cfrac{\cfrac{[\mathbb{T}:\mathcal{U}] \quad \cfrac{[\mathbb{S}:\mathcal{U}]}{[x:A(\mathbb{T})(\mathbb{S})]}}{\cfrac{\varepsilon W(x,\mathbb{T}):\Diamond A(\mathbb{S})}{\Lambda \mathbb{S}\varepsilon W(x,\mathbb{T}):\Box\Diamond A}}(*)}{\lambda x \Lambda \mathbb{S}\varepsilon W(x,\mathbb{T}):A \to \Box\Diamond A} ,$$

where the world variable '$\mathbb{T}$' is still free. In order to close the proof, we need to bind that free occurrence of '$\mathbb{T}$' and this means adding an antecedent saying that the universe '$\mathcal{U}$' is non-empty:

$$\mathcal{U} \to (A \to \Box\Diamond A).$$

If the universe '$\mathcal{U}$' of worlds is not empty, then the B-axiom schema holds.

We notice that, similarly to the non-intuitionistic existential quantifier (cf. [de Queiroz and Gabbay (1995)], the ε-abstraction made at step marked '$(*)$' is a vacuous abstraction — the term 'x' did not have any free occurrence of the world variable '$\mathbb{W}$'. Thus, as we would expect, this axiom schema is not valid in S4 (the 'modal' counterpart to intuitionistic logic).[10]

---

[10]In [de Queiroz and Gabbay (1995)], we show how to deal with (weakened versions of) classical first-order theorems which are not provable in first-order intuitionistic logic. This is done by allowing vacuous ε-abstraction in the rule of $\exists$-*introduction*. E.g.:

$$(\forall x^D F(x) \to C) \to \exists y^D (F(y) \to C) \text{ (classical)}$$

is proved as:

$$\cfrac{[t:D] \quad \cfrac{[t:D] \quad \cfrac{\cfrac{[f(t):F(t)]}{\Lambda t f(t):\forall t^D F(t)} \quad [z:\forall x^D F(x) \to C]}{APP(z,\Lambda t f(t)):C}}{\cfrac{\lambda f APP(z,\Lambda t f(t)):F(t) \to C}{\varepsilon y.(\lambda f.APP(z,\Lambda y.f(y)),t):\exists y^D.(F(y) \to C)}}(*)}{\cfrac{\lambda z \varepsilon y(\lambda f APP(z,\Lambda y f(y)),t):(\forall x^D F(x) \to C) \to \exists y^D (F(y) \to C)}{\lambda t \lambda z \varepsilon y(\lambda f APP(z,\Lambda y f(y)),t):D \to ((\forall x^D F(x) \to C) \to \exists y^D (F(y) \to C))}} .$$

7.2.6   *Modal logic S4 and its parallel with intuitionistic logic*

In [Gabbay and de Queiroz (1992)], in order to obtain intuitionistic implication
from relevant implication we had to introduce the device of 'vacuous' abstraction.
By doing this we were then able to construct a derivation of the axiom '$A \to
(B \to A)$' which is provable in intuitionistic logic but it is not provable in relevant
logic.

Now, the parallel between intuitionistic logic and modal logic S4 appears
rather naturally in the framework of the functional interpretation: in order to obtain S4 from K (which is characterised by an axiom schema similar to the one
which characterises relevant implication), we shall allow vacuous Λ-abstractions
on world-variables, i.e. $\Lambda\mathbb{S}$-like abstractions over terms which may not contain '$\mathbb{S}$'
free. By extending the Λ-abstraction discipline on world-variables in this way we
can construct a derivation of the axiom scheme which is valid in S4 but is not
valid in K, namely,

$$. \quad 4. \quad \Box A \to \Box\Box A$$

and the proof-tree will look like:

$$
\cfrac{
\cfrac{
\cfrac{
[\mathbb{S} : \mathcal{U}]
}{
[l : \Box A]
}
}{
\Lambda\mathbb{S}.l : \Box\Box A
}(*)
}{
\lambda l.\Lambda\mathbb{S}.l : \Box A \to \Box\Box A
}'
$$

which involves a proof step where a vacuous $\Lambda\mathbb{S}$-abstraction is being made, namely
the step marked with '$(*)$' where $\mathbb{S}$ does not occur free in the term 'l' over which
the $\Lambda\mathbb{S}$-abstraction is being performed.

7.2.7   *Grzegorczyk's extension of S4*

In [Grzegorczyk (1967)], an extension of S4 which has as theorems precisely the
sentences valid on all finite partially ordered frames, is characterised by the introduction of the following schema to the axioms of S4:

$$\text{Grz.} \quad \Box(\Box(A \to \Box A) \to A) \to A,$$

whose proof construction is the following:

---

and the step '$(*)$' is made by means of an εy-abstraction over a term which does not contain any free
occurrence of the variable 'y'.

Note that here we have a weakened version of the classical case where the full classical version is
conditional to the domain 'D' being non-empty.

Now, the interesting characteristic of this extension of S4 is that it differs from all the previous notions of necessity (even S5) in the sense that it admits a 'modal *reductio ad absurdum*', namely, the inference step marked with '$(*)$' where both the world variable '$\mathbb{S}$' and the label 'u' were bound because they were used (in conjunction) as ticket and minor: they appear in the term both in the function place and in the argument place, characterising a restricted self-application (a higher-order parallel to our generalised propositional *reductio ad absurdum* defined in [Gabbay and de Queiroz (1992)]). The formula '$(\Box(A \rightarrow \Box A))(\mathbb{S})$' was used both as minor (in the left hand corner of the proof tree) in an $\rightarrow$-*elimination* (*modus ponens*), and as ticket (in the right hand corner) in an $\Box$-*elimination* (universal *extraction*). And the modal counterpart to our extended propositional *reductio ad absurdum*

$$[x : A \rightarrow B]$$

$$\frac{b(x, \ldots, x) : B}{\lambda x.b(x, \ldots, x) : A} \boxed{\begin{array}{l} \text{minor} + \\ \text{ticket} \end{array}}$$

introduced in [Gabbay and de Queiroz (1992)], appears to reveal the parallel between the propositional and the modal versions of the device of *reductio ad absurdum*. The idea that the rule gives us classical positive implication finds support in the fact that there is a kind of restricted self-application of 'x' to itself, given that it labels a formula that was used both as minor premise and as ticket (major premise) of an application of *modus ponens*.[11]

The modal counterpart would then be introduced as

$$[\mathbb{S} : \mathcal{U}]$$

$$[u : (\Box(A \rightarrow \Box A))(\mathbb{S})]$$

$$\frac{F(G(\mathbb{S}, u), \ldots, H(\mathbb{S}, u)) : \Box A(\mathbb{S})}{\Lambda\mathbb{S}\lambda u F(G(\mathbb{S}, u), \ldots, H(\mathbb{S}, u)) : A} \boxed{\begin{array}{l} \text{minor} + \\ \text{ticket} \end{array}}.$$

Note that the term '$APP(\mathcal{EXTR}(u, \mathbb{S}), APP(\mathcal{EXTR}(l, \mathbb{S}), u))$' in the step '$(*)$', like $F(G(\mathbb{S}, u), \ldots, H(\mathbb{S}, u))$ has both '$\mathbb{S}$' and 'u' involved in a sort of

---

[11]With that special proviso we can prove *Peirce's axiom* in the following way:

$$\frac{[y : A \rightarrow B] \qquad [x : (A \rightarrow B) \rightarrow A]}{\dfrac{\dfrac{APP(x, y) : A}{APP(y, APP(x, y)) : B} \qquad [y : A \rightarrow B]}{\dfrac{\lambda y APP(y, APP(x, y)) : A}{\lambda x \lambda y APP(y, APP(x, y)) : ((A \rightarrow B) \rightarrow A) \rightarrow A}}} (*)$$

where in the step marked with '$(*)$' we have applied our *generalised reductio ad absurdum*. In the resulting term '$\lambda x \lambda y APP(y, APP(x, y))$' notice the (restricted) self-application which is being uncovered: 'y' is being applied to the result of an application of 'x' to 'y' itself.

restricted self-application: the functional expression '$\mathcal{EXTR}(u, \mathbb{S})$' is being applied to another functional expression where '$\mathbb{S}$' and 'u' are themselves the arguments.

We recall that our propositional *RAA* makes sure that for any propositional formulas 'A' and 'B', either '$(A \to B) \to A$' or '$A \to B$' is provable. Now, the modal logic S4Grz characterises the set of modal formulas which are valid on all finite partially ordered frames, meaning that for any frame, and for any formula 'A' in that frame it must be the case that either 'A' is provable, or that '$A \to B$', for some 'B', is provable.

7.2.8   *Modal logic S5 and its parallel with classical logic*

Usually, the modal logic S5 is said to differ from the modal logic S4 at least in the sense that the connective of *possibility* is defined as:

$$\Diamond A \equiv \neg\Box\neg A$$

for any formula A.

Sometimes S5 is also characterised an extension of S4 where negations of modal formulas can be introduced and later discharged as assumptions in a proof of a modal theorem.[12] In terms of axiom schema this is reflected in that the axiom characterising S4, namely:

$$\Box A \to \Box\Box A$$

becomes

$$\Box\neg B \to \Box\Box\neg B$$

for any formula B, which can be rewritten to the more commonly used axiom schema:

$$\Diamond C \to \Box\Diamond C$$

by putting $\Box \equiv \neg\Diamond\neg$ and $\neg\Diamond B \equiv \Diamond C$.

Here we want to see S5 through its parallel with first-order classical logic, and for that we want to work with the following equivalent axiom schema:

$$5. \quad \Diamond P \to \Box\neg\Box\neg P \qquad\qquad (*)$$

whose translation into a formula with quantification over world variables gives us

$$\exists \mathbb{S}^{\mathcal{U}}.P(\mathbb{S}) \to \forall \mathbb{T}^{\mathcal{U}}.(\forall \mathbb{W}^{\mathcal{U}}.(P(\mathbb{W}) \to \mathcal{F}) \to \mathcal{F})$$

---

[12]Here we are referring to Prawitz's proviso that 'an instance of the deduction rule for NI [Necessity Introduction] in S4 (resp. S5) is to have the form $<< \Gamma, A >, < \Gamma, NA >>$, where every formula of Γ is modal (resp. modal or *the negation of a modal formula*)' [Prawitz (1965)] (p. 74).

where $\neg P(\mathbb{S})$ is written as $P(\mathbb{S}) \to \mathcal{F}$, '$\mathcal{F}$' being the *falsum*.

The proof-tree can be constructed in the following way:

$$\cfrac{\cfrac{[W:\mathcal{U}]}{[u:\Diamond P]} \quad \cfrac{\cfrac{\cfrac{\cfrac{[\mathbb{T}:\mathcal{U}]}{[G(\mathbb{T}):P(\mathbb{T})]}}{\Lambda\mathbb{T}.G(\mathbb{T}):\Box P} \quad [z:\Box P \to \mathcal{F}]}{APP(z,\Lambda\mathbb{T}.G(\mathbb{T})):\mathcal{F}}}{\lambda z APP(z,\Lambda\mathbb{T}G(\mathbb{T})):(\Box P \to \mathcal{F}) \to \mathcal{F}}}{\cfrac{\cfrac{INST(u,\acute{G}\acute{\mathbb{T}}\lambda z APP(z,\Lambda\mathbb{T}.G(\mathbb{T}))):(\Box P \to \mathcal{F}) \to \mathcal{F}}{\Lambda W.INST(u,\acute{G}\acute{\mathbb{T}}\lambda z APP(z,\Lambda\mathbb{T}.G(\mathbb{T}))):\Box((\Box P \to \mathcal{F}) \to \mathcal{F})}}{\lambda u\Lambda WINST(u,\acute{G}\acute{\mathbb{T}}\lambda z APP(z,\Lambda\mathbb{T}G(\mathbb{T}))):\Diamond P \to \Box((\Box P \to \mathcal{F}) \to \mathcal{F})}}$$

and here we note that the assumption '$z : \Box P \to \mathcal{F}$' is not a (fully) modalised formula but the negation of a fully modalised formula.

7.2.9 *Non-normal modal logics*

A *normal* modal logic is usually defined as the modal logic which validates:

(1) every tautology;
(2) all formulas of the form $\Box(A \to B) \to (\Box A \to \Box B)$;
(3) $\Box A$ whenever it validates A.

Observe that the rule of *necessitation*, namely:

$$\frac{\vdash A}{\Box A}$$

is left vague as to what '$\vdash A$' is to mean. The formula may be 'provable' simply because it is a theorem of the logic, regardless of the assumptions (hypotheses) involved in the reasoning. It can also be a 'legitimately context-dependent' theorem.

The definition of *non-normal* modal logics is then given by replacing the necessitation rule by a weaker rule, usually the rule of 'regularity', which is defined in [Chellas (1980)] as

$$\text{RM} \quad \frac{\vdash A \to B}{\Box A \to \Box B}$$

As pointed out by various people, when studying the semantics of these logics, it is difficult to avoid the rule of necessitation because its soundness is 'virtually built into the definition of possible world models' [Fitting (1993)]. Kripke's own solution to such semantic difficulty was to account for exceptional ('queer') worlds which would not play by the usual rules [Kripke (1965)].

What would be a reasonable approach to non-normality in our labelled proof theory? In order to answer this question we need, first of all, to consider what is at stake here, namely the weakening of the standard notion of necessity, in order to make it, in some sense, *relevant*. The intended application of a formal characterisation of the non-normal modal logics is the formulation of a logic for notions such as knowledge and belief. In dealing with concepts like these one often needs to make use of the not-so-declarative notion of 'explicitness'.

As we have pointed out earlier on in this chapter, when formulated in the pattern of our labelled natural deduction, where names of possible worlds are introduced explicitly in the labels, the rule of necessitation

$$[\mathbb{W} : \mathcal{U}]$$
$$\vdots$$
$$\frac{F(\mathbb{W}) : A(\mathbb{W})}{\Lambda \mathbb{W} F(\mathbb{W}) : \Box A}$$

will give us a more refined mechanism with which to 'parameterise' our notion of necessity. In a similar way to the characterisation of relevant implication, we will be able to impose a 'relevant' discipline of Λ-abstraction over the label-expression alongside the 'provable A' formula, such it requires that the arbitrary world must have been used in order for the formula to be considered 'necessary'.

7.2.10 *Problematic cases: The scope of 'necessity'*

Here we shall attempt to demonstrate that the aspect of the methodology enforced by the functional interpretation, namely that which involves introducing names for 'worlds' (world variables), might help throwing some light on issues which have been traditionally looked at from the denotational semantics point of view.

The Barcan formula. The Barcan formula is said to be true in the denotational semantics approach when the domain of individuals is either constant or decreasing. We can perhaps understand this interpretation by constructing a proof-tree as in the previous case.

The Barcan formula is the following:

$$\forall x_1^D \Box F(x) \rightarrow \Box \forall x_2^D F(x)$$

which can be proved using our methodology as follows:

$$\cfrac{\cfrac{[\mathbb{S}:\mathcal{U}] \quad \cfrac{[t:D_1] \quad [u:\forall x_1^D \Box F(x)]}{EXTR(u,t):\Box F(t)}}{\cfrac{\mathcal{EXTR}(EXTR(u,t),\mathbb{S}):F(t)(\mathbb{S})}{\Lambda t\mathcal{EXTR}(EXTR(u,t),\mathbb{S}):(\forall t_2^D F(t))(\mathbb{S})}(*)}{\cfrac{\Lambda\mathbb{S}\Lambda t\mathcal{EXTR}(EXTR(u,t),\mathbb{S}):\Box\forall t_2^D F(t)}{\lambda u\Lambda\mathbb{S}\Lambda t\mathcal{EXTR}(EXTR(u,t),\mathbb{S}):\forall x_1^D\Box F(x) \to \Box\forall t_2^D\Box F(t)}}$$

and, similarly to the case below, the universal quantification (step '$(*)$') over the domain of individuals 'D_2' can only be safely made if this domain is contained in the domain 'D_1'. Unlike the previous case, the domain of individuals must either remain constant or else decrease as one moves from the 'current' world to any next possible world.

The Converse of the Barcan formula. Another example is the so-called Converse of the Barcan formula, namely,

$$\Box\forall x_1^D F(x) \to \forall x_2^D\Box F(x),$$

where the debate concerns the relation between D_1 and D_2, the domains of individuals.

Let us construct a proof tree for the Converse of the Barcan formula using the proof methodology advocated in this chapter:

$$\cfrac{\cfrac{[t:D_1(\mathbb{S})] \quad \cfrac{[\mathbb{S}:\mathcal{U}] \quad [l:\Box\forall x_1^D.F(x)]}{\mathcal{EXTR}(l,\mathbb{S}):(\forall x_1^D.F(x)(\mathbb{S})}}{\cfrac{EXTR(\mathcal{EXTR}(l,\mathbb{S}),t):F(t)(\mathbb{S})}{\cfrac{\Lambda\mathbb{S}EXTR(\mathcal{EXTR}(l,\mathbb{S}),t):\Box F(t)}{\Lambda t\Lambda\mathbb{S}EXTR(\mathcal{EXTR}(l,\mathbb{S}),t):\forall t_2^D\Box F(t)}(*)}}{\lambda l\Lambda t\Lambda\mathbb{S}EXTR(\mathcal{EXTR}(l,\mathbb{S}),t):\Box\forall x_1^D F(x) \to \forall t_2^D\Box F(t)},$$

where the in the step marked with '$(*)$' it is clear that the universal quantification over D_2' can only be made if all elements arbitrarily chosen from 'D_1' also inhabit 'D_2'. In other words, for the Converse of the Barcan formula to be true, the domain of individuals must be either constant or increase as one moves from the 'current' world to any next possible world.

7.3 Finale

Parallel to the functional interpretation of the (first-order) existential quantifier (which is developed in [de Queiroz and Gabbay (1995)]), we are also investigating how to characterise the functional interpretation of the modal connective of *possibility.*[13] (In fact, we have already given the reader some idea of how the parallel between the $\exists$ and the $\Diamond$ is dealt with when we discussed the modal logic S5 above.) By introducing names for worlds in conjunction with procedures for skolemising on world names, we shall also be able to demonstrate how to obtain natural extensions of fundamental theorems, such as Skolem's and Herbrand's theorems, to modal logics.

---

[13] In [Prawitz (1965)] it is already suggested the parallel between the *introduction* and *elimination* rules for the *possibility* connective and those for the first-order existential quantifier:
"One may add an additional modal operator, $\Diamond$ expressing *possibility*, and inference rules indicated by the figures:

$$\Diamond\text{I)} \quad \frac{A}{\Diamond A} \qquad\qquad \Diamond\text{E)} \quad \frac{\Diamond A \qquad \overset{(A)}{B}}{B}.$$

[Prawitz (1965)] (p. 75)

By using a similar pattern to the presentation of our $\exists$-type of ε-terms developed in [de Queiroz and Gabbay (1995)], and considering that now we are dealing with names of 'worlds' instead of names of (first-order) individuals, we put:

$$\Diamond A \equiv \exists \mathbb{W}^{\mathcal{U}}.A(\mathbb{W})$$

and we have the following counterparts to Prawitz' rules for *possibility*:

$\Diamond$-introduction $\qquad\qquad\qquad$ $\Diamond$-elimination

$$\frac{\mathbb{S}:\mathcal{U} \qquad F(\mathbb{S}):A(\mathbb{S})}{\varepsilon\mathbb{W}(F(\mathbb{W}),\mathbb{S}):\Diamond A} \qquad \frac{e:\Diamond A \qquad \overset{[\mathbb{T}:\mathcal{U},\,G(\mathbb{T}):A(\mathbb{T})]}{D(G(\mathbb{T}),\mathbb{T}):C}}{\mathcal{INST}(e,\acute{G}\grave{T}D(G(\mathbb{T}),\mathbb{T})):C}.$$

Chapter 8

Meaning and Proofs: A Reflection on Proof-Theoretic Semantics

8.1 Proof-theoretic semantics

In a recent special issue of *Synthese*, edited by R. Kahle & P. Schroeder-Heister, (Volume 148, Number 3/February, 2006), the possibility of a "proof-theoretic semantics" is put forward in the Editorial:

> "According to the model-theoretic view, which still prevails in logic, semantics is primarily denotational. Meanings are denotations of linguistic entities. The denotations of individual expressions are objects, those of predicate signs are sets, and those of sentences are truth values. The meaning of an atomic sentence is determined by the meanings of the individual and predicate expressions this sentence is composed of, and the meaning of a complex sentence is determined by the meanings of its constituents. A consequence is logically valid if it transmits truth from its premises to its conclusion, with respect to all interpretations. Proof systems are shown to be correct by demonstrating that the consequences they generate are logically valid. This basic conception also underlies most alternative logics such as intensional or partial logics. In these logics, the notion of a model is more involved than in the classical case, but the view of proofs as entities which are semantically dependent on denotational meanings remains unchanged.

> Proof-theoretic semantics proceeds the other way round, assigning proofs or deductions an autonomous semantic role from the very onset, rather than explaining this role in terms of truth transmission. In proof-theoretic semantics, proofs are not merely treated as syntactic objects as in Hilbert's formalist philosophy of mathematics, but as entities in terms of which meaning and logical consequence can be explained."

Still in the same Editorial the origins of the term is explained as:

> "The programme of proof-theoretic semantics can be traced back to Gentzen (1934). Seminal papers by Tait, Martin-Löf, Girard and Prawitz were published in 1967 and 1971. An explicit formulation of a semantic validity notion for generalized deductions with respect to arbitrary justifications was given by Prawitz (1973). Much of the philosophical groundwork for proof-theoretic semantics was

laid by Dummett from the 1970s on, culminating in Dummett (1991). Martin-Löf's type theory, whose philosophical foundation is proof-theoretic semantics, became a full-fledged theory in the 1970s as well (see Martin-Löf 1975b, 1982). The term "proof-theoretic semantics" was proposed by the second editor in a lecture in Stockholm in 1987."

Gentzen's analysis of logical deduction gave yet one more argument in favor of the view that the proof conditions determine the meaning of a proposition:

"The introductions represent, as it were, the 'definitions' of the symbols concerned, and the eliminations are no more, in the final analysis, than the consequences of these definitions." ([Gentzen (1935)], p. 80 of the English translation in [Szabo (1969)].).

In a series of lectures entitled 'On the meaning of logical constants and the justification of the logical laws' (1985), Martin-Löf presents philosophical explanations concerning the rôle of the definition of direct/canonical proofs in the proof-based account of meaning:

"The intuitionists explain the notion of proposition, not by saying that a proposition is the expression of its truth conditions, but rather by saying, in Heyting's words, that a proposition expresses an expectation or an intention, and you may ask, An expectation or an intention of what? The answer is that it is an expectation or an intention of a proof of that proposition. And Kolmogorov phrased essentially the same explanation by saying that a proposition expresses a problem or task (Ger. *Aufgabe*). Soon afterwards, there appeared yet another explanation, namely, the one given by Gentzen, who suggested that the introduction rules for the logical constants ought to be considered as so to say the definitions of the constants in question, that is, as what gives the constants in question their meaning. What I would like to make clear is that these four seemingly different explanations actually all amount to the same, that is, they are not only compatible with each other but they are just different ways of phrasing one and the same explanation." [Martin-Löf (1987)] (p. 410).

The aforementioned Editorial goes on to call Gentzen's remarks 'the most specific root' of the proposed account of proof-theoretic semantics:

"Proof-theoretic semantics has several roots, the most specific one being Gentzen's (1934) remarks that the introduction rules in his calculus of natural deduction define the meanings of logical constants while the elimination rules can be obtained as a consequence of this definition. More broadly, it belongs to the tradition according to which the meaning of a term has to be explained by reference to the way it is used in our language."

As a matter of fact, the idea that the meaning of a proposition is given by what counts as a proof of it was put forward by Dummett and Prawitz in the early 1970's, and the whole programme became known as the *Dummett-Prawitz argument*. First Dummett:

"the meaning of each [logical] constant is to be given by specifying, for any sentence in which that constant is the main operator, what is to count as a proof of that sentence, it being assumed that we already know what is to count as a proof of any of the constituents."

(p. 12, Michael Dummett in *Elements of Intuitionism* [Dummett (1977)])

Then Prawitz:

"As pointed out by Dummett, this whole way of arguing with its stress on communication and the role of the language of mathematics is inspired by ideas of Wittgenstein and is very different from Brouwer's rather solipsistic view of mathematics as a languageless activity. Nevertheless, as it seems, it constitutes the best possible argument for some of Brouwer's conclusions. (...)

I have furthermore argued that the rejection of the platonistic theory of meaning depends, in order to be conclusive, on the development of an adequate theory of meaning along the lines suggested in the above discussion of the principles concerning meaning and use. Even if such a Wittgensteinian theory did not lead to the rejection of classical logic, it would be of great interest in itself."

(p. 18, Dag Prawitz 'Meaning and proofs: on the conflict between classical and intuitionistic logic' [Prawitz (1977)])

So strong was Prawitz's confidence in the argument that he proposed to give a kind of 'technical' counterpart to Gentzen's dictum that the so-called *introduction* (proof) rules give the meaning to the logical constants. A general schema for obtaining the elimination rules from the introduction rules of any logical constant was proposed in a paper published in 1978:

"Gentzen suggested that 'it should be possible to display the elimination rules as unique functions of the corresponding introduction rules on the basis of certain requirements.' One has indeed a strong feeling that the elimination rules are obtained from the corresponding introduction in a uniform way. For the introduction rules of the schematic type, we can easily describe how the elimination rules are obtained uniformly from the corresponding introduction rule."

(p. 36f of 'Proofs and the Meaning and Completeness of the Logical Constants', [Prawitz (1978)])

As a result, a set of 'generalized elimination rules', following the pattern of the elimination of disjunction, was put forward. The resulting rule of elimination of implication, however, turned out to be rather different from the usual *modus ponens*, and seemed to carry certain degree of circularity, as in:

$$\frac{A \to B \quad \begin{array}{c}[A \vdots B]\\ C\end{array}}{C}$$

Note that the rule involves an assumption which is itself a hypothetical deduction, and which is discharged by the application of the rule. A further development by P.

Schroeder-Heister called 'A Natural Extension to Natural Deduction' [Schroeder-Heister (1984)] argues for the adoption of such a general account of hypothetical reasoning where rules themselves may be introduced as assumptions (and later discharged).

In spite of Shcroeder-Heister's proposed way out of the circularity shown in Prawitz's general schema for obtaining/deriving the elimination rules from the introduction rules, there was an attempt at a revision of the whole enterprise along the lines of the proposal that the elimination rules should be seen, not as *derivable from*, but rather as *justified by* the introduction rules. In his 1991 book *The Logical Basis of Metaphysics*, Dummett says:

> "Gerhard Gentzen, who, by inventing both natural deduction and the sequent calculus, first taught us how logic should be formalised, gave a hint how to do this, remarking without elaboration that 'an introduction rule gives, so to say, a definition of the constant in question', by which he meant that it fixes its meaning, and that the elimination rule is, in the final analysis, no more than a consequence of this definition. (...) Plainly, the elimination rules are not consequences of the introduction rules in the straightforward sense of being derivable from them; Gentzen must therefore have had in mind some more powerful means of drawing consequences. He was also implicitly claiming that the introduction rules are, in our terminology, self-justifying.
> (...)
> It is confusing to speak of the elimination rules as *consequences* of the introduction rules: it is better to speak of them as being *justified* by reference to them."
> (Michael Dummett in *The Logical Basis of Metaphysics* [Dummett (1991)], p. 251f)

He goes further to suggest an opposing view, namely the one where it is the elimination rule which fixes the meaning of the logical constants:

> "Intuitively, Gentzen's suggestion that the introduction rules be viewed as fixing the meanings of the logical constants has no more force than the converse suggestion, that they are fixed by the elimination rules; intuitive plausibility oscillates between these opposing suggestions as we move from one logical constant to another. Per Martin-Löf has, indeed, constructed an entire meaning-theory for the language of mathematics on the basis of the assumption that it is the elimination rules that determine meaning. The underlying idea is that the content of a statement is what you can *do* with it if you accept it—what difference learning that it is true will, or at least may, make to you. This is, of course, the guiding idea of a pragmatist meaning-theory. When applied to the logical constants, the immediate consequences of any logically complex statement are taken as drawn by means of an application of one of the relevant elimination rules."
> (Ibid., p. 280)

This seems to be a clear departure, on the part of Dummett, from the shared vision that the Gentzen's dictum was the ultimate justification for a theory of meaning for the language of intuitionistic mathematics. And indeed, in a paper published in 1998, Dummett clearly steps out of the joint programme with Prawitz, this time acknowledging explicitly that Prawitz's perspective is different from (and rival to) Wittgenstein's own view on the connections between meaning and use:

> "Prawitz's view is a genuine rival to Wittgenstein's only if we do understand the notion of truth as having metaphysical resonance. If we do, then we need to think very hard about the basis of the constructivist view of meaning. I presented it as deduced from a principle and an ancillary thesis. The principle was that the meaning of any given form of statement is encapsulated in what we learn when we learn to use statements of that form; and this principle must surely be correct. How could we come to attach a meaning to any such statements if a grasp of it were not intrinsic to the use we make of them? The ancillary thesis, as I stated it, was crude: it was to the effect that all that we learn, when we learn the use of a given form of statement, was what constitutes the verification of a statement of that form, or what establishes its truth and hence conclusively warrants an assertion of it. I already acknowledged that, as applied to empirical statements, this is too restrictive; grounds that fall short of being conclusive may also be true in virtue of a past, but no longer present, possibility of verifying it presents a different reason for qualifying the ancillary thesis. I regret that I do not feel able at present to offer a clear formulation of the necessary qualification; but it surely involves a fairly substantial concession to realism."
>
> (p. 137f of 'Truth from the constructivist standpoint', [Dummett (1998)])

In a paper published in 1998 Prawitz remains confident that Gentzen's claim should be understood in the sense that the elimination rules are *justified* on the basis of the meaning given to the logical constants by the introduction rules:

> "(...) This may sound as a formalist position: the meaning of the logical constants are given by inference rules that are just taken for granted, in other words, their validity is not thought of as depending on anything. However, Gentzen's point was that only inference rules of a certain kind, that is, the introduction rules, were given as valid in the sense that they just state what we mean by the sentences that occur as conclusions of the inferences in question. Other inference rules were to be justified on the basis of the meaning given to the logical constants by the introduction rules."
>
> (p. 42 of 'Truth and objectivity from a verificationist point of view', [Prawitz (1998)])

> "A way of formulating Gentzen's idea more systematically in semantic terms is to start from the idea that the meaning of a statement is determined by what counts as a proof of the statement, an idea formulated already before Gentzen by Heyting in his account of intuitionism."
>
> (Ibid., p. 44)

> "That the meaning of a statement is determined by what counts as its canonical proof or verification is thus the essence of the verificationism that I have in mind."
>
> (Ibid., p. 45)

But in a paper published in 2002, Prawitz shows a certain degree of flexibility:

"As Michael Dummett has put it, the rules that govern the use of expressions may be divided into two kinds: the ones that tell on what conditions it is appropriate to utter them, and the ones that concern the appropriate reactions to the utterance. In the case of assertive sentences, the latter rules concern the expectations we have when trusting an assertion, or, in other words, the conclusions that can be rightly inferred from the asserted sentence. The latter aspect is an aspect of the use of sentences in another sense of "use" again, namely what such sentences are used for, what they are good for. With Dummett we may call a theory of meaning pragmatic when it is based on this other aspect of the use of expressions. What are the prospects for such a theory? Is it more promising?

(...)

No one has really tried to construe a comprehensive semantics built on the idea that the meaning of a sentence is given by what counts as direct consequences of the sentence. It seems easy enough to work this idea for the language of predicate logic. Unfortunately, it seems less easy to do the same for mathematics."

(p. 92 of 'Problems of a Generalization of a Verificationist Theory of Meaning', [Prawitz (2002)])

The prospect for an alternative account for the relation between proofs and meaning which departs from the original formulation of the so-called 'Dummett-Prawitz argument' is also mentioned in a more recent publication:

"The term proof-theoretic semantics would have sounded like a *contradictio in adjecto* to most logicians and philosophers half a century ago, when proof theory was looked upon as a part of syntax, and model theory was seen as the adequate tool for semantics. Michael Dummett is one of the earliest and strongest critics of the idea that meaning could fruitfully be approached via model theory, the objection being that the concept of meaning arrived at by model theory is not easily connected with our speech behaviour so as to elucidate the phenomenon of language. Dummett pointed out at an early stage that Tarski's T-sentences, i.e. the various clauses in Tarski's definition of truth, cannot simultaneously serve to determine both the concept of truth and the meaning of the sentences involved. Either one must take the meaning as already given, which is what Tarski did, or one has to take truth as already understood, which is the classical approach from Frege onwards.

This latter alternative amounts to an account of meaning in terms of truth conditions depending on a tacit understanding of truth. In the case of a construed formal language, the T-sentences become postulated semantic rules that are supposed to give the formulas a meaning (a representative presentation of this view is in *Introduction to Mathematical Logic* by Alonzo Church (1956)). If the T-sentences are to succeed in conferring meaning to sentences, this must be because of some properties of the notion of truth. A person not familiar with the notion of truth would obviously not learn the meaning of a sentence by being told what its truth condition is. It therefore remains to state what it is about truth that makes the semantic rules function as genuine meaning explanations. the semantics has to be embedded in a meaning theory as Dummett puts it.

In the case of an already given natural language, the T-sentences become instead hypotheses, which must somehow be connected with speech behaviour. Here one may follow Donald Davidson's suggestion which may roughly be put: if ..A is true iff C.. is a correct T-sentence for the sentence A in a language L, then a speaker of L who asserts A normally believes that the truth condition C is satisfied; cases when a speaker is noticed both to observe that C is satisfied and to assert A therefore constitute data supporting the T-sentence. By making this connection between T-sentences and speech behaviour for at least observation sentences, one begins spelling out the concept of truth, which is needed to support the claim that the T-sentences give the meaning of the sentences of a language. However, as argued by Dummett (e.g. in Dummett 1983), it is only a beginning, because the assertion of sentences is only one aspect of their use. If the T-sentences are really to be credited with ascribing meaning to sentences, they must be connected with all aspects of the use of sentences that do depend on meaning. In other words, there are further ingredients in the concept of truth that must be made explicit, if the truth condition of a sentence is to become connected with all features of the use of the sentence that do depend on meaning. One such feature is the use of sentences as premisses of inferences. When asserting a sentence we are not only expected to have grounds for the assertion, we also become committed to certain conclusions that can be drawn from the assertion taken as a premiss.

I shall leave the prospects of rightly connecting the meanings of expressions with our use of them within a theory of meaning developed along these lines, and shall instead review some approaches to meaning that are based on how we use sentences in proofs. One advantage of such an approach is that from the beginning meaning is connected with aspects of linguistic use."

(p. 507f of 'Meaning approached via proofs', [Prawitz (2006)])

Indeed, since Dummett's attempt to give a philosophical justification of intuitionistic logic from a 'linguistic' perspective, drawing on Wittgenstein's dictum 'meaning-is-use', the idea of giving an account of the semantics of the language of mathematics which accounts for the relation between proof, inference rule, and meaning has motivated a number of developments, most, if not all, in line with Gentzen's suggestion that the meaning of a logical constant is fully specified by the corresponding conditions for assertion. In a recent article, Schroeder-Heister says:

Proof-theoretic semantics is an alternative to truth-condition semantics. It is based on the fundamental assumption that the central notion in terms of which meanings can be assigned to expressions of our language, in particular to logical constants, is that of *proof* rather than *truth*. In this sense proof-theoretic semantics is inherently inferential in spirit, as it is the inferential activity of human beings which manifests itself in proofs.

Proof-theoretic semantics has several roots, the most specific one being Gentzen's (1934) remarks that the introduction rules in his calculus of natural deduction define the meanings of logical constants, while the elimination rules can be obtained as a consequence of this definition. More broadly, it is part of the tradition according to which the meaning of a term should be explained by reference to the way it is *used* in our language.

Although the "*meaning as use*" approach has been quite prominent for half a century now and has provided one of the cornerstones of the philosophy of language, in particular of ordinary language philosophy, it has never prevailed in the formal semantics of artificial and natural languages. In formal semantics, the denotational approach, which starts with interpretations of singular terms and predicates, then fixes the meaning of sentences in terms of truth conditions, and finally defines logical consequence as truth preservation under all interpretations, has always been dominant. The main reason for this, as I see it, is the fact that from the very beginning, denotational semantics received an authoritative rendering in Tarski's (1933) theory of truth, which combined philosophical claims with a sophisticated technical exposition and, at the same time, laid the ground for model theory as a mathematical discipline. Compared to this development, the "meaning as use" idea was a slogan supported by strong philosophical arguments, but without much formal underpinning."

(p. 525f of 'Validity Concepts in Proof-Theoretic Semantics', [Schroeder-Heister (2006)])

It is exactly with the intention of giving a formal underpinning to the idead of 'meaning-is-use' that I develop below an alternative account of proof-theoretic semantics based on the idea that the meaning of logical constants are given by the explanation of immediate consequences, which in formalistic terms means the effect of elimination rules on the result of introduction rules, i.e. the so-called reduction rules.

8.2   Meaning, use and consequences

Further to the connections between *meaning* and *use*, it seems useful to consider the (explanation of the immediate) *consequences* one is allowed to draw from a proposition as something directly related to its *meaning/use*. And indeed, Wittgenstein's references to the connections between *meaning* and the *consequences*, as well as between *use* and *consequences* are sometimes as explicit as his celebrated 'definition' of *meaning* as *use* given in the *Investigations*. Here we attempt to collect some of these references, discussing how an intuitive basis for the construction of a more convincing proof-theoretic semantics (than, say, assertability conditions semantics) for the mathematical language can arise out of this connection *meaning/use/*(explanation of the immediate) *consequences.*[1]

---

[1] An earlier version of parts of this chapter was presented at the *Fourteenth International Wittgenstein Symposium (Centenary Celebration)*, Kirchberg/Wechsel, 14–21 August 1989, and accepted for publication in the *Reports* of the Symposium. Due to length restrictions for the papers to go into the

8.3 Meaning and purpose

The profound insights into the connections between *meaning* and *use*, *usefulness*, *purpose* and *function* provided by Wittgenstein's later writings, such as the remark: "Meaning, function, purpose, usefulness — interconnected concepts"[2] serve as a remarkably intuitive basis on which a proof-theoretic semantics for the mathematical language can be constructed. Not just another semantics but one that proves more convincing than many 'verificationistic' semantics because, in addition to prescribing that 'the meaning of a proposition is determined by its method of verification', no canonical value or mathematical construction is said to have meaning in itself, assertability conditions do not suffice, and no dialogical notions such as game or dialogue need to be introduced. Accepting the explanation of the *consequences* one is allowed to draw from a proposition as the explanation of its *meaning*, the explanation of how it is to be *used*, one may reasonably take that *meaning*, *use* and *consequences* are indeed interconnected concepts. Further to the connections between *meaning* and *use*, it seems useful to consider the *consequences* one is allowed to draw from a proposition as something directly related to its *meaning/use*. And indeed, Wittgenstein's references to the connections between *meaning* and *consequences*, as well as between *use* and *consequences* are sometimes as explicit as his celebrated 'definition' of *meaning* as *use* given in the *Investigations*. The resulting 'picture' is a triangle-like diagram with vertices on meaning-use-consequences, which highlights the rôle of the explanation of the *consequences* one can draw from a proposition for the settlement of its meaning as a fundamental ingredient of Wittgenstein's development of his celebrated semantical paradigm *meaning-is-use*. Here we attempt to collect some of these references, discussing how an intuitive basis for the construction of a more convincing proof-theoretic semantics for the mathematical language can arise out of this connection *meaning/use/consequences*. The main point here is to demonstrate how a particular formal counterpart to the explanation of (immediate) consequences, namely the proof-theoretic device of normalisation, can be used to explain some fundamental concepts of the logical foundations of computation. (E.g. the λ-calculus and the rule for explaining the *meaning/use/usefulness/consequences* of a λ-term: the β-rule of normalisation.)

Despite the technical presentation given at the final sections here, the chief motivation arises from an attempt to understand the question as to what extent Wittgenstein's philosophies might be seen as natural developments of a (roughly)

Reports, the material was not published. The material was then reviewed for publication in *Studia Logica*, and ended up published in the *Logic Journal of the IGPL*.

[2] §291, p. 41^e of the *Last Writings on the Philosophy of Psychology*, Vol. I.

uniform intuition concerning the relationship between philosophy and critique of language, and its implications to the foundations of mathematical logic. Recent studies based on the catalogued *Nachlass* have helped to uncover many misconceptions generated by the rather fragmented character of Wittgenstein's philosophical writings. In the light of the recent findings, it seems reasonable to argue, for example, that the logical atomism of the *Tractatus* was rejected by Wittgenstein in so far as it assumed that a name denoted an object, and the object was its 'meaning' (i.e. *denotation*). The opening lines of the *Investigations* (and beginning of: MS 111, TS 213, *Blue Book*, *Brown Book*) reveals one of the main criticisms against the *Tractatus'* chemical-like analysis: the assumption that names denote objects which are their meanings. In the later phase, whilst that assumption is dropped, the atomism idea is retained, leaving a sort of 'functional atomism', where words are taken to be individually identifiable devices with a meaning (function, purpose, usefulness) in the calculus of language. It can be seen as a revision of his early ('official') *strict atomism* into a kind of *manageable holism*, which accounts for both the continuity (i.e. atomism, with a 'functional' ('manageably holistic') ingredient, as D. Pears confirms: "the idea that a sentence can only be explained by their places in systems ... is one of the central points of the picture theory"[3] and "Everyone is aware of the holistic character of Wittgenstein's later philosophy, but it is not so well known that it was already beginning to establish itself in the *Tractatus*."[4]) and the discontinuity (i.e. 'meaning as denotation' vs. 'meaning as use') of his main philosophies.

As an attempt to provide a concrete example of how such a functional atomism might be used to construct a framework where the logical connectives have a meaning (function, purpose, usefulness) in the calculus of language which is made clear by the explanation of the (immediate) consequences one is supposed to draw from the corresponding proposition, we discuss the so-called functional interpretation of logics. The (Curry–Howard) functional interpretation of logical connectives is an 'enriched' system of Natural Deduction, in the sense that terms representing proof-constructions are carried alongside the formulas. It can be used to formulate various logical calculi in an operational manner, where the way one handles assumptions determines the logic. It is naturally atomistic, it is verificationist, and it is functional in the sense that the main semantical device is that of normalisation of proofs (not assertability conditions), finding a parallel in the dialogue/game-like semantics of Lorenzen (and Hintikka): canonical terms (*introduction*) stand for the 'Assertion' (resp. 'Myself') moves, and noncanonical terms (*elimination*) stand for the 'Attack' (resp. 'Nature') moves. It seems to comply

---

[3] footnote 34, p. 121 of [Pears (1987)].
[4] Opening lines of [Pears (1990)].

Meaning and Proofs   235

with Wittgenstein's 'functional atomism': it accounts for the early atomism (already with holistic tendencies according to D. Pears [Pears (1990)]) in a functional fashion, abiding by 'words are tools' and 'language is a calculus' (which, S. Hilmy verifies, "was already in place in early 1930s, and remained at least until 1946"[5]

It seems as though signs of Wittgenstein's attempt to make the connections between explaining the *meaning* of a proposition, demonstrating how to *use* it, and showing the (immediate) *consequences* one is allowed to draw from it, can be found in many different presentations. One finds many relevant remarks on the topic scattered among his writings, irrespective of phases. (The earliest reference to that particular aspect is perhaps in a letter sent to Russell in 1912, as we have previously pointed out.[6]) In the later writings, however, the remarks of that nature become much more frequent, and significantly more incisive than those of the early phase. In a series of *Lectures on the Foundations of Mathematics* given in Cambridge in 1939, for example, he explains how the meaning of a proposition can be changed by changing the way it is used, by changing the consequences that can be drawn from it. The core of the discussion can be summarised in the following:

> *Wittgenstein:* Yes; and let us take another example; the use of "all". "If all the chairs in this room were bought at Eaden Lilley's then this one was. $(x).fx$ entails fa." Suppose I ask, "Are you sure fa follows from $(x).fx$? Can we assume that it does not follow? What would go wrong if we did assume that?"
>
> *Wisdom:* One reply which might be given is that it is impossible to make such an assumption.
>
> *Wittgenstein:* Yes. But let us look into this, because such things as "Let us assume that $(x).fx$ does not entail fa" have been said. Now the reply you suggested did

---

[5]Cf. Hilmy [Hilmy (1987)]. Hilmy's extensive study of the *Nachlass* has greatly helped removing classical misconceptions such as Hintikka's claim that "Wittgenstein in the *Philosophical Investigations* almost completely gave up the calculus analogy.", p. 14 of [Hintikka and Hintikka (1986)].

Hilmy points out that even in the *Investigations* one finds the use of the calculus/game paradigm to the understanding of language, such as "im Läufe des Kalküls" (Part I, §559) (in the English translation: "in operating with the word"), and "it plays a different part in the calculus" (footnote of p. 14). Hilmy also quotes from a late (1946) unpublished manuscript (MS 130) "this sentence has use in the calculus of language".), and is compatible with "asking whether and how a proposition can be verified is only a particular way of asking "How d'you mean?"" (*Investigations*, Part I, §353).

[6]As early as 1912, in a letter to Russell, dated 1.7.12, he writes (our *emphasis*):

> Will you think that I have gone mad if I make the following suggestion?: The sign $(x).\varphi x$ is not a complete symbol but has meaning only in an inference of the kind: *from* $\vdash \varphi x \supset_x \psi x.\varphi(a)$ *follows* $\psi(a)$. Or more generally: *from* $\vdash (x).\varphi x.\varepsilon_0(a)$ *follows* $\varphi(a)$. I am—of course—most uncertain about the matter but something of the sort might really be true.

Letter *R.3*, p. 12 of Wittgenstein, L.: 1974, *Letters to Russell, Keynes and Moore*, Ed. with an Introd. by G. H. von Wright, (assisted by B. F. McGuinness), Basil Blackwell, Oxford, 190pp.

not mean that it is psychologically impossible to assume that; for if it did, one might say that although Wisdom cannot do it yet perhaps other people can.

In what way is it impossible to assume $(x).fx$ does not entail fa?

Wisdom: Isn't the assumption like saying "Couldn't we have a zebra without stripes?"

Wittgenstein: Yes. It would be said that the meaning of '$(x).fx$' had been changed.

What then would go wrong if someone assumes that $(x).fx$ does not entail fa? I would say that all I am assuming is a different use of "all", and there is *nothing* wrong in this.

If I stick to saying that the meaning is given by the use, then I cannot use an expression in a different way without changing the meaning. But it is then misleading to say, "The expression *must* have a different meaning if used differently." It is merely that it *has* a different meaning—the different use *is* the different meaning.

And if one says, "If one assumes fa does not follow from $(x).fx$, one must use $(x).fx$ in a different way"—we reply, in assuming this one *does* it in a different way.—But if we make this assumption, nothing goes wrong.[7]

From the above, it would not be unreasonable to say, for example, that not only the meaning of a proposition changes when the way it is used changes, when what can be deduced from it changes, but that the conclusions one can draw from a proposition *determine* its meaning. This view becomes much sharper and the evidences for it much clearer in Wittgenstein's very late writings, and that is what we shall attempt to demonstrate below. (And it is for that reason that we attempt to provide historical information such as dates, typescripts and manuscripts numbers whenever appropriate, etc.).

What we do here in trying to relate those three concepts of *meaning*, *use* and *consequences* by means of Wittgenstein's (mainly later) writings is in effect an attempt to show that a semantical standpoint based on the paradigm 'meaning is use' can be not just practically useful, as far as a basis for operational semantics of formalised languages is concerned, but it can also be very natural for a semantical standpoint supporting logic and computing science as a whole where meaning and rules have a close interplay. In order to define a logical sign by means of deduction/computation rules what one does, besides giving the 'grammatical' rules describing the assertability conditions, is essentially to demonstrate what can be obtained from the 'term' having the sign as its major one after *deducing, eliminating* the logical sign.

---

[7] p. 192 of C. Diamond's *Wittgenstein's Lectures on the Foundations of Mathematics,* Cambridge, 1939.

8.4 Meaning and use

There has been a number of different readings of Wittgenstein's 'definition' of meaning as given in "the meaning of a word is its use in the language.[8] Much has already been said on the different interpretations one can give to such a 'definition'. *Use* being such a vague concept as it is in ordinary language, enough room is left for distinct and perhaps uncongenial readings of Wittgenstein's semantical paradigm. There are readings discussed in the literature that lend too vague a character to Wittgenstein's use-based semantical paradigm, some attempting to discuss use in the sense of 'usage', some in the sense of 'practice', etc. The general picture which remains is that of vagueness, usually in the least positive sense. As a general rule, one's first reaction to the adoption of a use-based semantics in the analysis of formalised languages is that of scepticism with respect to the feasibility of the enterprise. One is induced to think that the enterprise may turn out to be unworkable, because when describing the use of a word or sentence one has to refer to the whole context within which it is embedded, and therefore, the holistic character of language is definitely 'unmanageable'. And from this, one is almost inevitably led to conclude that, according to Wittgenstein, 'language is really unanalysable' (*not* Wittgenstein's own words!). One might then justify the arrival at such a negative conclusion by saying that if one has to deal with meaning as functionality, purpose, usefulness, in a calculus/game/system of logical signs, without having the support of some 'universe' of values (cf. interpretations-based semantical theories), then one makes the semantical standpoint sound inadequate for linguistic analysis through formalised description and reasoning mechanisms, given the inadequacy of a holistic approach to an enterprise of that sort.

Now, what we find remarkable about Lorenzen's and Hintikka's game-based approach to the characterisation of logical connectives is that a (semi-)formal account of Wittgenstein's "the *use* of the logical sign in the language-game" for the case of the language of mathematics is given, and yet no prohibitively insurmountable difficulty is dealt with by handling language in a manageably holistic fashion. It actually makes the explanation of use of logical constants 'atomistic', in the sense that each individual connective is treated separately, and 'functional' in the sense that each connective gets its meaning from the explanation of the 'rules of the game'. Moreover, each connective is explained by stating in each case what can be deduced from the propositions with those logical signs as the major connective, and the game follows after the *elimination* step is made. In other words, the rules are laid down by means of explanations of what (immediate) consequences one is allowed to draw

---

[8]§43, Part I, p. 20e of the *Philosophical Investigations*.

from the corresponding propositions. It is then left implicit (as in the context of a proper game where moves other than those stated are simply invalid, nonsensical) that any consequence beyond the one(s) indicated cannot be drawn from the proposition(s), whatever context they may be uttered in (wherever the game is played!). And this seems to be a very reasonable rationale to be followed by any theory of meaning which, whereas taking into account contexts (language-games) as frames of reference, is meant to be placing the semantical emphasis on the use and functionality of logical signs ("All signals get their sense from circumstances, it is true; but the way they function in the circumstances is different."[9]).

As a matter of fact, *use* (as *usefulness*) is not just a concept arbitrarily chosen by Wittgenstein as the basis for his semantical investigations, for it entails the crucial concept of *intentionality*. And, as a matter of course, there is a great deal of *intentional* ingredients in Wittgenstein's concept of meaning as use. That is what makes his semantic theory fundamentally different from many of the previous accounts of the relation between language and the world, which were mostly extensional and followed Augustine's tradition "as I heard words repeatedly used in their proper places in various sentences, I gradually learnt to understand what objects they signified". In such an *intentional* view of language as Wittgenstein's, there is no place for interpretations as endpoints for settling meaning. Moreover, in the light of such an alternative account of the links between language and reality, one can say with Wittgenstein that the phrase "the interpretation of signs" is misleading, and that one ought to say "the use of signs" instead,[10] because the former suggests that a sign is to be understood by perhaps another sign which does not need any further interpretation, whereas the latter conveys a much broader conceptual basis which can accommodate intentionality: the kind of 'manageable' holism, which we are here referring to as 'functional atomism' to emphasise that the words of a language are identified as individuals which have the meaning determined by their function/use/rôle/usefulness in the calculus of language.

In the *Investigations* Wittgenstein speaks of 'interpretations hanging in the air along with what they interpret':

"But how can a rule shew me what I have to do at *this* point? Whatever I do is, on some interpretation in accord with the rule." — That is not what we ought to say, but rather: any interpretation still hangs in the air along with what it interprets, and cannot give it any support. Interpretations by themselves do not determine meaning.[11]

---

[9]p. 76 of Geach, P. T. (ed.): 1988, *Wittgenstein's Lectures on Philosophical Psychology 1946–47*, (Notes by P. T. Geach, K. J. Shah & A. C. Jackson), Harvester-Wheatsheaf, Hemel Hempstead, xv+348pp.
[10]§32, Part III, p. 70 of the *Philosophical Remarks*.
[11]§198, Part I, p. 80[e] of the *Investigations*, op. cit.

The pictorial device of 'hanging in the air' is sometimes replaced by the figure of a 'rider' which stands there to help giving the sentence its meaning. This is the particular presentation of the shortcomings of interpretations as semantical devices that appears as early as in *Philosophical Grammar*, which collects writings from the early thirties:

> But an *interpretation* is something that is given in signs. It is this interpretation as opposed to a different one (running differently). — So when we wanted to say "Any sentence still stands in need of an interpretation", that meant: no sentence can be understood without a rider.[12]

The 'rider' of a sentence can only disappear when the element of *intentionality* is acknowledged, and the intention is uncovered by a demonstration of the conclusions one can draw from the sentence. The intentional aspect of a sentence is not made clear until one makes explicit where one intends to arrive at by asserting it. Compare, e.g. 'what do you mean?' with 'what are you trying to say?', 'what conclusion(s) do you want one to draw from what you have just said?', 'what is your intention when asserting the sentence?', etc.. And it is 'intention' which is going to play a crucial rôle in breaking the apparently infinite chain of interpretations. And indeed, in another remark from the same *Philosophical Grammar*, he expresses the view that *intention* can be a useful concept in the breaking of such a chain of interpretations, but does not sound too satisfied when saying that such a concept is of a 'psychological' rather than 'logical' nature:

> By "intention" I mean here what uses a sign in a thought. The intention seems to interpret, to give the final interpretation; which is not a further sign or picture, but something else — the thing that cannot be further interpreted. But what we have reached is a psychological, not a logical terminus.[13]

In attempting to extract from Wittgenstein's early as well as later writings a certain uniformity with respect to the connections between the meaning of a proposition and the (immediate) consequences one is allowed to draw from it, it is also our endeavour to demonstrate that a logical counterpart to such a 'psychological' terminus, namely *intention*, is already present in his semantic theory, in spite of an apparent lack of the explicit and definite acknowledgement that this is the case by Wittgenstein himself.

---

[12] §9, p. 47 of *Philosophical Grammar*, Part I.
[13] Ibid., §98, p. 145.

8.5   Meaning and the explanation of consequences

By contrast with the wide popularity enjoyed by the connections between *meaning* and *use* that Wittgenstein's (especially later) writings convey, the position on the connections between the *meaning* of an assertion and the explanation of (immediate) *consequences* one is allowed to draw from it is hardly emphasised in most of the relevant literature on Wittgenstein's semantic theory. Nevertheless, we find that without proper attention to this particular aspect of his writings on the notion of meaning, the understanding of Wittgenstein's use-based semantic theory must remain incomplete. There are a great many explicit references to such a crucial semantical standpoint especially, as we have said previously, throughout the writings of his later phase.

For example, in a typescript dictated in late autumn 1947 (TS 229), whose underlying manuscripts stem from the period between 10 May 1946 and 11 October 1947, published as *Remarks on the Philosophy of Psychology*, Vol. I, one can find a remark relating what an utterance means and what one is permitted to draw from it (our *emphasis*):

> ... someone who does philosophy or psychology will perhaps say "*I* feel that I think in my head". But what that means he won't be able to say. For he will not be able to say *what* kind of feeling that is; but merely to use the expression that he 'feels'; as if he were saying "*I* feel this stitch *here*". Thus he is unaware that *it remains to be investigated what his expression "I feel" means here, that is to say: what consequences [Konsequenzen] we are permitted to draw from this utterance.* Whether we may draw the same ones as we would from the utterance "I feel a stitch here".[14]

Here he uses the German '*Konsequenz*', as opposed to '*Folges*' which is used more frequently in his attempts to relate the consequences one can draw from an assertion and its meaning.

From another typescript dictated in early autumn 1948 (TS 232), bearing remarks which stem from manuscripts written between November 1947 and August 25th 1948, also published as *Remarks on the Philosophy of Psychology*, but this time Vol. II, one can see remarkably direct and explicit references to the connections between the meaning of a statement and what can be deduced from it.

> For the question is not, 'What am I doing when . . .?' (for this could only be a psychological question) — but rather, 'What meaning does the statement have, what can be deduced from it, *what consequences does it have*?'[15]

---

[14] §350, p. 69ᵉ of the *Remarks on the Philosophy of Psychology*, Vol. I.
[15] §38, p. 8ᵉ of the *Remarks on the Philosophy of Psychology*, Vol. II.

In other notebooks from a particularly late period of his later phase, namely the two last years of his life, which are collected in the book published under the title of *On Certainty* as a direct reference to the main subject of the remarks, one can find clear traces of what one is tempted to call one of the canons of L. Wittgenstein's philosophical writings. An answer to the question 'what follows from your statement' plays the rôle of settling its meaning, giving its final interpretation, making the intention explicit (our *emphasis*):

> I know that this is my foot. I could not accept any experience as proof to the contrary. — That may be an exclamation; *but what follows from it?* At least that I shall act with a certainty that knows no doubt, in accordance with my belief.[16]

Another passage from the same *On Certainty*, this time dated in the original as 27.3.51, again reflects his position with respect to the fundamental rôle of the demonstration of the consequences of an assertion in settling its meaning. It says (our *emphasis*):

> One is often bewitched by a word. For example, by the word "know".
>
> Is God bound by our knowledge? Are a lot of our statements *incapable* of falsehood? For that is what we want to say.
>
> I am inclined to say: "That *cannot* be false." That is interesting; *but what consequences has it?*[17]

8.6   Use and the explanation of consequences

Similarly to the previous case, it does not seem to be widely acknowledged the attempt Wittgenstein has made in demonstrating the close links between the *use* of a proposition and the *consequences* one can draw from it. Nevertheless, one can demonstrate that such an attempt is made very clear in later writings, by contrast with the case of *meaning* and *consequences*, which although less emphasised in the literature, yet can be traced back in early remarks as we will see in the sequel.

From the notes of a series of *Lectures on Religious Belief*, given in Cambridge sometime around the summer of 1938, taken by Y. Smythies and published in *Lectures and Conversations on Aesthetics, Psychology and Religious Belief*, one can choose a passage where the links between *use* and *consequences* are made very explicit (our *emphasis*):

> Yes, this might be a disagreement—*if he himself were to use the word in a way in*

---

[16]§360, p. 47ᵉ of *On Certainty/Über Gewissheit*.
[17]¹⁶ Ibid., §§435–7, p. 57ᵉ.

which I did not expect, or were to draw conclusions I did not expect him to draw. (...)[18]

In the first volume of the *Last Writings on the Philosophy of Psychology*, which is not based on a typescript as is the case for the previously published *Remarks on the Philosophy of Psychology*, but on manuscript writings dating from the period between 22 October 1948 and 22 March 1949 except for the last remark dated 20 May (second half of MS 137 and the whole of MS 138), one can find examples of very explicit references to the connections between the use of words and the consequences of the corresponding utterance. Presented sometimes in a positive way such as in (our *emphasis*):

> What are you telling me when you *use* the words . . .? What can I do with this utterance? *What consequences does it have?*[19]

and in (our *emphasis*):

> What does anyone tell me by saying "Now I see it as . . ."? *What consequences has this information?* What can I do with it?[20]

(which also appears in the *Investigations*, Part II, section xi, p. 202^e) or in a negative manner such as in (our *emphasis*):

> The report "The word . . . was crammed full of its meaning" *is used quite differently, has quite different consequences*, from "It had the meaning . . .".[21]

his remarks on the connections between the use of a sentence and the consequences one is allowed to draw from it constitute a key ingredient in his investigations on the concept of meaning.

Considering that his main focus of attention in the later phase is indeed on the 'logic of ordinary language', rather than on the technicalities of Frege–Russell's logical calculus, it is not quite surprising that Wittgenstein's very late writings on meaning and use, strikes one as so intuitively appealing. By contrast, in early as well as not-so-late writings, in spite of a clear attempt to show that 'meaning' is better understood as 'use', generally speaking there is no clear-cut, incisive account of what should be taken as the logical counterpart of the latter. There are, however, a few references to meaning-use-consequences scattered amongst early writings, as we have attempted to demonstrate in a previous opportunity.[22]

---

[18] p. 71 of the *Lectures and Conversations on Aesthetics, Psychology and Religious Belief*.
[19] §624, p. 80^e of *Last Writings I*, op. cit.
[20] Id. Ibid., §630.
[21] Ibid., §785, p. 100^e.
[22] Cf. [de Queiroz (1989)].

Unfortunately, the remaining general picture of the transitional period suggests a certain hesitation which is revealed quite clearly when the rôle of proofs in mathematics is analysed. In the writings of the early period of his later phase, documented in the notebooks containing remarks mainly on the foundations of mathematics now published as *Philosophical Remarks* and *Philosophical Grammar*, there is a clear shift in the focus of attention from truth-values and truth-functions which were characteristic of the *Tractatus*, to mathematical proofs. The concept of mathematical proof is thoroughly investigated, and in some passages it is even placed in a rather privileged position with respect to the semantics of the mathematical language. One of the lessons taught by those 'transitional' writings seems to be that (using Wittgenstein's own words) "when one wants to know the meaning, the sense, of a proposition, one has to look at what its proof proves".

It does not take too long, however, for a revision of the position on the connections between meaning and proofs to be made, and it is in the *Remarks on the Foundations of Mathematics*, written between 1941 and 1944, that the revision is made explicit:

> I once said: 'If you want to know what a mathematical proposition says, look at what its proof proves'.[23] Now is there not both truth and falsehood in this? For is the sense, the point, of a mathematical proposition really clear as soon as we can follow the proof?[24]

A fair number of remarks which reveal a much less explicit statement of the revision made with respect to the connections between the meaning of a proposition and its proof appear in later stages of the later period. One of them, from the same *Remarks on the Foundations of Mathematics* and written around 1943/1944, says:

> The proof of a proposition certainly does not mention, certainly does not describe, the whole system of calculation that stands behind the proposition and gives it its sense.[25]

which reveals his concern on the insufficiency of the explanation of what counts as a proof (i.e. explanation of assertability conditions) with respect to the explanation of meaning. The apparently unworkable holism is shown perfectly feasible to the formulation of a theory of meaning for the language of mathematics by his suggestions on the connections between the meaning of a proposition and the explanation of the immediate consequences one can draw from it (Cf. "One learns

---

[23] Cf. *Philosophical Grammar*, p. 369; *Philosophical Remarks*, op. cit., pp. 183–4.
[24] §10, Part VII, p. 367 of the *Remarks on the Foundations of Mathematics*.
[25] Ibid., §11, Part VI, p. 313.

the meaning [*Bedeutung*] of 'all' by learning that 'fa' follows from '$(x).fx$'."
and "And the meaning [*Sinn*] of '$(x).fx$' is made clear by our insisting on 'fa's
following from it."[26]). It is as if the parts of language are understood only in
relation to other neighbouring parts, as D. Pears points out.[27]

His insistence on the idea that "the meaning of a word is determined by its
rôle in the language" would be understood in the following way: by showing how
to 'eliminate', 'get rid of' the logical sign, demonstrating the conclusions one can
draw from a proposition which has it as its major connective, one is demonstrating
the rôle it plays in the calculus/system of propositions.

Unfortunately, his acknowledgement of the inappropriateness of using mathematical proofs and assertability conditions as semantical notions does not seem to
have been taken into consideration by some 'verificationistic' theories of meaning
which still look for a conceptual basis in Wittgenstein's semantic theory and suggest that 'meaning is use' is their underlying semantical paradigm. Wittgenstein's
clear emphasis on the *deductions* that can be made from a proposition as explanations of its meaning does not appear to have been given the attention it would
seem to deserve by those who advocate a use-based semantic theory but still insist
on the primacy of assertability conditions. Neither there seems to be much concern about this particular aspect of Wittgenstein's very late writings in most of the
literature on Wittgenstein's philosophy, to the best of our knowledge.

8.7   Early signs of 'meaning–use/usefulness–consequences'

The intuition regarding the connections between meaning and use (as usefulness),
which appears explicitly in very late remarks such as the one shown in the opening
lines as well as in the title of this chapter, and underlies the "words are tools"[28]
conceptual link made mainly in Part I of the *Investigations*, seems to find its roots
in very early remarks documented in early writings such as the *Notebooks* and the
Tractatus. If the former shows the early signs of the connections between meaning
and use — "The way in which language signifies is mirrored in its use"[29] —, in
the latter one can find remarks such as:

---

[26] Ibid., Part I, §10, p. 41 and §11, p. 42, respectively.

[27] "The new idea [of W.'s later theory of language] was that we experience each part of language in its
relations to other, neighbouring parts.", p. 171 of [Pears (1990)].

[28] "Think of the tools in a tool-box: there is a hammer, pliers, a saw, a screw-driver, a rule, a glue-pot,
glue, nails and screws. — The functions of words are as diverse as the functions of these objects. (And
in both cases there are similarities.)", §11, Part I, p. 6[e] of the *Investigations*, op. cit.

[29] dated 11.9.16, p. 82[e] of the *Notebooks 1914–1916*.

If a sign is *useless*, it is meaningless. That is the point of Occam's maxim.[30]

and 'useless' here is to be understood as 'purposeless', as it becomes clear later in the same *Tractatus*:

> Occam's maxim is, of course, not an arbitrary rule, not one that is justified by its success in practice: its point is that *unnecessary* units in a sign-language mean nothing.
>
> Signs that serve *one* purpose are logically equivalent, and signs that serve *none* are logically meaningless.[31]

In the former, the question whether the logical constants exist is not even considered to be relevant, given that they can 'disappear'. And again, the position can be seen to be very reasonable and coherent with the one expressed in the remark above if one looks at such 'constants' as devices, tools, which find meaning in their *use/usefulness*. In this particular respect, Wittgenstein's words in the *Notebooks* are put in the following way:

> With the logical constants one need never ask whether they exist, for they can even *vanish!*[32]

In the same *Notebooks* the (introduction of the) sign '0' is suggested to obtain its meaning from its usefulness in making the decimal notation possible:

> Einführung des Zeichens "0" um die Dezimalnotation möglich zu machen: Die logische Bedeutung dieses Vorgehens.[33]

(translated as "The introduction of the sign "0" in order to make the decimal notation possible: the logical significance of this procedure.")

If meaning is to be understood as use/usefulness, logical connectives should be better seen as *instruments*, rather than as *constants*, and that is what seems to be suggested by Wittgenstein even in the *Tractatus* where the truth-values based semantical view inherited from Frege and Russell characterises the 'early Wittgenstein' as distinct from the 'later Wittgenstein and the use-based semantical view'. And one can see his rather original point of view on logic and language *versus* reality, when he suggests that logical connectives are *tools*, therefore the idea that 'logical connectives are *constants*' and the viewpoint that 'there are logical constants, namely "∨", "⊃", "." ("∧"), etc.' are hardly tenable. He makes the point by recalling the so-called interdefinability of such 'constants':

---

[30] §3.328, p. 16 of the *Tractatus Logico-Philosophicus*.

[31] Ibid., §5.47321, p. 48.

[32] dated 25.10.14, p. 19e of the *Notebooks*, op. cit.

[33] dated 16.11.14, p. 31 of the *Notebooks*, op. cit.

At this point it becomes manifest that there are no 'logical objects' or 'logical constants' (in Frege's and Russell's sense).[34]

Under such a viewpoint, it seems rather clear that insofar as being understood as useful *instruments*, the logical connectives would not have to be looked at as 'primitive' signs. This appears to be suggested by Wittgenstein in (again the *Tractatus*):

It is self-evident that $\vee$, $\supset$, etc. are not relations in the sense in which right and left etc. are relations.

The interdefinability of Frege's and Russell's 'primitive signs' of logic is enough to show that they are not primitive signs, still less signs for relations.[35]

Furthermore, in spite of resorting to truth-values and truth-functions when providing a precise account of meaning for logical propositions by means of his "a proposition is a truth-function of elementary propositions. (an elementary proposition is a truth-function of itself.),"[36] Wittgenstein did not seem to be too convinced that the concept of truth-function could work equally well for all logical concepts and connectives of Frege-Russell's logical calculus. For example, the generality concept in which the universal quantifier was embedded was explicitly dissociated from truth-functions in:

I dissociate the concept *all* from truth-functions.

Frege and Russell introduced generality in association with logical product or logical sum. This made it difficult to understand the propositions '$(\exists x).fx$' and '$(x).fx$', in which both ideas are embedded.[37]

---

[34] §5.4, p. 44 of the *Tractatus*, op. cit.

[35] Ibid., §5.42, p. 44.

[36] Ibid., §5, p. 36.

[37] Ibid., §5.521, p. 51. The use of 'dissociate' in the translation by D. Pears and B. McGuinness appears to reinforce Wittgenstein's deliberate intention in "Ich trenne den Begriff *Alle* von der Wahrheitsfunktion" to disconnect one from another. This is clear from the correspondence between Wittgenstein and the first (English) translator C. K. Ogden. In the 'Comments II on Separate Sheets', from *Letters to C. K. Ogden*, p. 61, he says:

> 5.521 I mean "separate" and *not* "derive". They were connected and I separate them.

and the comments by the Editor, p. 65:

> 5.521 The passage under discussion runs: "I separate the concept *all* from the truth-function."
> Ogden had queried whether "separate" should not be changed to "derive".

From this it should be clear that it is unreasonable to say that Wittgenstein in the *Tractatus* "construes quantified sentences as conjunctions and disjunctions" (p. 110 of [Hintikka and Hintikka (1986)],

For the particular case of the logical notion of generality and the corresponding connective for universal quantification, the notion of 'following from' was preferred to the more *Tractatus*-characteristic truth-function:

(Daß man aus (x).fx auf fa schließen kann, das zeigt, daß die Allgemeinheit auch im Symbol ≫(x).fx≪ vorhanden ist.)

(translated as 'The possibility of inference from $(x).fx$ to fa shows that the symbol $(x).fx$ itself has generality in it.'[38])

The inference demonstrating the (immediate) consequences one can draw from the proposition '$(x).fx$' is actually making clear that the distinguishing characteristic of the major logical sign (namely '$(x).$', *for all*) is that of generality. From this standpoint, one can read many of the remarks contained in the *Tractatus* itself in a way which has, to the best of our knowledge, hardly been done in the literature on the early Wittgenstein, and place them still in line with the remarks from the later Wittgenstein using the same connections between *meaning*, *use/usefulness* and *consequences*. For example, there is a remark concerning the view that in logic the deductions ('*Folgern*') happen a priori, and it seems that one can understand this in the light of, e.g. Hintikka's *semantical games*, a (semi-)formal counterpart to later Wittgenstein's 'the meaning of a sign is its use in the language-game', where the logical connectives are given meaning by means of rules of 'one-step' deduction. And indeed, if "before a proposition can have a sense, it must be completely settled what propositions follow from it"[39], according to the

---

Investigating Wittgenstein). Hintikka and Hintikka's conclusion seems to draw on Moore's saying that Wittgenstein in 1930–33, "went on to say that one great mistake which he made in the *Tractatus* was that of supposing that in the case of *all* classes "defined by grammar", general propositions were identical with logical products or with logical sums (meaning by this logical products or sums of the propositions which are values of fx) as, according to him, they really are in the case of the class of "primary colours". He said that, when he wrote the *Tractatus*, he had supposed that *all* such general propositions were "truth-functions"; (...) He said that, when he wrote the *Tractatus*, he would have defended the mistaken view which he then took by asking the question: How can $(x).fx$ possibly entail fa, if $(x).fx$ is not a logical product? And he said that the answer to this question is that where $(x).fx$ is not a logical product, the proposition "$(x).fx$ entails fa" *is* "taken as a primary proposition"*, whereas it is a logical product this proposition is deduced from other primary propositions.", (our *emphasis*), p. 3 of Moore, G. E.: 1955, 'Wittgenstein's Lectures in 1930–33', in *Mind*, vol. 54, n. 253, pp. 1–27. Wittgenstein might well have said that, given that he was very critical, almost ruthless, with himself and he often insisted on the big mistakes he made in the *Tractatus*. For example, he says in a section entitled '*Criticism of my former view of generality*' of the *Philosophical Grammar*, p. 268, that his "view about general propositions was that $(\exists x).\varphi x$ is a logical sum". Nonetheless, nowhere in the *Tractatus* the existential quantifier is treated as a logical sum. Moreover, as we shall see below, early remarks bear witness of the different, 'non-truth-functional' treatment he gave to the universal quantifier.

[38] Ibid., §5.1311, p. 38.

[39] §3.20103, p. 65 of the *Prototractatus*. An almost identical remark already appears in the *Notebooks 1914–16*: "We might demand definiteness in this way too: if a proposition is to make sense then the

Prototractatus, then:

   Alles Folgern geschieht a priori.

(translated as 'All deductions are made a priori.'[40]) according to the *Tractatus*. It does not seem unreasonable to say that these remarks can be understood along the same lines advocated here by simply acknowledging that to show the consequences which are allowed to be drawn from a proposition is to demonstrate what can be done with it, the deductions happening a priori as part of the 'definition'. Moreover, in the light of this same conceptual basis one has to say that (using early Wittgenstein's own words), "logic must look after itself"[41] just because for a proposition to be defined and have a sense in logic, the rule showing the consequences (deductions) one can draw from the proposition must be given a priori. Therefore, "we cannot give a sign the wrong sense"[42], and so, the (one-step) 'deductions' are made a priori simply because they *determine* the meaning of the proposition.

   Here it is relevant to reemphasise that one can find examples of remarks suggesting the links between the meaning of an assertion and what can be deduced from it, even in very early writings, such as: "A proposition affirms every proposition that follows from it."[43], and: " '$p.q$' is one of the propositions that affirm 'p' and at the same time one of the propositions that affirm 'q'."[44] of the *Tractatus*, and, e.g.:

   That the *proposition* "ϕa" can be inferred from the *proposition* "$(x).\phi x$" shews how generality is present even in the *sign* "$(x).\phi x$".[45]

of the *Notebooks*.

---

syntactical employment of each of its parts must be settled in advance. — It is, e.g. not possible *only subsequently to come upon* the fact that a proposition follows from it. But, e.g. *what propositions follow from a proposition must be completely settled before that proposition can have a sense!*", *Notebooks*, dated 18.6.15, p. 64ᵉ. (Our *emphasis*)

[40] §5.133, p. 39, of the *Tractatus*, op. cit.

[41] Ibid., §5.473, p. 47.

[42] Id. Ibid., §5.4732.

[43] Ibid., §5.124, p. 38.

[44] Id. Ibid., §5.1241.

[45] dated 24.11.14, p. 32ᵉ of the *Notebooks*, op. cit.

8.8 Normalisation of proofs: the explanation of the consequences

We have mentioned in the opening section that the formal counterpart to the informal notion of 'explanation of the (immediate) consequences' would seem to find in the proof-theoretic device of normalisation a good candidate. In this section we would like to pursue this a little further, and for that we need to refer to the framework of the functional interpretation of logics, where normalisation plays a definite semantical role.

When it comes to the quantifiers, observe that in the presentation of both '∀' and '∃' the canonical proof involves a 'course-of-values' term, a 'Λ'-term for the universal quantifier, and an 'ε'-term for the existential quantifier. The fundamental difference comes with 'choice' of witnesses: the 'Λ'-term ('$\Lambda x.f(x)$') does not carry a particular witness with it, whereas the 'ε'-term does ('$\varepsilon x.(f(x), a)$'). This crucial difference in the 'choice' of witnesses becomes clear when the normalisation (reduction) rules are laid down: the '*EXTR*action' eliminatory operator, or 'destructor', introduces a new open term ('a') which is not part of the canonical term ('$\Lambda x.f(x)$'), indicating the 'arbitrariness' of the choice of the witness; the '*INST*antiation' destructor introduces only closed terms. The open term which plays the rôle of witness, namely 'a', is already part of the canonical term ('$\varepsilon x.(f(x), a)$'), therefore chosen at the time of assertion. This situation, where the name for the individual is bound in the introduction of a universal quantifier, and remains free after the introduction of an existential quantifier is reflected in the meta-language distinction between an 'arbitrary' and a 'specific' individual, as in Martin-Löf's [Martin-Löf (1987)] (p. 411) explanation of the inference rules for the quantifiers:

$$\frac{A(a) \text{ is true}}{\exists x.A(x)} \qquad (\text{`}A(a)\text{ is true of a } \textit{specific } \text{individual'})$$

$$\frac{A(x) \text{ is true}}{\forall x.A(x)} \qquad (\text{`}A(x)\text{ is true of an } \textit{arbitrary } \text{individual'})$$

Note that the specific versus arbitrary is reflected not simply by means of the typographical convention (initial versus terminal roman letters), but rather in the individual's name remaining unbound or becoming bound in the logical formula. As for the 'functional' side, namely the functional calculus on the labels, the same guidelines should apply, and that is why we have a Λ-term in one case and an ε-term in the other. Only, in our framework we need to make explicit the 'arbitrary' quality of the individual by writing out the assumption 'suppose x is an arbitrary element from the domain...' as in:

$$[\text{Suppose } x \text{ is in } D]$$

$$\vdots$$

$$\frac{A(x) \text{ is true}}{\forall x^D.A(x)}$$

and the three dots indicate that some inference steps were made between the supposition that x was an arbitrary element from domain D, and the premise of the inference rule, which says that $A(x)$ is true. And the fact that the introduction of the universal quantifier involves the discharge of assumptions characterises the $\forall$-*introduction* as an improper inference rule (to use a terminology from [Prawitz (1965)], similarly to the introduction of implication ($\rightarrow$-*introduction*). The introduction of the existential quantifier, however, does not assume that the individual was an arbitrary element from the domain concerned, thus it does not involve any assumption discharge.

Similar treatments of the difference between the two quantifiers appear in other formulations of the semantics of logical connectives based on the explanation of the (immediate) consequences of the corresponding proposition, such as Lorenzen's [Lorenzen (1969)] *dialogical games* [Lorenzen (1969)] (p. 25) and Hintikka's *semantical games* [Hintikka (1974)] (p. 156); [Hintikka (1983)] (p. 3). The general underlying principle is that of the logical principle of inversion, uncovered by Gentzen, and later by Lorenzen [Lorenzen (1955)]: the elimination procedure is the exact inverse of the introduction, therefore all that can be asked from an assertion is what is indicated in the explanation of elimination procedure. We can look at Lorenzen's dialogical games and try to show where the principle of inversion resides in the game-approach by comparing it to the normalisation-approach (read '$\triangleright_\beta$' as 'β-normalises to'):

Assertion/Introd. (Myself/	Attack/Elim. Nature/	$\triangleright_\beta$	Defense Game continues)
Conjunction ('$\wedge$'):			
$A_1 \wedge A_2$	$L?$		A_1
$A_1 \wedge A_2$	$R?$		A_2
$\langle a_1, a_2 \rangle : A_1 \wedge A_2$	$\text{FST}(\langle a_1, a_2 \rangle)$	$\triangleright_\beta$	$a_1 : A_1$
$\langle a_1, a_2 \rangle : A_1 \wedge A_2$	$\text{SND}(\langle a_1, a_2 \rangle)$	$\triangleright_\beta$	$a_2 : A_2$

Disjunction ('∨'):

$A_1 \vee A_2$? A_1
$A_1 \vee A_2$? A_2

$\text{inl}(a_1) : A_1 \vee A_2$ $\text{CASE}(\text{inl}(a_1), \acute{s_1}d(s_1), \acute{s_2}e(s_2))$ $\rhd_\beta$ $d(a_1/s_1) : C$
$\text{inr}(a_2) : A_1 \vee A_2$ $\text{CASE}(\text{inr}(a_2), \acute{s_1}d(s_1), \acute{s_2}e(s_2))$ $\rhd_\beta$ $e(a_2/s_2) : C$

If-then ('→'):

$A \rightarrow B$ A   ?   B

$\lambda x.b(x)$ $\text{APP}(\lambda x.b(x), a)$ $\rhd_\beta$ $b(a/x) : B$

Universal Quantifier ('∀'):

$\forall x^D.F(x)$ $a : D$? $F(a)$

$\Lambda x.f(x)$ $\text{EXTR}(\Lambda x.f(x), a)$ $\rhd_\beta$ $f(a/x) : F(a)$

Existential Quantifier ('∃'):

$\exists x^D.F(x)$? $t : D, F(t)$

$\varepsilon x.(f(x), a)$ $\text{INST}(\varepsilon x.(f(x), a), \acute{g}\acute{t}d(g, t))$ $\rhd_\beta$ $d(f/g, a/t) : C$

If we look at the conclusion of *reduction* inference rules, and we take the DESTRUCTOR as being the Attack (or 'Nature', its counterpart in the terminology of Hintikka's Game-Theoretical Semantics), and the constructor as being the Assertion (or Hintikka's 'Myself'), it is straightforward to see that the game-theoretic explanations of logical connectives find a counterpart in the functional interpretation with the semantics of convertibility [de Queiroz (1988)].

It would not seem unreasonable to say that the parallels mentioned above are to be attributed to the 'universal validity' of the underlying logical principle of inversion. (Recall that the cut-rule, sometimes said to be a fundamental rule for any calculus to have in order to be considered a logic (cf. [Hacking (1979)]), can be seen as an instance of the principle of inversion.) The principle of inversion and the 'naturally' opposite sides appear in many different foundational frameworks:

- Proof Theory (Gentzen): Introduction/Elimination
- λ-Calculus (Church): Canonical Terms/Noncanonical Terms
- Dialogue Games (Lorenzen): Assertion/Attack
- Semantical Games (Hintikka): Myself/Nature

8.9 Concluding remarks

By attempting to fill in the two lesser known sides of the semantic triangle-like diagram with vertices in *meaning, use* and (the explanation of the immediate) *consequences* we have demonstrated how one can find a formal counterpart to Wittgenstein's 'meaning-is-use' in the proof-theoretic semantics based on convertibility (normalisation). In [de Queiroz (1990)] there was already an attempt at presenting a reformulated type theory and its underlying 'semantics of use', demonstrating why its appropriateness in studying the logical foundations of computation. We also attempt to show that a number of ideas arising out of a semantical reformulation of P. Martin-Löf's *Intuitionistic Type Theory* where the main semantical device is not that of canonical proofs but that of normalisation, find themselves useful in the understanding of some of the issues dealt with in theories of specification of abstract data types, a major topic of modern theoretical computing science.

Concerning the use of such a framework for the development of new foundational basis for proof-theoretic semantics, we have been engaged in the development of what we call *Labelled Natural Deduction*, which provides the basis for

(1) the study of various notions of implication by taking β-normalisation on $\rightarrow$-types as the main semantical device, and varying disciplines of assumption discharge and corresponding λ-abstractions [Gabbay and de Queiroz (1991, 1992)];

(2) the reassessment of Skolem and Herbrand fundamental results, as well as the various formulations of the notion of singular terms (including Hilbert's ε-calculus) via the functional interpretation [de Queiroz and Gabbay (1995)]; and

(3) the study of modal logics via the use of an extension of the variable-binding technique to deal with arbitrary 'structured' collections of formulas and domains ('worlds'): the procedure is similar to the one followed in (1) where the β-normalisation on terms containing 'world-variables' is kept fixed, while the disciplines of discharging assumptions of the form 'let $\mathbb{S}$ name an arbitrary world' (and corresponding Λ-abstraction disciplines on terms containing $\mathbb{S}$) are modified to suit the particular kind of necessity [de Queiroz and Gabbay (1997)].

Bibliography

Abramsky, S. (1990). Computational Interpretations of Linear Logic. *Theoretical Computer Science* **111**(1–2), 3–57.

Aczel, P. H. G. (1980). Frege Structures and the Notions of Proposition, Truth and Set, in J. Barwise, H.-J. Keisler and K. Kunen (eds.), *The Kleene Symposium, Studies in Logic and The Foundations of Mathematics*, Vol. 101 (North-Holland Publishing Co., Amsterdam, xx+425pp), pp. 31–59, Proceedings of the Symposium held in June 18–24, 1978, at Madison, Wisconsin, USA.

Aczel, P. H. G. (1991). Term Declaration Logic and Generalised Composita, in *Sixth Annual IEEE Symposium on Logic in Computer Science (LICS '91)* (IEEE Press), pp. 22–30, Proceedings of the Symposium held July 15–18 1991, in Amsterdam, The Netherlands.

Anderson, A. R. and Belnap Jr., N. D. (1975). *Entailment. The Logic of Relevance and Necessity I* (Princeton University Press, Princeton, NJ, xxxii+541pp), with contributions by J. Michael Dunn and Robert K. Meyer.

Artemov, S. (1998). Logic of Proofs: A Unified Semantics for Modality and λ-terms, Tech Report, Department of Mathematics, Cornell University, Ithaca, NY.

Avron, A. (1988). The semantics and proof theory of linear logic, *Theoretical Computer Science* **57**, 161–184.

Barendregt, H. P. (1981). *The Lambda Calculus. Its Syntax and Semantics, Studies in Logic and The Foundations of Mathematics*, Vol. 103, revised 1984 edn. (North-Holland, Amsterdam, xv+621pp).

Barwise, J. (1989). *The Situation in Logic, CSLI Lecture Notes*, Vol. 17 (Center for the Study of Language and Information, Stanford, CA, xvi+327pp).

Bishop, E. (1967). *Foundations of Constructive Analysis, McGraw-Hill Series in Higher Mathematics* (McGraw-Hill Book Company, New York, xiv+371pp).

Buchberger, B. (1987). History and basic features of the critical-pair/completion procedure, *Journal of Symbolic Computation* **3**, 3–38.

Chellas, B. F. (1980). *Modal Logic. An Introduction* (Cambridge University Press, Cambridge, xii+295pp).

Chenadec, P. L. (1989). On the logic of unification, *Journal of Symbolic Computation* **8**(1 and 2), 141–199.

Church, A. (1941). *The Calculi of Lambda-Conversion*, second printing (1951) edn., no. 6 in Annals of Mathematics Studies (Princeton University Press, Princeton, NJ, ii+82pp).

Copi, I. (1954). *Symbolic Logic*, fifth (1979) edn. (Macmillan Pub. Co., New York, xiii+398pp).

Curry, H. B. (1931). The universal quantifier in combinatory logic, *Annals of Mathematics* **32**, 154–180.

Curry, H. B. (1934). Functionality in combinatory logic, *Proceedings of the National Academy of Sciences of USA* **20**, 584–590.

Curry, H. B. (1942). The combinatory foundations of mathematical logic, *Journal of Symbolic Logic* **7**, 49–64.

Curry, H. B. (1950). *A Theory of Formal Deducibility*, *Notre Dame Mathematical Lectures*, Vol. 6, third printing (1966) of second (1957) edn. (Notre Dame University Press, Notre Dame, IN, xi+129pp).

Curry, H. B. (1963). *Foundations of Mathematical Logic*, reprinted with corrections (1977) edn. (Dover Publications, New York, viii+408pp).

Curry, H. B. and Feys, R. (1958). *Combinatory Logic, Vol. I, Series Studies in Logic and The Foundations of Mathematics* (North-Holland, Amsterdam, xvi+417pp).

Curry, H. B., Hindley, J. R. and Seldin, J. P. (1972). *Combinatory Logic, Vol. II, Studies in Logic and The Foundations of Mathematics* (North-Holland, Amsterdam, xiv+520pp).

Davis, M. and Fechter, R. (1991). A free variable version of the first-order predicate calculus, *Journal of Logic and Computation* **1**(4), 431–451.

de Oliveira, A. G. (1995). *Proof Transformations for Labelled Natural Deduction via Term Rewriting. (In Portuguese)*, Ph.D. thesis, Master's thesis, Depto. de Informática, Universidade Federal de Pernambuco, Recife, Brasil.

de Oliveira, A. G. and de Queiroz, R. J. (1994). Term Rewriting Systems with *LDS*, in T. Pequeno and F. Carvalho (eds.), *Proceedings of the Brazilian Symposium of Artificial Intelligence (SBIA'94)* (Fortaleza, Oct 1994), pp. 425–439.

de Oliveira, A. G. and de Queiroz, R. J. G. B. (1995). A new basic set of proof transformations (abstract), *Bulletin of Symbolic Logic* **3**, pp. 124–126, abstract of a paper accepted for presentation at *Logic Colloquium '95* held in Haifa, Israel, August 1995.

de Oliveira, A. G. and de Queiroz, R. J. G. B. (1999). A normalization procedure for the equational fragment of labelled natural deduction, *Logic Journal of the Interest Group in Pure and Applied Logics* **7**(2), 173–215.

de Oliveira, A. G. and de Queiroz, R. J. G. B. (2005). A new basic set of proof transformations, in S. Artemov, H. Barringer, A. Garcez, L. Lamb and J. Woods (eds.), *We Will Show Them: Essays in Honour of Dov Gabbay*. Vol. 2 (College Publications, London, ISBN 1-904987-26-5), pp. 499–528.

de Queiroz, R. J. G. B. (1987). Note on Frege's notions of definition and the relationship proof theory vs. recursion theory (extended abstract), in *Abstracts of the VIIIth International Congress of Logic, Methodology and Philosophy of Science. Vol. 5, Part I* (Institute of Philosophy of the Academy of Sciences of the USSR, Moscow), pp. 69–73.

de Queiroz, R. J. G. B. (1988). A proof-theoretic account of programming and the rôle of reduction rules, *Dialectica* **42**(4), 265–282.

de Queiroz, R. J. G. B. (1989). The mathematical language and its semantics: to show the consequences of a proposition is to give its meaning, in P. Weingartner and G. Schurz (eds.), *Reports of the Thirteenth International Wittgenstein Symposium 1988, Schriftenreihe der Wittgenstein-Gesellschaft*, Vol. 18 (Hölder-Pichler-Tempsky,

Vienna, 304pp), pp. 259–266, Symposium held in Kirchberg/Wechsel, Austria, 14–21 August 1988.

de Queiroz, R. J. G. B. (1990). *Proof Theory and Computer Programming. The Logical Foundations of Computation*, Ph.D. thesis, Department of Computing, Imperial College, University of London.

de Queiroz, R. J. G. B. (1991). Meaning as grammar *plus* consequences, *Dialectica* **45**(1), 83–86.

de Queiroz, R. J. G. B. (1992). *Grundgesetze* alongside *Begriffsschrift* (abstract), in *Abstracts of Fifteenth International Wittgenstein Symposium*, pp. 15–16, Symposium held in Kirchberg/Wechsel, 16–23 August 1992.

de Queiroz, R. J. G. B. (1994). Normalisation and language-games, *Dialectica* **48**, pp. 2, 83–123, early version presented at Logic Colloquium '88, Padova. Abstract in *JSL* **55**, 425, 1990.

de Queiroz, R. J. G. B. (2001). Meaning, function, purpose, usefulness, *consequences* — interconnected concepts, *Logic Journal of the Interest Group in Pure and Applied Logics* **9**(5), 693–734.

de Queiroz, R. J. G. B. (2008). On reduction rules, meaning-as-use, and proof-theoretic semantics, *Studia Logica* **90**(2), pp. 211–247.

de Queiroz, R. J. G. B. and de Oliveira, A. G. (2011). Natural deduction for equality: The missing entity, in E. Haeusler, L. Pereira and V. de Paiva (eds.), *Advances in Natural Deduction*, Trends in Logic (Kluwer), pp. 244–273, to appear. Invited talk at *Natural Deduction — Rio 2001*.

de Queiroz, R. J. G. B. and Gabbay, D. M. (1992). The Functional Interpretation of Propositional Equality, Tech. Report, Department of Computing, Imperial College.

de Queiroz, R. J. G. B. and Gabbay, D. M. (1995). The functional interpretation of the existential quantifier, *Bulletin of the Interest Group in Pure and Applied Logics* **3**(2–3), 243–290, presented at *Logic Colloquium '91*, Uppsala, 9–16 August 1991. Abstract *JSL* **58**(2):753–754, 1993.

de Queiroz, R. J. G. B. and Gabbay, D. M. (1997). The functional interpretation of modal necessity, in M. de Rijke (ed.), *Advances in Intensional Logic*, Applied Logic Series (Kluwer, Dordrecht), pp. 61–91.

de Queiroz, R. J. G. B. and Maibaum, T. S. E. (1990). Proof theory and computer programming, *Zeitschrift für mathematische Logik und Grundlagen der Mathematik* **36**, 389–414.

de Queiroz, R. J. G. B. and Maibaum, T. S. E. (1991). Abstract data types and type theory: Theories as types, *Zeitschrift für mathematische Logik und Grundlagen der Mathematik* **37**, 149–166.

de Queiroz, R. J. G. B. and Smyth, M. B. (1989). Induction Rules for Non-Inductive Types in Type Theory, Tech. Report, Department of Computing, Imperial College, presented at the *Fifth British Colloquium for Theoretical Computer Science*, Royal Holloway and Bedford New College, Egham, Surrey, UK, 11–13 April 1989. Abstract in the *Bulletin of European Association for Theoretical Computer Science (EATCS)* 37:220.

Dershowitz, N. (1979). Orderings for term-rewriting systems, in *Proceedings of the Twentieth IEEE Symposium on Foundations of Computer Science* (IEEE Press, San Juan, PR), pp. 123–131.

Dershowitz, N. (1982). Ordering for term-rewriting systems, *Theoretical Computer Science* **17**, 279–301.

Dershowitz, N. (1987). Termination of rewriting, *Journal of Symbolic Computation* **2**, 69–116.

Dershowitz, N. and Jouannaud, J.-P. (1990). Rewrite systems, in J. van Leeuwen (ed.), *Handbook of Theoretical Computer Science* Vol. B, pp. 243–320 (Elsevier Science Publishers B.V., Amsterdam).

Došen, K. (1988). Sequent systems and grupoid models, *Studia Logica* **47**, 353–385.

Došen, K. (1989). Sequent systems and grupoid models ii, *Studia Logica* **48**, 41–65.

Dummett, M. A. E. (1959). Truth, *Proceedings of the Aristotelian Society (new series)* **59**, 141–162.

Dummett, M. A. E. (1973). *Frege: Philosophy of Language*, second (1981) edn. (Duckworth, London, xliii+708pp).

Dummett, M. A. E. (1975). The philosophical basis of intuitionistic logic, in H. E. Rose and J. C. Shepherdson (eds.), *Logic Colloquium '73, Studies in Logic and The Foundations of Mathematics*, Vol. 80 (North-Holland, Amsterdam, viii+513pp), pp. 5–40, Proceedings of the Colloquium held in Bristol, UK, 1973.

Dummett, M. A. E. (1977). *Elements of Intuitionism*, reprinted (1985) edn., Series *Oxford Logic Guides* (Clarendon Press, Oxford, xii+467pp), with the assistance of Roberto Minio.

Dummett, M. A. E. (1980). Comments on Professor Prawitz's paper, in G. H. von Wright (ed.), *Logic and Philosophy*, Series *Entretiens of the International Institute of Philosophy* (Martinus Nijhoff Publishers, The Hague, viii+84pp), pp. 11–18, Symposium held in Düsseldorf, August 27–1 September 1978.

Dummett, M. A. E. (1991a). *Frege: Philosophy of Mathematics* (Duckworth, London, xiii+331pp).

Dummett, M. A. E. (1991b). *The Logical Basis of Metaphysics* (Duckworth, London, xi+355p), revised account (1989) of *The William James Lectures* given at Harvard University in 1976.

Dummett, M. A. E. (1998). Truth from the constructivist standpoint, *Theoria* **LXIV**, 122–138, Special issue on the Philosophy of Dag Prawitz, P. Pagin (ed.).

Fariñas del Cerro, L. and Herzig, A. (1990). Modal deduction with applications in epistemic and temporal logics, Research Report, LSI-IRIT, Toulouse, p. 60.

Fenstad, J. E. (ed.) (1971). *Proceedings of the Second Scandinavian Logic Symposium*, Studies in Logic and The Foundations of Mathematics, Vol. 63 (North-Holland, Amsterdam, viii+405pp), Proceedings of the Symposium held in Oslo, 18–20 June 1970.

Fine, K. (1985). *Reasoning with Arbitrary Objects*, Aristotelian Society series, Vol. 3 (Basil Blackwell, Oxford, viii+220pp).

Fitch, F. B. (1952). *Symbolic Logic. An Introduction* (The Ronald Press Company, New York, x+238pp).

Fitch, F. B. (1966a). Natural deduction rules for obligation, *American Philosophical Quarterly* **3**, 27–38.

Fitch, F. B. (1966b). Tree proofs in modal logic, *Journal of Symbolic Logic* **31**, 152, Abstract of a paper presented at a meeting of the Association for Symbolic Logic in conjunction with the American Philosophical Association, at Chicago, Illinois, 29–30 April 1965.

Fitting, M. (1972). An epsilon-calculus system for first-order S4, in Wilfred Hodges (ed.), *Conference in Mathematical Logic, London '70*, Series *Lecture Notes in Mathematics* (Springer-Verlag), pp. 103–110.

Fitting, M. (1975). A Modal logic epsilon-calculus, *Notre Dame Journal of Formal Logic* **16**, 1–16.

Fitting, M. (1981). *Proof Methods for Modal and Intuitionistic Logics, Synthese Library. Studies in Epistemology, Logic, Methodology and Philosophy of Science*, Vol. 169 (D. Reidel, Dordrecht, viii+555pp).

Fitting, M. (1989). Modal logic should say more than it does, in J.-L. Lassez and G. Plotkin (eds.), *Computational Logic. Essays in Honor of Alan Robinson* (MIT Press, Cambridge, MA), pp. 113–135.

Fitting, M. (1993). Modal logic, in D. Gabbay, C. Hogger and J. Robinson (eds.), *Handbook of Logic in Artificial Intelligence and Logic Programming. Vol. I: Logical Foundations* (Oxford University Press, Oxford), pp. 365–448.

Frege, G. (1879). *Begriffsschrift, eine der arithmetischen nachgebildete Formelsprache des reinen Denkens* (Verlag von Louis Nebert, Halle), english translation 'Begriffsschrift, A Formula Language, Modeled upon that of Arithmetic, for Pure Thought' in [van Heijenoort (1967)], pp. 1–82.

Frege, G. (1891). Funktion und Begriff, *Proceedings of the Jena Medical and Scientific Society* English translation 'Function and Concept' (by Peter Geach) published in [McGuinness (1984)], pp. 137–156.

Frege, G. (1893). *Grundgesetze der Arithmetik. Begriffsschriftlich abgeleitet. I* (Verlag von Hermann Pohle, Jena), reprinted in Vol. 32 of *Olms Paperbacks* (Georg Olms Verlagsbuchhandlung, Hildesheim, 1966, XXXII+254pp). Partial English translation in [Furth (1964)].

Frege, G. (1903). *Grundgesetze der Arithmetik. Begriffsschriftlich abgeleitet. II* (Verlag von Hermann Pohle, Jena), reprinted in Vol. 32 of *Olms Paperbacks* (Georg Olms Verlagsbuchhandlung, Hildesheim, 1966, XVI+266pp). Partial English translation in [Geach and Black (1952)].

Frege, G. (1914). Logic in mathematics, In [Hermes *et al.* (1979)], pp. 203–250.

Furth, M. (ed.) (1964). *The Basic Laws of Arithmetic. Exposition of the System* (University of California Press, Berkeley and Los Angeles, lxiv+143pp). Partial English translation of Gottlob Frege's *Grundgesetze der Arithmetik*.

Gabbay, D. and Hodkinson, I. (1990). An axiomatization of the temporal logic with until and since over the real numbers, *Journal of Logic and Computation* **1**(2), 229–259.

Gabbay, D. M. (1994). *Labelled Deductive Systems, Vol. I — Foundations.* (Oxford University Press. First Draft 1989. Current Draft, 465pp., May 1994), published as MPI-I-94-223, Max-Planck-Institut für Informatik, Im Stadtwald D 663123 Saarbrücken, Germany.

Gabbay, D. M. and de Queiroz, R. J. G. B. (1991). Extending the Curry–Howard–Tait intepretation to linear, relevant and other resource logics (abstract), *Journal of Symbolic Logic* **56**(3), 1139–1140, presented at the *Logic Colloquium '90*, Helsinki, 15–22, July 1990.

Gabbay, D. M. and de Queiroz, R. J. G. B. (1992). Extending the Curry–Howard intepretation to linear, relevant and other resource logics, *Journal of Symbolic Logic* **57**(4), pp. 1319–1365.

Gabbay, D. M. and Kempson, R. M. (1991). Labelled abduction and relevance reasoning, in *Proceedings of the Workshop on Non-Standard Queries and Non-Standard Answers* (Toulouse).

Gabbay, D. M. and Kempson, R. M. (1992). Natural-language content and information flow: A proof-theoretic perspective, in P. Dekker (ed.), *Proceedings of the 8th Amsterdam Colloquium on Formal Semantics*.

Geach, P. and Black, M. (eds.) (1952). *Translations from the Philosophical Writings of Gottlob Frege*, third edn. (Basil Blackwell, Oxford, x+228pp).

Gentzen, G. (1935). Untersuchungen Über das logische Schliessen, *Mathematische Zeitschrift* **39**, pp. 176–210 and 405–431, English translation of 'Investigations into Logical Deduction' in [Szabo (1969)], pp. 68–131.

Girard, J.-Y. (1971). Une Extension de l'Interpretation de Gödel à l'Analyse, et son Application à l'Elimination des Coupures dans l'Analyse et la Théorie des Types, in [Fenstad (1971)], pp. 63–92.

Girard, J.-Y. (1987a). Linear logic, *Theoretical Computer Science* **50**, 1–102.

Girard, J.-Y. (1987b). *Proof Theory and Logical Complexity*, *Studies in Proof Theory*, Vol. I (Bibliopolis, Napoli, 503pp).

Girard, J.-Y. (1989). Towards a geometry of interaction, in J. W. Gray and A. Scedrov (eds.), *Category Theory in Computer Science and Logic*, *Contemporary Mathematics*, Vol. 92 (American Mathematical Society), pp. 69–108, Proceedings of the Symposium held in 1987, Boulder, Colorado.

Girard, J.-Y. (1991). A new constructive logic: Classical logic, *Mathematical Structures in Computer Science* **1**, 255–296.

Girard, J.-Y., Lafont, Y. and Taylor, P. (1989). *Proofs and Types*, *Cambridge Tracts in Theoretical Computer Science*, Vol. 7, reprinted with minor corrections (1990) edn. (Cambridge University Press, Cambridge, xii+175pp).

Gödel, K. (1933). Eine intepretation des intuitionistischen aussagenkalküls, *Ergebnisse eines mathematischen Kolloquiums* **4**, 39–40, English translation 'An interpretation of the intuitionistic propositional calculus' in J. Hintikka (ed.), *The Philosophy of Mathematics* (Oxford University Press, 1969).

Gödel, K. (1958). Über eine bisher noch nicht benützte Erweiterung des finiten Standpunktes, *Dialectica* **12**, 280–287, English translation 'On a hitherto unexploited extension of the finitary standpoint' in *Journal of Philosophical Logic* **9**, pp. 133–142, 1980. Reprinted with facing English translation, in S. Fefermal, *et al.* (eds.) *Kur Gödel's Collected Works, Vol. II: Publications 1938–1974* (Oxford University Press, New York, 1990), pp. 240–251.

Gödel, K. (1995). Lecture at zilsel's, in S. F. *et al.* (ed.), *Kurt Gödel Collected Works. Vol. III* (Oxford University Press, Oxford), pp. 86–113.

Goodman, N. D. (1970). A theory of constructions equivalent to arithmetic, in A. Kino, J. Myhill and R. E. Vesley. (eds.), *Intuitionism and Proof Theory*, Series *Studies in Logic and The Foundations of Mathematics* (North-Holland, Amsterdam, vii+516pp), pp. 101–120, Proceedings of the *Summer Conference* at Buffalo, New York, 1968.

Grzegorczyk, A. (1967). Some relational systems and the associated topological spaces, *Fundamenta Mathematicae* **60**, 223–231.

Guttag, J. (1977). Abstract data types and the development of data structures, *Communications of the ACM* **20**(6), 396–404.

Hacking, I. (1979). What is logic? *Journal of Philosophy* **LXXVI**, 6, 285–319.

Hailperin, T. (1957). A theory of restricted quantification, *Journal of Symbolic Logic* **22**, 19–35 and 113–129.

Helman, G. H. (1977). *Restricted Lambda-Abstraction and the Interpretation of Some Non-classical Logics*, Ph.D. thesis, University of Pittsburgh.

Hermes, H., Kambartel, F. and Kaulbach, F. (eds.) (1979). *Gottlob Frege. Posthumous Writings*, (Basil Blackwell, Oxford, XIII+288pp), translated by Peter Long and Roger White.

Heyting, A. (1930). Die formale Regeln der intuitionistische Logik, *Sitzungsberichte der preuszischen Akademie von Wissenschaften (physicalischmathematische Klasse)*, pp. 42–56.

Heyting, A. (1946). On weakened quantification, *Journal of Symbolic Logic* **11**, 119–121.

Heyting, A. (1956). *Intuitionism. An Introduction*, Series *Studies in Logic and the Foundations of Mathematics* (North-Holland, Amsterdam, viii+133pp).

Hilbert, D. and Ackermann, W. (1938). *Grundzüge der Theoretischen Logik*, 2nd edn., English translation *Principles of Mathematical Logic* (Chelsea, New York, 1950).

Hilbert, D. and Bernays, P. (1934). *Grundlagen der Mathematik I, Die Grundlehren der mathematischen Wissenschaften*, Vol. XL (Verlag von Julius Springer, Berlin, XII+471pp), reprinted by Edwards Brothers, Ann Arbor, Michigan, 1944.

Hilbert, D. and Bernays, P. (1939). *Grundlagen der Mathematik II, Die Grundlehren der mathematischen Wissenschaften*, Vol. L (Verlag von Julius Springer, Berlin, XII+498pp), reprinted by Edwards Brothers, Ann Arbor, Michigan, 1944.

Hilmy, S. S. (1987). *The Later Wittgenstein. The Emergence of a New Philosophical Method* (Basil Blackwell, Oxford, x+340pp).

Hindley, J. R. and Meredith, D. (1990). Principal type-schemes and condensed detachment, *Journal of Symbolic Logic* **55**(1), pp. 90–105.

Hindley, J. R. and Seldin, J. P. (1986). *Introduction to Combinators and λ-Calculus, London Mathematical Society Student Texts*, Vol. 1 (Cambridge University Press, Cambridge, vi+360pp).

Hindley, J. R. and Seldin, J. P. (2008). *Lambda Calculus and Combinators: An Introduction* (Cambridge University Press, Cambridge, xi+345pp).

Hintikka, K. J. (1974). Quantifiers vs. quantification theory, *Linguistic Inquiry* **5**, 153–177.

Hintikka, K. J. (1983). Game-theoretical semantics: Insights and prospects, in J. Hintikka and J. Kulas (eds.), *The Game of Language, Synthese Language Library*, Vol. 22 (D. Reidel, Dordrecht, xii+319pp), pp. 1–31.

Hintikka, M. B. and Hintikka, J. (1986). *Investigating Wittgenstein* (Basil Blackwell, Oxford, xx+326pp).

Howard, W. (1991). Review of Girard, Lafont & Taylor's proofs and types, *Journal of Symbolic Logic* **56**(2) pp. 760–761.

Howard, W. A. (1980). The formulae-as-types notion of construction, in J. P. Seldin and J. R. Hindley (eds.), *To H. B. Curry: Essays on Combinatory Logic, Lambda Calculus and Formalism* (Academic Press, London, xxv̂+606pp), pp. 479–490, privately circulated notes, 1969, only later published in Curry's *Festschrift*.

Hughes, G. E. and Cresswell, M. J. (1968). *An Introduction to Modal Logic*, reprinted (1982) edn. (Methuen and Co., xii+388pp).

Jaśkowski, S. (1934). On the rules of suppositions in formal logic, *Studia Logica* **1**, 5–32, reprinted in [McCall (1967)], pp. 232–258.

Johansson, I. (1936). Der minimalkalkül, ein reduzierter intuitionistischer formalismus, *Compositio Mathematica* **4**, 119–136.

Kalish, D., Montague, R. and Marr, G. (1964). *Logic: Techniques of Formal Reasoning*, second (1980) edn. (Harcourt Brace Jovanovich, New York, xvi+520pp).

Kamp, H. and Reyle, U. (1993). *From Discourse to Logic* (Kluwer, Dordrecht).

Kleene, S. C. (1945). On the interpretation of intuitionistic number theory, *Journal of Symbolic Logic* **10**, 109–124.

Kleene, S. C. (1967). *Mathematical Logic* (Wiley Interscience, New York).

Klop, J. W. (1990). Term rewriting systems, in S. Abramsky, D. Gabbay and T. Maibaum (eds.), *Handbook of Logic in Computer Science,* Vol. 2 (Oxford University Press, Oxford), pp. 1–116.

Knuth, D. and Bendix, P. (1970). Simple word problems in universal algebras, in J. Leech (ed.), *Computational Problems in Abstract Algebra* (Ed. J. Leech), pp. 263–297.

Komori, Y. (1983). The variety generated by BCC-algebras is finitely based, *Reports of Faculty of Science, Shizuoka University* **17**, 13–16.

Komori, Y. (1989). Illative combinatory logic based on *bck*-logic, *Mathematica Japonica* **4**, 585–596.

Komori, Y. (1990). On bb'i logic, bb'ik logic and bb'iw logic, Handwritten memo, 27 February 1989, 8pp. It should be obtainable from the author at Department of Mathematics, Faculty of Science, Shizuoka University, Ohya, Shizuoka 422, Japan.

Kreisel, G. and Tait, W. (1961). Finite definability of number theoretic functions and parametric completeness of equational calculi, *Zeitschrift für Mathematische Logik und Grundalgen der Mathematik* **7**, 28–38.

Kripke, S. A. (1959). A completeness theorem in modal logic, *Journal of Symbolic Logic* **24**, 1–14.

Kripke, S. A. (1963). Semantic analysis of modal logic. I: Normal propositional calculi, *Zeitschrift für mathematische Logik und Grundlagen der Mathematik* **9**, 67–96.

Kripke, S. A. (1965). Semantical analysis of modal logic II: Non-normal modal propositional calculi, in J. W. Addison, L. Henkin and A. Tarski (eds.), *The Theory of Models* (North-Holland, Amsterdam), pp. 206–220.

Lambek, J. (1958). The mathematics of sentence structure, *American Mathematical Monthly* **65**, 154–170.

Lambek, J. (1980). From λ-calculus to cartesian closed categories, in J. P. Seldin and J. R. Hindley (eds.), *To H. B. Curry: Essays on Combinatory Logic, Lambda Calculus and Formalism* (Academic Press, London, xxv+606pp), pp. 375–402.

Lambek, J. (1989). Multicategories revisited, *Contemporary Mathematics* **92**, 217–239.

Lambek, J. and Scott, P. J. (1986). *Introduction to Higher Order Categorical Logic*, *Cambridge Studies in Advanced Mathematics*, Vol. 7 (Cambridge University Press, Cambridge, ix+293pp).

Läuchli, H. (1965). Intuitionistic propositional calculus and definably non-empty terms (abstract), *Journal of Symbolic Logic* **30**, 263.

Leisenring, A. C. (1969). *Mathematical Logic and Hilbert's ε-Symbol*, A volume of *University Mathematical Series* (MacDonald Technical and Scientific, London, ix+142pp).

Lewis, C. I. and Langford, C. H. (1932). *Symbolic Logic*, second (with Dover, New York, 1959) edn. (The Century Co., New York).

Löb, M. H. (1955). Solution of a problem of Leon Henkin, *Journal of Symbolic Logic* **20**, 115–118.

Lorenzen, P. (1955). *Einführung in die operative Logik und Mathematik, Die Grundlehren der mathematischen Wissenschaften*, Vol. LXXVIII (Springer-Verlag, Berlin, iv+298pp).

Lorenzen, P. (1961). Ein dialogisches konstruktivitätskriterium, in *Infinitistic Methods* (Pergamon Press, Oxford, 362pp), Proceedings of the Symposium on the Foundations of Mathematics (International Mathematical Union and Mathematical Institute of the Polish Academy of Sciences) held in Warsaw, 2–9 September 1959.

Lorenzen, P. (1969). *Normative Logic and Ethics*, B.I-Hochschultaschenbücher. *Systematische Philosophie*, Vol. 236* (Bibliographisches Institut, Mannheim/Zürich), 89pp.

Martin-Löf, P. (1971). A Theory of Types, Report 71-3, Department of Mathematics, University of Stockholm, 57pp. February 1971, revised October 1971.

Martin-Löf, P. (1972). Infinite terms and a system of natural deduction, *Compositio Mathematica* **24**, 93–103.

Martin-Löf, P. (1975a). About models for intuitionistic type theories and the notion of definitional equality, in S. Kanger (ed.), *Proceedings of the Third Scandinavian Logic Symposium*, Series *Studies in Logic and The Foundations of Mathematics* (North-Holland, Amsterdam), pp. 81–109, Symposium held in 1973.

Martin-Löf, P. (1975b). An intuitionistic theory of types: Predicative part, in H. E. Rose and J. C. Shepherdson (eds.), *Logic Colloquium '73, Studies in Logic and The Foundations of Mathematics*, Vol. 80 (North-Holland, Amsterdam, viii+513pp), pp. 73–118, Proceedings of the Colloquium held in Bristol, UK, in 1973.

Martin-Löf, P. (1982). Constructive mathematics and computer programming, in L. J. Cohen, J. Los, H. Pfeiffer and K.-P. Podewski (eds.), *Logic, Methodology and Philosophy of Science VI*, Series *Studies in Logic and The Foundations of Mathematics* (North-Holland, Amsterdam, xiii+738pp), pp. 153–175, Proceedings of the International Congress held in Hannover, 22–29 August 1979.

Martin-Löf, P. (1984). *Intuitionistic Type Theory*, Series *Studies in Proof Theory* (Bibliopolis, Naples, iv+91pp), notes by Giovanni Sambin of a series of lectures given in Padova, June 1980.

Martin-Löf, P. (1985). On the meanings of the logical constants and the justifications of the logical laws, in C. Bernardi and P. Pagli (eds.), *Atti degli incontri di logica matematica*. Vol. 2, Series *Scuola di Specializzazione in Logica Matematica* (Dipartimento di Matematica, Università di Siena), pp. 203–281.

Martin-Löf, P. (1987). Truth of a proposition, evidence of a judgement, validity of a proof, *Synthese* **73**, 407–420, Special issue on *Theories of Meaning*, Guest Editor: Maria Luisa Dalla Chiara, collecting articles originally presented as contributions to the conference "Theories of Meaning", organised by the Florence Center for the History and Philosophy of Science, Firenze, Villa di Mondeggi, June 1985.

McCall, S. (ed.) (1967). *Polish Logic 1920–1939* (Clarendon Press, Oxford, viii+406pp).

McGuinness, B. (ed.) (1984). *Gottlob Frege Collected Papers on Mathematics, Logic and Philosophy* (Basil Blackwell, Oxford, viii+412pp), translated by Max Black, V. H. Dudman, Peter Geach, Hans Kaal, E.-H. W. Kludge, Brian McGuinness and R. H. Stoothoff.

Mitchell, J. C. and Scedrov, A. (1993). Notes on sconing and relators, in E. B. *et al.* (ed.), *Computer Science Logic '92, Selected Papers*, LNCS 702 (Springer LNCS 702), pp. 352–378.

Montague, R. (1970). Universal grammar, *Theoria* **36**, 373–398, reprinted in [Thomason (1974)], pp. 222–246.

Newman, M. H. A. (1942). On theories with a combinatorial definition of equivalence, *Annals of Mathematics* **42**(2), 223–243.

Nordström, B., Petersson, K. and Smith, J. M. (1990). *Programming in Martin-Löf's Type Theory. An Introduction, The International Series of Monographs on Computer Science*, Vol. 7 (Clarendon Press, Oxford, x+221pp).

Ohlbach, H. J. (1991). Semantics based translation methods for modal logics, *Journal of Logic and Computation* **1**(5), 691–746.

Ono, H. (1988). Structural rules and a logical hierarchy, Preprint, Faculty of Integrated Arts and Sciences, Hiroshima University, Higashisendamachi, Hiroshima 730, Japan, to appear in the Proceedings of the Summer School and Conference on Mathematical Logic *Heyting '88*, held at Varna, Bulgaria. 11pp.

Ono, H. and Komori, Y. (1985). Logics without the contraction rule, *Journal of Symbolic Logic* **50**, 169–201.

Peano, G. (1889). *Arithmetices principia, nova methodo exposita* (Turin), english translation *The Principles of Arithmetic, Presented by a New Method* published in [van Heijenoort (1967)], pp. 83–97.

Pears, D. F. (1987). *The False Prison. A Study of the Development of Wittgenstein's Philosophy*. Vol. I (Clarendon Press, Oxford, xii+202pp).

Pears, D. F. (1990). Wittgenstein's holism, *Dialectica* **44**, 165–173.

Peterson, G. E. and Stickel, M. E. (1981). Complete sets of reductions for some equational theories, *Journal of the ACM* **28**(2), 233–264.

Plaisted, D. A. (1994). Equational reasoning and term rewriting systems, in D. Gabbay, C. Hogger and J. A. Robinson (eds.), *Handbook of Logic in Artificial Intelligence*. Vol. I (Oxford University Press, Oxford), pp. 273–364.

Poigné, A. (1992). Basic Category Theory, in S. Abramsky, D. Gabbay and T. Maibaum (eds.), *Handbook of Logic in Computer Science*. Vol. I (Oxford University Press, Oxford).

Prawitz, D. (1965). *Natural Deduction. A Proof-Theoretical Study, Acta Universitatis Stockholmiensis. Stockholm Studies in Philosophy*, Vol. 3 (Almqvist & Wiksell, Stockholm), 113pp.

Prawitz, D. (1971). Ideas and results in proof theory, In [Fenstad (1971)], 235–307.

Prawitz, D. (1977). Meaning and proofs: On the conflict between classical and intuitionistic logic, *Theoria* **XLIII**, 2–40.

Prawitz, D. (1978). Proofs and the meaning and completeness of the logical constants, in J. Hintikka, I. Niiniluoto and E. Saarinen (eds.), *Essays on Mathematical and Philosophical Logic, Synthese Library*, Vol. 122 (D. Reidel Publishing Company, Dordrecht, viii+462pp), pp. 25–40, Proceedings of the Fourth Scandinavian Logic Symposium and of the First Soviet-Finnish Logic Conference, Jyväskylä, Finland, June 29–July 6 1976.

Prawitz, D. (1980). Intuitionistic logic: A philosophical challenge, in G. H. von Wright (ed.), *Logic and Philosophy*, Series *Entretiens of the International Institute of Philosophy* (Martinus Nijhoff Publishers, The Hague, viii+84pp), pp. 1–10, Proceedings of the Symposium held in Düsseldorf, August 27–1 September 1978.

Prawitz, D. (1998). Truth and objectivity from a verificationist point of view, in H. G. Dales and G. Olivieri (eds.), *Truth in Mathematics* (Clarendon Press, Oxford), pp. 41–51.

Prawitz, D. (2002). Problems of a generalization of a verificationist theory of meaning, *Topoi* **21**, pp. 87–92, Special issue dedicated to *Justification and Meaning*, held at the University of Siena, Certosa di Pontignano, on 15–17, June 2000.

Prawitz, D. (2006). Meaning approached via proofs, *Synthese* **148**(3), 507–524, Special issue on *Proof-Theoretic Semantics*, R. Kahle and P. Schroeder-Heister (eds.), collecting contributions to a conference with that title held at the University of Tübingen in January 1999.

Quine, W. V. O. (1950). *Methods of Logic*, fourth (1982) edn. (Harvard University Press, Cambridge, Massachussetts, x+333pp).

Reynolds, J. C. (1974). Towards a theory of type structure, in B. Robinet (ed.), *Programming Symposium. Colloque sur la Programmation, Lecture Notes in Computer Science*, Vol. 19 (Springer-Verlag, Berlin/New York, 425pp), pp. 408–425, held in Paris, 9–11 April 1974.

Rose, A. (1956). Formalisation du calcul propositionnel implicatif à *m* valeurs de łukasiewicz, *Comptes rendus hebdomadaires des séances de l'Académie des Sciences* **243**, 1263–1264.

Sahlqvist, H. (1975). Completeness and correspondence in the first and second order semantics for modal logic, in S. Kanger (ed.), *Proceedings of the Third Scandinavian Logic Symposium* (North-Holland), pp. 110–143, Symposium held in Uppsala, 1973.

Schönfinkel, M. (1924). Über die bausteine der mathematischen logik, *Mathematische Annalen* **92**, pp. 305–316, English translation 'On the building blocks of mathematical logic' in [van Heijenoort (1967)], pp. 355–366.

Schroeder-Heister, P. (1984). A natural extension of natural deduction, *Journal of Symbolic Logic* **49**(4), 1284–1300.

Schroeder-Heister, P. (2006). Validity concepts in proof-theoretic semantics, *Synthese* **148**(3), 525–571, Special issue on *Proof-Theoretic Semantics*, Reinhard Kahle and Peter Schroeder-Heister (eds.), collecting contributions to a conference with that title held at the University of Tübingen in January 1999.

Scott, D. S. (1970). Constructive Validity, in M. Laudet, D. Lacombe, L. Nolin and M. Schützenberger (eds.), *Symposium on Automated Deduction, Lecture Notes in Mathematics*, Vol. 125 (Springer-Verlag, Berlin, v+310pp), pp. 237–275, Proceedings of the Symposium held in Versailles, December 1968.

Seldin, J. P. (1989). Normalization and Excluded Middle. i, *Studia Logica* **48**, 193–217.

Shoenfield, J. (1967). *Mathematical Logic* (Addison-Wesley).

Sluga, H. D. (1980). *Gottlob Frege Series, The Arguments of the Philosophers* (Routledge & Kegan Paul, London, xi+203pp).

Snyder, W. (1991). *A Proof Theory for General Unification*, Progress in Computer Science and Applied Logic (Birkhäuser, Boston, MA).

Statman, R. (1977). Herbrand's theorem and gentzen's notion of a direct proof, in J. Barwise (ed.), *Handbook of Mathematical Logic*, Studies in Logic and The Foundations of Mathematics (North-Holland, Amsterdam), pp. 897–912.

Statman, R. (1978). Bounds for proof-search and speed-up in the predicate calculus, *Annals of Mathematical Logic* **15**, 225–287.

Szabo, M. E. (ed.) (1969). *The Collected Papers of Gerhard Gentzen*, Series *Studies in Logic and The Foundations of Mathematics* (North-Holland, Amsterdam, xiv+338pp).

Tait, W. W. (1965). Infinitely long terms of transfinite type, in J. N. Crossley and M. A. E. Dummett (eds.), *Formal Systems and Recursive Functions*, Series *Studies in Logic and The Foundations of Mathematics* (North-Holland, Amsterdam, 320pp), pp. 176–185, Proceedings of the *Logic Colloquium '63*, held in Oxford, UK.

Tait, W. W. (1967). Intensional interpretations of functionals of finite type I, *Journal of Symbolic Logic* **32**, 198–212.

Tait, W. W. (1983). Against intuitionism: Constructive mathematics is part of classical mathematics, *Journal of Philosophical Logic* **12**, 173–195.

Thomason, R. (ed.) (1974). *Formal Philosophy. Selected Papers of Richard Montague.* (Yale University Press, New Haven, CT and London).

Thomason, R. and Stalnaker, R. (1968). Modality and reference, *Noûs* **2**, 359–372.

Troelstra, A. S. and van Dalen, D. (1988). *Constructivism in Mathematics: An Introduction. Vol. II*, *Studies in Logic and The Foundations of Mathematics*, Vol. 123 (North-Holland, Amsterdam, xvii+535pp).

Tuziak, R. (1988). An axiomatization of the finite-valued łukasiewicz calculus, *Studia Logica* **47**, 49–55.

van Benthem, J. (1989). Categorial grammar and type theory, *Journal of Philosophical Logic* **18**, 115–168.

van Benthem, J. (1990). Categorial grammar meets unification, in J. W. *et al.* (ed.), *Unification Formalisms: Syntax, Semantics and Implementation* (Kluwer, Dordrecht).

van Heijenoort, J. (ed.) (1967) *From Frege to Gödel: A Source Book in Mathematical Logic. 1879–1931*, Series *Source Books in the History of the Sciences* (Harvard University Press, Cambridge, xii+664pp).

Wansing, H. (1990). Formulas-as-types for a hierarchy of sublogics of intuitionistic propositional logic, Technical Report 9/90, Department of Philosophy, Freie Universität Berlin, 29pp.

Index